OPTICAL WDM NETWORKS
Principles and Practice

**THE KLUWER INTERNATIONAL SERIES
IN ENGINEERING AND COMPUTER SCIENCE**

OPTICAL WDM NETWORKS
Principles and Practice

edited by

KRISHNA M. SIVALINGAM
Washington State University

SURESH SUBRAMANIAM
George Washington University

Kluwer Academic Publishers
Boston//London/Dordrecht

Distributors for North, Central and South America:
Kluwer Academic Publishers
101 Philip Drive
Assinippi Park
Norwell, Massachusetts 02061 USA
Telephone (781) 871-6600
Fax (781) 871-6528
E-Mail <kluwer@wkap.com>

Distributors for all other countries:
Kluwer Academic Publishers Group
Distribution Centre
Post Office Box 322
3300 AH Dordrecht, THE NETHERLANDS
Telephone 31 78 6392 392
Fax 31 78 6546 474
E-Mail services@wkap.nl>

 Electronic Services <http://www.wkap.nl>

Library of Congress Cataloging-in-Publication

Optical WDM networks : principles and practice / edited by Krishna M. Sivalingam,
Suresh Subramaniam.
    p. cm. -- (The Kluwer international series in engineering and computer science ; SECS 554)
    Includes bibliographical references and index.
  ISBN 0-7923-7825-3
    1. Optical communications. 2. Multiplexing. 3. Fiber optics. 4. Routers (Computer
networks) I. Sivalingam, Krishna M. II. Subramaniam, Suresh.

  TK5103.59. O686 2000
  621.382'7--dc21

                          00-026008

Printed on acid-free paper.

Printed in the United States of America

Contents

Preface

Optical Wavelength Division Multiplexed (WDM) networking technology is spearheading a bandwidth revolution in the infrastructure of the next generation Internet and beyond. Demand for bandwidth is rapidly increasing with the possibility of new kinds of applications such as electronic commerce, video on demand, and global cooperative work. Optical WDM networks offer tremendous promise in meeting this demand.

The basic concept of WDM technology is the ability to simultaneously transmit data on multiple wavelengths on a single fiber. WDM provides a practical solution to the *opto-electronic* speed mismatch problem. This mismatch arises since the theoretical capacity of fiber optics is close to 75 Terabits-per-second (Tbps), while current electronic processing is limited to a few Gigabits-per-second (Gbps). With WDM, several independent channels each operating at a few Gbps are created – a speed that is within electronic processing limits.

The theoretical upper limit on the number of channels is close to one thousand which has been recently achieved in laboratory demonstrations at Lucent Technologies. Individual bit-streams operating at 160 Gbps have also been recently demonstrated, again by Lucent Technologies. At present, available commercial technology provides up to 128 channels each at 2.5 Gbps (OC-48).

Rapid advances in optical components such as tunable lasers and filters have enabled the transition from WDM point-to-point links to WDM networks. Early WDM networking systems used a star coupler as a multi-channel broadcast network. The relatively recent inventions of the optical amplifier and wavelength crossconnect have combined to enable new scalable network architectures with exciting possibilities. A single, core, nation-wide optical network carrying traffic transparently and all-optically from a variety of sources has been envisioned. However, a number of challenges remain for that vision to be realized.

The purpose of this book is to present a collection of excellent papers from leading researchers on various aspects of WDM networks. We realize that technology is so rapidly changing that it is nearly impossible to keep up with the changes. However, we have endeavored to capture a substantial subset of the key problems and known solutions to these problems. Some of the papers present a survey of past research on important problems while others deal with specific problems and solutions.

The audience for this book are practitioners who are interested in recent research work that has not appeared in textbooks; those who are interested in survey articles on specific topics without having to pore over larger textbooks; and graduate students and others who are starting research in this field.

Our hope is that the reader gains valuable insight into the mainstream ideas behind the technology and is inspired to go forth and innovate new ideas and technologies.

All the articles in this book have been written by leading researchers who have made significant contributions to this field over the past decade and longer. The submitted articles were closely reviewed by the editors and their graduate students. The published articles are the revised versions based on these comments.

The book is organized into seven parts. Part I contains Biswanath Mukherjee's paper on a survey of the state-of-the-art in WDM technology.

Part II deals with the physical components that make WDM fiber-optic networks possible. The first paper presents an introduction to fibers, amplifiers, lasers, and receivers. The second paper discusses wavelength routers, switches and filters. Both the papers are written by Byrav Ramamurthy, with Jason Jue co-authoring the first article.

Part III contains three papers on wavelength routed networks. The first paper is a survey on logical topology design, and is written by Rudra Dutta and George Rouskas. The second paper by Chunming Qiao and Myungsik Yoo presents a survey of optical switching techniques in general, and discusses optical burst switching in detail. This is followed by Galen Sasaki's paper on wavelength assignment in networks with limited or no wavelength conversion.

Part IV deals with the passive broadcast star architecture for local area networks. A survey of medium access control (MAC) protocols for star networks is presented by Bo Li, Maode Ma, and Mounir Hamdi. This is followed by a survey of scheduling algorithms for star networks written by George Rouskas. The next paper by Krishna Sivalingam presents the design and analysis of a packet switched MAC protocol that has been implemented in the LIGHTNING testbed. The final paper in this part is by Mike Borella and Biswanath Mukherjee on multicasting in passive star coupled networks.

Part V is dedicated to the performance evaluation of wavelength-routing networks. The first paper by Suresh Subramaniam presents an analytical model for evaluating the blocking performance of wavelength routing networks using a statistical traffic model for state independent routing. The next paper by Ling Li and Arun Somani presents new routing techniques, and analyzes their performance.

Part VI deals with miscellaneous topics. The first paper by Eytan Modiano describes the ONRamp testbed at MIT for providing ultra high speed access networks. This is followed by Alan McGuire's paper on network management issues – a critical area in future networks. The last paper deals with the reconfiguration of a WDM optical network in response to traffic changes in an overlaid ATM network. The paper is written by Georgios Ellinas, Krishna Bala, and Chien-Ming Yu.

Part VII contains Imrich Chlamtac and Jason Jue's outlook into the future of WDM networks and where WDM networking might be headed.

We would like to acknowledge the help of Nilesh Bhide (whose help was exceptionally valuable), Sasidhar Reddy, Ramakrishna Shenai, Stephanie Lindsey, and Bo Wen, graduate students at Washington State University; and Dongyun Zhou, graduate student at the George Washington University. We also gratefully acknowledge our research sponsors – Cisco Systems, National Science Foundation, Alcatel Packet Engines, Telcordia Technologies, and Washington Technology Center. We are also grateful to Alex Greene and Patricia Lincoln at Kluwer Academic for their help and patience, without which this book would not have been possible.

Krishna Sivalingam
Boeing Associate Professor
of Computer Science
Washington State University, Pullman
krishna@eecs.wsu.edu

September 2000

Suresh Subramaniam
Assistant Professor
George Washington University
suresh@seas.gwu.edu

This book is dedicated to our families.

Contributing Authors

Dr. Krishna Bala, *Tellium Inc.*

Dr. Mike Borella, *3Com Corp.*

Prof. Imrich Chlamtac, *University of Texas at Dallas*

Mr. Rudra Dutta, *North Carolina State University*

Dr. Georgios Ellinas, *Telcordia Technologies*

Prof. Mounir Hamdi, *Hong Kong University of Science and Technology*

Prof. Jason P. Jue, *University of Texas at Dallas*

Prof. Bo Li, *Hong Kong University of Science and Technology*

Mr. Ling Li, *Iowa State University*

Mr. Maode Ma, *Hong Kong University of Science and Technology*

Mr. Alan McGuire, *British Telecom*

Prof. Eytan Modiano, *Massachusetts Institute of Technology*

Prof. Biswanath Mukherjee, *University of California at Davis*

Prof. Chunming Qiao, *SUNY at Buffalo*

Prof. Byrav Ramamurthy, *University of Nebraska-Lincoln*

Prof. George N. Rouskas, *North Carolina State University*

Prof. Galen Sasaki, *University of Hawaii*

Prof. Krishna M. Sivalingam, *Washington State University*

Prof. Arun K. Somani, *Iowa State University*

Prof. Suresh Subramaniam, *The George Washington University*

Mr. Myungsik Yoo, *SUNY at Buffalo*

Chien-Ming Yu, *Columbia University*

Ms. Hui Zang, *University of California at Davis*

I

INTRODUCTION

Chapter 1

SURVEY OF STATE-OF-THE-ART

Biswanath Mukherjee
University of California, Davis
mukherje@cs.ucdavis.edu

Hui Zang
University of California, Davis
zang@cs.ucdavis.edu

Abstract   Although optical networking is a relatively young topic, this field has experienced
                  explosive growth over the past few years. This growth has been fueled mainly by
                  the demands for enormous bandwidth on our data networks, namely the Internet,
                  to satisfy the high-speed applications of the network users. Wavelength-division
                  multiplexing (WDM) is the current favorite technology for building optical com-
                  munication networks since all of the end-user equipment needs to operate only
                  at the bit rate of a WDM channel, which can be chosen arbitrarily, e.g., peak
                  electronic processing speed. The state-of-the-art in designing optical WDM net-
                  works is reviewed in this chapter, starting with WDM networking evolution, and
                  covering new WDM technologies, how to construct WDM networks, followed
                  by recent trends in WDM research.

1.   INTRODUCTION

We are moving towards a society which requires that we have access to
information at our finger tips *when we need it, where we need it, and in
whatever format we need it*. The information is provided to us through our
global mesh of communication networks, whose current implementations, e.g.,
today's Internet and asynchronous transfer mode (ATM) networks, do not have
the capacity to support the foreseeable bandwidth demands.

Fiber-optic technology can be considered our saviour for meeting our above
need because of its potentially limitless capabilities [Mukherjee, 1997],
[Cochrane, 1995]: huge bandwidth (nearly 50 terabits per second (Tbps)), low

signal attenuation (as low as 0.2 dB/km), low signal distortion, low power requirement, low material usage, small space requirement, and low cost. Our challenge is to turn the promise of fiber optics to reality to meet our information networking demands of the next decade (and well into the next century!).

Thus, the basic premise of the subject on optical wavelength-division multiplexing (WDM) networks is that, as more and more users start to use our data networks, and as their usage patterns evolve to include more and more bandwidth-intensive networking applications such as data browsing on the world wide web (WWW), java applications, video conferencing, etc., there emerges an acute need for very high-bandwidth transport network facilities, whose capabilities are much beyond those that current high-speed (ATM) networks can provide. There is just not enough bandwidth in our networks today to support the exponential growth in user traffic!

Given that a single-mode fiber's potential bandwidth is nearly 50 Tbps, which is nearly four orders of magnitude higher than electronic data rates of a few gigabits per second (Gbps), every effort should be made to tap into this huge opto-electronic bandwidth mismatch. Realizing that the maximum rate at which an end-user, which can be a workstation or a gateway that interfaces with lower-speed subnetworks, can access the network is limited by electronic speed (to a few Gbps), the key in designing optical communication networks in order to exploit the fiber's huge bandwidth is to introduce concurrency among multiple user transmissions into the network architectures and protocols. In an optical communication network, this concurrency may be provided according to either wavelength or frequency [wavelength-division multiplexing (WDM)], time slots [time-division multiplexing (TDM)], or wave shape [spread spectrum, code-division multiplexing (CDM)].

Optical TDM and CDM are somewhat futuristic technologies today. Under (optical) TDM, each end-user should be able to synchronize to within one time slot. The optical TDM bit rate is the aggregate rate over all TDM channels in the system, while the optical CDM chip rate may be much higher than each user's data rate. As a result, both the TDM bit rate and the CDM chip rate may be much higher than electronic processing speed, i.e., some part of an end user's network interface must operate at a rate higher than electronic speed. Thus, TDM and CDM are relatively less attractive than WDM, since WDM – unlike TDM or CDM – has no such requirement.

Specifically, WDM is the current favorite multiplexing technology for optical communication networks since all of the end-user equipment needs to operate only at the bit rate of a WDM channel, which can be chosen arbitrarily, e.g., peak electronic processing speed. Hence, the major carriers today all devote significant effort to developing and applying WDM technologies in their businesses.

2.   WAVELENGTH-DIVISION MULTIPLEXING (WDM)

Wavelength-division multiplexing (WDM) is an approach that can exploit the huge opto-electronic bandwidth mismatch by requiring that each end-user's equipment operate only at electronic rate, but multiple WDM channels from different end-users may be multiplexed on the same fiber. Under WDM, the optical transmission spectrum (see Fig. 1.1) is carved up into a number of non-overlapping wavelength (or frequency) bands, with each wavelength

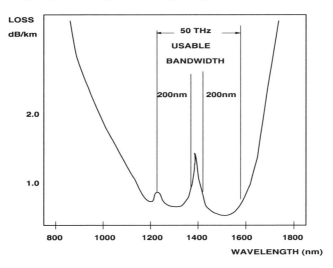

Figure 1.1   The low-attenuation regions of an optical fiber.

supporting a single communication channel operating at whatever rate one desires, e.g., peak electronic speed. Thus, by allowing multiple WDM channels to coexist on a single fiber, one can tap into the huge fiber bandwidth, with the corresponding challenges being the design and development of appropriate network architectures, protocols, and algorithms. Also, WDM devices are easier to implement since, generally, all components in a WDM device need to operate only at electronic speed; as a result, several WDM devices are available in the marketplace today, and more are emerging.

Research and development on optical WDM networks have matured considerably over the past few years, and they seem to have suddenly taken on an explosive form, as evidenced by recent publications [Mukherjee, 1997; JSAC98, 1998; JLT96, 1996; JSAC96, 1996; JLT93, 1993; JSAC90, 1990; JHSN95, 1995] on this topic as well as overwhelming attendance and enthusiasm at the WDM workshops during recent conferences: Optical Fiber Communications conference and IEEE International Conference on Communications. A number of experimental prototypes have been and are currently being deployed and tested mainly by telecommunication providers in the U.S., Europe, and Japan. It is anticipated that the next generation of the Internet will employ WDM-based optical backbones.

Current development activities indicate that this sort of WDM network will be deployed mainly as a backbone network for large regions, e.g., for nationwide or global coverage. (However, optical WDM networks for local applications are also being researched and prototyped, e.g., Rainbow, STARNET, etc.) End-users, to whom the architecture and operation of the backbone will be transparent except for significantly improved response times, will attach to the network through a wavelength-sensitive switching/routing node. An end-user in this context need not necessarily be a terminal equipment but the aggregate activity from a collection of terminals – including those that may possibly be feeding in from other regional and/or local subnetworks – so that the end-user's aggregate activity on any of its transmitters is close to the peak electronic transmission rate.

3.   WDM NETWORKING EVOLUTION

3.1.   POINT-TO-POINT WDM SYSTEMS

WDM technology is being deployed by several telecommunication companies for point-to-point communications. This deployment is being driven by the increasing demands on communication bandwidth. When the demand exceeds the capacity in existing fibers, WDM is turning out to be a more cost-effective alternative compared to laying more fibers. A study [Melle et al., 1995] compared the relative costs of upgrading the transmission capacity of a point-to-point transmission link from OC-48 (2.5 Gbps) to OC-192 (10 Gbps) via three possible solutions. (The terminology OC-n is a widely used telecommunications jargon. "OC" stands for "optical channel" and it specifies electronic data rates. "OC-n" stands for a data rate of $n \times 51.84$ megabits per second (Mbps) approximately; so OC-48 and OC-192 correspond to approximate data rates of 2.5 Gbps and 10 Gbps, respectively. OC-768 (40 Gbps) is the next milestone in highest achievable electronic communication speed.)

1 installation/burial of additional fibers and terminating equipment (the "multi-fiber" solution);

2 a four-channel "WDM solution" (see Fig. 1.2) where a WDM multiplexer (mux) combines four independent data streams, each on a unique wavelength, and sends them on a fiber; and a demultiplexer (demux) at the fiber's receiving end separates out these data streams; and

3 OC-192, a "higher-electronic-speed" solution.

The analysis in [Melle et al., 1995] shows that, for distances lower than 50 km for the transmission link, the "multi-fiber" solution is the least expensive; but for distances longer than 50 km, the "WDM" solution's cost is the least with the cost of the "higher-electronic-speed" solution not that far behind.

WDM mux/demux in point-to-point links is now available in product form from several vendors such as IBM, Pirelli, and AT&T [Green, 1996]. Among these products, the maximum number of channels is 64 today, but this number is expected to increase soon.

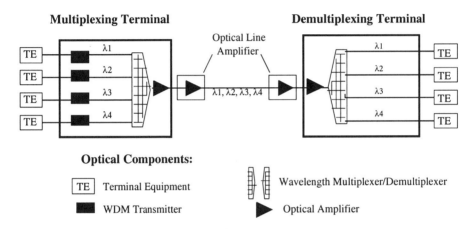

Figure 1.2   A four-channel point-to-point WDM transmission system with amplifiers.

3.2.   WAVELENGTH ADD/DROP MULTIPLEXER (WADM)

A Wavelength Add/Drop Multiplexer (WADM) is shown in Fig. 1.3. It consists of a demux, followed by a set of 2×2 switches, one switch per wavelength, followed by a mux. The WADM can be essentially "inserted" on a physical fiber link. If all of the 2×2 switches are in the "bar" state, then all of the wavelengths flow through the WADM "undisturbed." However, if one of the 2×2 switches is configured into the "cross" state (as is the case for the λ_i switch in Fig. 1.3) via electronic control (not shown in Fig. 1.3), then the signal on the corresponding wavelength is "dropped" locally, and a new data stream can be "added" on to the same wavelength at this WADM location. More than one wavelength can be "dropped and added" if the WADM interface has the necessary hardware and processing capability.

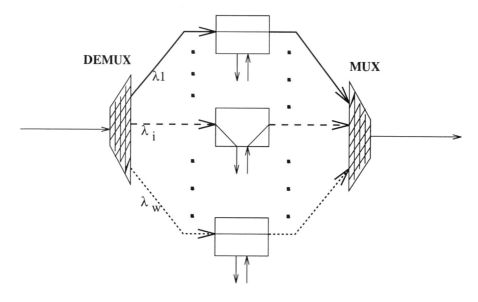

Figure 1.3   A Wavelength Add/Drop Multiplexer (WADM).

3.3.   FIBER AND WAVELENGTH CROSSCONNECTS – PASSIVE STAR, PASSIVE ROUTER, AND ACTIVE SWITCH

In order to have a "network" of multiwavelength fiber links, we need appropriate fiber interconnection devices. These devices fall under three broad categories:

- passive star (see Fig. 1.4),

- passive router (see Fig. 1.5), and

- active switch (see Fig. 1.6).

The *passive star* is a "broadcast" device, so a signal that is inserted on a given wavelength from an input fiber port will have its power equally divided among (and appear on the same wavelength on) all output ports. As an example, in Fig. 1.4, a signal on wavelength λ_1 from Input Fiber 1 and another on wavelength λ_4 from Input Fiber 4 are broadcast to all output ports. A "collision" will occur when two or more signals from the input fibers are simultaneously launched into the star on the same wavelength. Assuming as many wavelengths as there are fiber ports, an $N \times N$ passive star can route N simultaneous connections through itself.

A *passive router* can separately route each of several wavelengths incident on an input fiber to the same wavelength on separate output fibers, e.g., wavelengths λ_1, λ_2, λ_3, and λ_4 incident on Input Fiber 1 are routed to the same

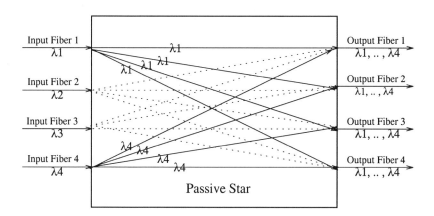

Figure 1.4   A 4 × 4 passive star.

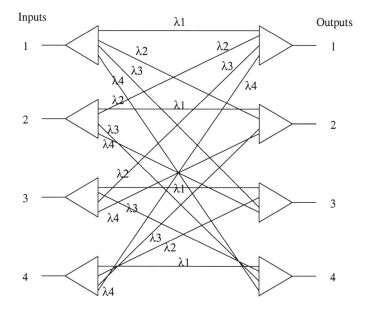

Figure 1.5   A 4 × 4 passive router (four wavelengths).

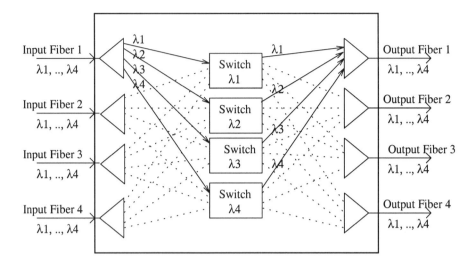

Figure 1.6   A 4 × 4 active switch (four wavelengths).

corresponding wavelengths to Output Fibers 1, 2, 3, and 4, respectively, in Fig. 1.5. Observe that this device allows *wavelength reuse*, i.e., the same wavelength may be spatially reused to carry multiple connections through the router. The wavelength on which an input port gets routed to an output port depends on a "routing matrix" characterizing the router; this matrix is determined by the internal "connections" between the demux and mux stages inside the router (see Fig. 1.5). The routing matrix is "fixed" and cannot be changed. Such routers are commercially available and are also known as Latin routers, waveguide grating routers (WGRs), wavelength routers (WRs), etc. Again, assuming as many wavelengths as there are fiber ports, a $N \times N$ passive router can route N^2 simultaneous connections through itself (compared to only N for the passive star); however, it lacks the broadcast capability of the star.

The *active switch* also allows *wavelength reuse*, and it can support N^2 simultaneous connections through itself (like the passive router). However, the active star has a further enhancement over the passive router in that its "routing matrix" can be *reconfigured* on demand, under electronic control. However the "active switch" needs to be powered and is not as fault-tolerant as the passive star and the passive router which do not need to be powered. The active switch is also referred to as a *wavelength-routing switch (WRS)*, *wavelength selective crossconnect (WSXC)*, or just crossconnect for short. (We will refer to it as a WRS in the remainder of this chapter.)

The active switch can be enhanced with an additional capability, viz., a wavelength may be converted to another wavelength just before it enters the mux stage before the output fiber (see Fig. 1.6). A switch equipped with such a wavelength-conversion facility is more capable than a WRS, and it is referred to as a *wavelength-convertible switch, wavelength interchanging crossconnect (WIXC)*, etc.

The passive star is used to build local WDM networks, whereas the active switch is used for constructing wide-area wavelength-routed networks. The passive router has mainly found application as a mux/demux device.

4.   NEW WDM TECHNOLOGIES

Recent developments in fiber optics have further expanded the usable fiber bandwidth. A new type of fiber, called the allwave fiber, does not have the 1385 nm "water-peak window" (which the conventional fiber has) and therefore, it provides a more usable optical spectrum. Figure 1.7 shows the low-attenuation region of the allwave fiber when compared with that of the conventional fiber.

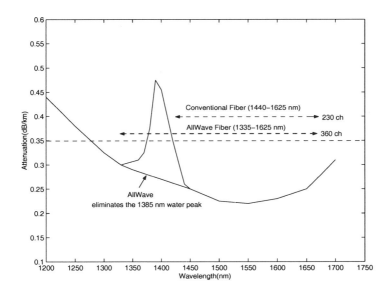

Figure 1.7   Low-attenuation region of allwave fiber vs. conventional fiber.

Another new and interesting technology is a new type of amplifier device which uses the erbium-doped fiber amplifier (EDFA) as a building block. The normal EDFA has a gain spectrum of 30-40 nm (typically in the 1530-1560 nm range). Thus, fibers employing EDFAs for long-haul communication would normally require that their carried signals be confined to the EDFA gain spectrum. However, by using the EDFA as an elementary gain building block, we can build a "circuit" of EDFAs configured such that the "circuit designer" has full control of the gain as well as gain bandwidth of the composite "circuit" (see Fig. 1.8 for an example) [Fan, 1998]. This "amplifier circuit" is referred to as an *Ultra Wide-Band EDFA*, which can fully exploit the expanded low-attenuation region of the "allwave fiber." As an analogy, if we think of the EDFA as equivalent to a "transistor," then the *Ultra Wide-Band EDFA* device would be analogous to an "integrated circuit" [Fan, 1998].

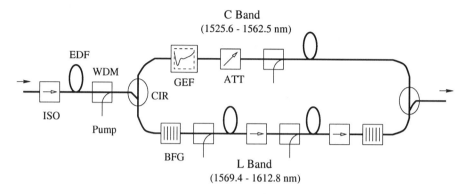

Figure 1.8   Ultra Wide-Band EDSFA.

5.   WDM NETWORK CONSTRUCTIONS

5.1.   BROADCAST-AND-SELECT (LOCAL) OPTICAL WDM NETWORK

A local WDM optical network may be constructed by connecting network nodes via two-way fibers to a passive star, as shown in Fig. 1.9. A node sends its transmission to the star on one available wavelength, using a laser which produces an optical information stream. The information streams from multiple sources are optically combined by the star and the signal power of each stream is equally split and forwarded to all of the nodes on their receive fibers. A node's receiver, using an optical filter, is tuned to only one of the wavelengths; hence it can receive the information stream. Communication between sources and receivers may follow one of two methods: (1) *single-hop* [Mukherjee, 1992a], or (2) *multi-hop* [Mukherjee, 1992b]. Also, note that, when a source transmits on a particular wavelength λ_1, more than one receiver can be tuned to wavelength λ_1, and all such receivers may pick up the information stream. Thus, the passive-star can support "*multicast*" services.

5.2.   WAVELENGTH-ROUTED (WIDE-AREA) OPTICAL NETWORK

A wavelength-routed (wide area) optical WDM network is shown in Fig. 1.10. The network consists of a *photonic switching fabric*, comprising "active swit-

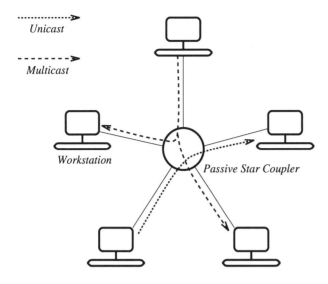

Figure 1.9 A passive-star-based local optical WDM network.

ches" connected by fiber links to form an arbitrary *physical topology*. Each end-user is connected to an active switch via a fiber link. The combination of an end-user and its corresponding switch is referred to as a network *node*.

Each node (at its access station) is equipped with a set of transmitters and receivers, both of which may be wavelength tunable. A transmitter at a node sends data into the network and a receiver receives data from the network.

The basic mechanism of communication in a wavelength-routed network is a *lightpath*. A *lightpath* is an all-optical communication channel between two nodes in the network, and it may span more than one fiber link. The intermediate nodes in the fiber path route the lightpath in the optical domain using their active switches. The end-nodes of the lightpath access the lightpath with transmitters and receivers that are tuned to the wavelength on which the lightpath operates. For example, in Fig. 1.10, lightpaths are established between nodes A and C on wavelength channel λ_1, between B and F on wavelength channel λ_2, and between H and G on wavelength channel λ_1. The lightpath between nodes A and C is routed via active switches 1, 6, and 7. (Note the wavelength reuse for λ_1.)

In the absence of any wavelength-conversion device, a lightpath is required to be on the same wavelength channel throughout its path in the network; this requirement is referred to as the *wavelength-continuity* property of the lightpath. This requirement may not be necessary if we also have wavelength converters in the network. For example, in Fig. 1.10, the lightpath between nodes D and E traverses the fiber link from node D to switch 10 on wavelength λ_1, gets converted to wavelength λ_2 at switch 10, traverses the fiber link between switch

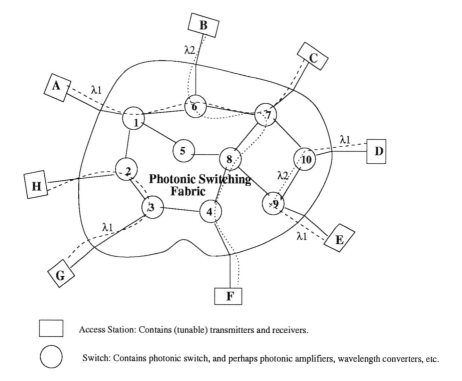

Access Station: Contains (tunable) transmitters and receivers.

Switch: Contains photonic switch, and perhaps photonic amplifiers, wavelength converters, etc.

Figure 1.10   A wavelength-routed (wide-area) optical WDM network.

10 and switch 9 on wavelength λ_2, gets converted back to wavelength λ_1 at switch 9, and traverses the fiber link from switch 9 to node E on wavelength λ_1.

A fundamental requirement in a wavelength-routed optical network is that two or more lightpaths traversing the same fiber link must be on different wavelength channels so that they do not interfere with one another.

5.3.   A SAMPLE WDM NETWORKING PROBLEM

As we have described in Section 5.2, end-users in a fiber-based WDM backbone network may communicate with one another via *all-optical (WDM) channels*, which are referred to as *lightpaths*. A *lightpath* may span multiple fiber links, e.g., to provide a "circuit-switched" interconnection between two nodes which may have a heavy traffic flow between them and which may be located "far" from each other in the physical fiber network topology. Each intermediate node in the lightpath essentially provides an all-optical bypass facility to support the lightpath.

In an N-node network, if each node is equipped with $N - 1$ transceivers [transmitters (lasers) and receivers (filters)] and if there are enough wavelengths on all fiber links, then every node pair could be connected by an all-optical lightpath, and there is no networking problem to solve. However, it should be noted that the network size (N) should be scalable, transceivers are expensive so that each node may be equipped with only a few of them, and technological constraints dictate that the number of WDM channels that can be supported in a fiber be limited to W (whose value is a few tens today, but is expected to improve with time and technological breakthroughs). Thus, only a limited number of lightpaths may be set up on the network.

Under such a network setting, a challenging networking problem is that, given a set of lightpaths that need to be established on the network, and given a constraint on the number of wavelengths, determine the routes over which these lightpaths should be set up and also determine the wavelengths that should be assigned to these lightpaths so that the maximum number of lightpaths may be established [Ramaswami and Sivarajan, 1995]. While shortest-path routes may be most preferable from the individual point of view of each lightpath, note that this choice may have to be sometimes sacrificed, in order to allow more lightpaths to be set up. Thus, one may allow several alternate routes for lightpaths to be established [Ramamurthy and Mukherjee, 1998]. Lightpaths that cannot be set up due to constraints on routes and wavelengths are said to be blocked, so the corresponding network optimization problem is to minimize this blocking probability.

In this regard, note that, normally, a lightpath operates on the same wavelength across all fiber links that it traverses, in which case the lightpath is said to satisfy the *wavelength-continuity constraint*. Thus, two lightpaths that share a common fiber link should not be assigned the same wavelength. However, if a switching/routing node is also equipped with a *wavelength-converter facility*, then the *wavelength-continuity constraints* disappear, and a lightpath may switch between different wavelengths on its route from its origin to its termination.

This particular problem, referred to as the Routing and Wavelength Assignment (RWA) problem, has been examined in detail in [Zang et al., 2000], whereas the general topic of *wavelength-routed networks* has been studied in [Mukherjee, 1997].

Returning to our sample networking problem, note that designers of next-generation lightwave networks must be aware of the properties and limitations of optical fiber and devices in order for their corresponding protocols and algorithms to take advantage of the full potential of WDM. Often a network designer may approach the WDM architectures and protocols from an overly simplified, ideal, or traditional-networking point of view. Unfortunately, this may lead an individual to make unrealistic assumptions about the properties of fiber and optical components and hence may result in an unrealizable or impractical design.

6.   RECENT TRENDS IN WDM RESEARCH

Recent research interests in WDM networks include network control and management, fault management, multicasting, physical-layer issues, IP over WDM, traffic grooming, and optical packet switching, just to name a few topics. We briefly examine a subset of these issues in the following subsections, namely network control and management, fault management, multicasting, and physical-layer issues.

6.1.   NETWORK CONTROL AND MANAGEMENT

In a wavelength-routed WDM networks, a control mechanism is needed to set up and take down all-optical connections (i.e., lightpaths) [Zang et al., 2000; Zang et al., 1999]. Upon the arrival of a connection request, this mechanism must be able to select a route, assign a wavelength to the connection, and configure the appropriate optical switches in the network. The mechanism must also be able to provide updates to reflect which wavelengths are currently being used on each fiber link so that nodes may make informed routing decisions. This control mechanism can either be centralized or distributed. Distributed systems are usually more robust than centralized systems so they are generally more preferred. The objectives of various research efforts on this subject are to minimize (1) the blocking probability of connection requests, (2) the connection setup delays, and (3) the bandwidth used for control messages as well as to maximize the scalability of such networks [Zang et al., 1999].

There are two distributed network control management schemes which have been examined in the literature. The first approach is proposed in [Ramaswami and Segall, 1997], and we refer to it as the "link-state approach" because it routes connections in a link-state fashion. The second approach is proposed by us in [Zang et al., 1999], and we refer to it as the "distributed routing approach" because it utilizes the distributed Bellman-Ford routing algorithm. We describe the two approaches below.

In the link-state approach, each node maintains the complete network topology, including information on which wavelengths are in use on each fiber link. Upon the arrival of a connection request, a node utilizes the topology information to select a route and a wavelength. Once the route and wavelength are selected, the node attempts to reserve the selected wavelength along each fiber link on the route by sending reservation requests to each node in the route. If an intermediate node is able to reserve the wavelength on the appropriate link, it sends an acknowledgement directly back to the source node. If all of the reservations are successful, then the source sends a SETUP message to each of the nodes. The appropriate switches are then configured at each node, and the connection is established. If even one of the reservations is not successful, then the call is blocked and the source node sends a TAKEDOWN message to each node on the route in order to release the reserved resources. When a connection is established or torn down, each node involved in the connection broadcasts a topology-update message which indicates any changes in the status of wavelengths being used on the node's outgoing links.

In the distributed-routing approach, routes are selected in a distributed fashion without knowledge of the overall network topology. Each node maintains a routing table which specifies the next hop and the cost associated with the shortest path to each destination on a given wavelength. The cost may reflect hop counts or actual fiber-link distances. The routing table is established by employing a distributed Bellman-Ford algorithm [Garcia-Luna-Aceves, 1992]. In the distributed-routing approach, upon receiving a connection request, a node will choose the wavelength which results in the shortest distance to the destination, and it will forward the connection request to the next node in the path. Thus, the connection request is routed one hop at a time, with each node along the route independently selecting the next hop based on routing information, and reserving the appropriate wavelength on the selected link. Once the request reaches the destination node, the destination node sends an acknowledgement back to the source node along the reverse path (i.e., it sends the acknowledgement back to the node from which it received the connection request). Upon receiving the ACK, each node along the reverse path configures its wavelength-routing switch. The source node begins transmitting data after it receives the acknowledgement. If a node along the path is unable to reserve the desired wavelength on a link, it will send a negative acknowledgement back to the source along the reverse path. The nodes on the reverse path will release the reserved wavelengths as they receive the negative acknowledgement. The source node may then re-attempt the connection on a different wavelength. If the source node is unable to establish the connection on any wavelength, the call is blocked. Once a connection is established, each node along the route sends to each of its neighbors an update message reflecting the status of the newly-occupied link and wavelength. Each node receiving an update message may then update its routing table. Similar updates occur when a connection is taken down.

Both schemes can be implemented using a separate channel, namely, a control channel in the WDM network. Besides supporting the signaling protocol and the network-topology and status update protocol, the control channel should also have the ability to discover and recover from faults, which we explain below.

6.2.   FAULT MANAGEMENT

In a wavelength-routed WDM network (as well as in other networks), the failure of a network element (e.g., fiber link, cross-connect, etc.) may cause the failure of several optical channels, thereby leading to large data (and revenue) losses. Studies have been conducted to examine different approaches to protect WDM optical networks from single fiber-link failures. Earlier studies were focused on ring topologies while recently mesh-based networks have been considered [Ramamurthy and Mukherjee, 1999a; Ramamurthy and Mukherjee, 1999b; Ramamurthy, 1998].

There are several approaches to ensure fiber-network survivability. Survivable network architectures are based either on dedicating backup resources in advance or on dynamic restoration. In dedicated-resource survivability (which includes Automatic Protection Switching and Self-Healing Rings [Wu, 1992; Grover, 1987]), the disrupted network service is restored by utilizing the dedicated network resources. In dynamic restoration, the spare-capacity available within the network is utilized for restoring services affected by a failure. Generally, dynamic restoration schemes are more efficient in utilizing capacity due to the multiplexing of the spare-capacity requirements, and they provide resilience against different kinds of failures, whereas dedicated restoration schemes have a faster restoration time and provide guarantees on the restoration ability.

We will examine three approaches to protecting against fiber-link failures in an optical network: (a) a dedicated-resource survivability approach called 1+1 protection, and (b) two dynamic approaches called link restoration and path restoration. Each of these approaches have different wavelength-capacity requirements and blocking performance. So far, research studies have focused on single-link failures, which are the most common form of failures in optical networks.

- 1+1 Protection: For each connection that needs to be 1+1 protected, a dedicated link-disjoint backup route and wavelength is set up in advance (at the time of connection setup). Upon a link failure on the primary path, the end-nodes of the connection start utilizing the backup route and wavelength.

- Link Restoration: In link restoration, all the connections that traverse the failed link are re-routed around that link. The source and destination nodes of the connections traversing the failed link are oblivious to the link failure. The end-nodes of the failed link dynamically discover a route around the link, for each wavelength that traverses the link. Upon a failure, the end-nodes of the failed link may participate in a distributed procedure to hunt for new paths for each active wavelength on the failed link. When a new route is discovered around the failed link for a wavelength channel, the end-nodes of the failed link re-configure their cross-connects to re-route that channel onto the new route. If no new routes are discovered for a wavelength channel, the connection that utilizes that wavelength is blocked.

- Path Restoration: In path restoration, when a link fails, the source node and the destination node of each connection that traverses the failed link are informed about the failure (possibly via messages from the nodes adjacent to the link failure). The source and the destination nodes of each connection independently discover a backup route on an end-to-end basis (such a backup path could be on a different wavelength channel). When a new route and wavelength channel is discovered for a connection, network elements such as wavelength cross-connects are reconfigured appropriately, and the connection switches to the new path. If no new routes (and associated wavelength) are discovered for a broken connection, that connection is blocked.

For a comprehensive review of the literature on the design of survivable optical networks, please see [Ramamurthy, 1998].

6.3. MULTICASTING – "LIGHT TREES"

Multicasting is the ability of a communication network to accept a single message from an application and to deliver copies of the message to multiple recipients at different locations. The connections (i.e., lightpaths) we examined earlier are point-to-point connections. However, we can extend the lightpath concept to point-to-multipoint, i.e., a multicast concept, which we call a "light tree" [Sahasrabuddhe and Mukherjee, 1999]. A light tree enables a transmitter at a node to have many more logical neighbors, thereby leading to a denser virtual interconnection diagram and lower hop distance. A collection of light trees embedded on an optical WDM backbone network can improve the performance of unicast, multicast, and broadcast traffic. However, the corresponding network will require multicast-capable optical switches and more power budget to combat the effect of power losses due to signal splitting. These tradeoffs, including an integer linear program (ILP) formulation of the problem, are examined in [Sahasrabuddhe and Mukherjee, 1999].

Many multicast applications exist, but their implementations are not necessarily efficient because today's networks were designed to mainly support point-to-point communication. Such applications include video conferencing, software/file distribution including file replication on mirrored sites, distributed games, Internet news distribution, e-mail mailing lists, etc. In the future, as multicast applications become more popular and bandwidth intensive, there emerges a pressing need to provide multicast support in the underlying communication network. A multicast-capable WDM WAN should not only support efficient routing for multicast traffic, but it may also enhance routing for unicast traffic by allowing more densely-connected virtual topologies. To realize multicast-capable WDM WANs, we need to develop novel multicast-capable WRS architectures and design multicast routing and wavelength assignment algorithms, as outlined below.

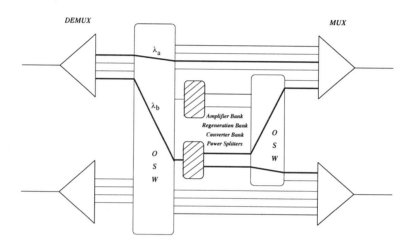

Figure 1.11   A multicast-capable WRS (MWRS).

6.3.1   Multicast-Capable WRS Architectures.

An optical splitter splits an input signal into multiple output signals, with the output powers being governed by the splitter's "splitting ratio." Figure 1.11 shows a 2×2 multicast-capable wavelength-routing switch (MWRS), which can support four wavelengths on each link. The information from each incoming fiber link is first demultiplexed (DEMUX) into separate wavelengths. Then, the separate signals, each on separate wavelengths, are switched by an optical switch (OSW). Unicast signals are sent directly to OSW ports corresponding to their output links, while those signals which need to be multicast are sent to an OSW port connected to a splitter bank. (*The splitter bank may be enhanced to provide optical signal amplification, wavelength conversion, and signal regeneration for multicast as well as unicast connections.*) For example, in Fig. 1.11, wavelength λ_a is an unicast signal, and λ_b is a multicast signal. The output of the splitter is connected to a smaller optical switch, which routes the multicast signals to their respective output fiber links. Since an optical splitter is a passive device, the power of the signals at each output of a n-way splitter (with equal splitting ratio) is $\frac{1}{n}$ times the input power. To be detected, the optical signal power needs to be higher than a threshold; hence, a MWRS may have a limited multicasting capability.

6.3.2   Multicast Routing and Wavelength Assignment.   The multi-casting problem in communication networks is often modeled as the Steiner tree problem in networks (SPN), which is defined as follows. Given

- a graph $G = (V, E)$,

- a cost function $C : E \to R^+$, and

- a set of nodes $D \subset V$,

find a subtree $T = (V_T, E_T)$ which spans D, such that its cost $C(T) = \sum_{e \in E_T} C(e)$ is minimized.

In [Karp, 1972], SPN was shown to be NP-complete. Heuristics are preferred in practice. Mechanisms described in Section 6.1 can be extended to route and assign wavelength to multicast traffic. See also [Sahasrabuddhe and Mukherjee, 1999] on how to set up "light trees" to improve the performance of unicast and broadcast traffic.

6.4.   PHYSICAL-LAYER ISSUES

Optics has many desirable characteristics, but it also possesses some not-so-desirable properties. It is beneficial to correct for several of these "mismatches" using intelligent network algorithms, one example of which is outlined below.

6.4.1   BER-Based Call Admission in Wavelength-Routed Optical Net-works.   In a wavelength-routed optical network, a new call can be admitted if an all-optical lightpath can be established between the call's source and destination nodes. Most previous networking studies have concentrated on the RWA problem to set up lightpaths while assuming an ideal physical layer, e.g., [Barry and Humblet, 1996; Birman, 1996; Ramaswami and Sivarajan, 1995]. It should, however, be noted that a signal degrades in quality due to physical-layer impairments as it proceeds through switches (picking up crosstalk) and EDFAs (picking up amplified spontaneous emission (ASE) noise). As a result, the bit-error rate (BER) at the receiving end of a lightpath may become unacceptably high.

The work in [Ramamurthy et al., 1998a] estimates the *on-line BER* on candidate routes and wavelengths before setting up a call. Thus, one approach would be to set up a call on a lightpath with minimum BER. Another approach would be to establish a call on any lightpath with a BER lower than a certain threshold, e.g., 10^{-12}; if no such lightpath is found, the call is blocked. It is feasible to develop network-layer solutions to combat the physical-layer impairments, including laser shift, dispersion in fiber, and also impairments that affect optical components such as wavelength converters, switch architectures, etc.

Other networking studies which attempt to incorporate physical-layer device characteristics while attempting to solve network-layer problems include (1) amplifier placement in wavelength-routed optical networks [Li et al., 1994; Ramamurthy et al., 1998b; Ramamurthy et al., 1998c] and (2) sparse optical regeneration in nationwide WDM networks [Ramamurthy et al., 1999].

References

Barry, R. A. and Humblet, P. A. (1996). Models of blocking probability in all-optical networks with and without wavelength changers. *IEEE J. Sel. Areas Comm.*, 14:858–867.

Birman, A. (1996). Computing approximate blocking probabilities for a class of all-optical networks. *IEEE J. Sel. Areas Comm.*, 14:852–857.

Cochrane, P. (1995). *Optical Network Technology – Foreword.* Chapman Hall.

Fan, C. (1998). Keynote address – optical networking: A paradigm shift. In *WDM Forum*, London.

Garcia-Luna-Aceves, J. J. (1992). Distributed routing with labeled distances. In *Proc. INFOCOM '92*, Florence, Italy.

Green, P. E. (1996). Optical networking update. *IEEE J. Sel. Areas Comm.*, 8:764–779.

Grover, W. D. (1987). The selfhealing network: A fast distributed restoration technique for networks using digital crossconnect machines. In *Proc. IEEE Globecom '87*, pages 28.2.1–28.2.6, Tokyo, Japan.

JHSN95 (1995). Special Issue on WDM Networks. *Journal of High-Speed Networks*, 4(1/2).

JLT93 (1993). Special Issue on Broad-Band Optical Networks. *IEEE/OSA J. Lightwave Tech.*, 11(5/6).

JLT96 (1996). Special Issue on Broad-Band Optical Networks. *IEEE/OSA J. Lightwave Tech.*, 14(6).

JSAC90 (1990). Special Issue on Dense Wavelength Division Multiplexing Techniques for High Capacity and Multiple Access Communication Systems. *IEEE J. Sel. Areas Comm.*, 8(6).

JSAC96 (1996). Special Issue on Optical Networks. *IEEE J. Sel. Areas Comm.*, 14(5).

JSAC98 (1998). Special Issue on High-Capacity Optical Transport Networks. *IEEE J. Sel. Areas Comm.*, 16(7).

Karp, R. M. (1972). Complexity of computer computations. chapter Reducibility among combinatorial problems, pages 85–104. Plenum Press.

Li, C. S. et al. (1994). Gain equalization in metropolitan and wide area optical networks using optical amplifiers. In *Proc. INFOCOM '94*, pages 130–137, Toronto, Canada.

Melle, S., Pfistner, C. P., and Diner, F. (1995). Amplifier and multiplexing technologies expand network capacity. *Lightwave Magazine*, pages 42–46.

Mukherjee, B. (1997). *Optical Communication Networks*. McGraw-Hill.

Mukherjee, B. (1992a). WDM-based local lightwave networks – part I: Single-hop systems. *IEEE Network Magazine*, 6(3):12–27.

Mukherjee, B. (1992b). WDM-based local lightwave networks – part II: Multihop systems. *IEEE Network Magazine*, 6(4):20–32.

Ramamurthy, B., Datta, D., Feng, H., Heritage, J., and Mukherjee, B. (1998a). Impact of transmission impairments on the teletraffic performance of wavelength routed optical networks. *IEEE/OSA J. Lightwave Tech.*, 17(10):1713–1723.

Ramamurthy, B., Datta, D., Feng, H., Heritage, J. P., and Mukherjee, B. (1999). Transparent vs. opaque vs. translucent wavelength-routed optical networks. In *Proc. OFC '97*, volume 1, pages 59–61, San Diego, CA.

Ramamurthy, B., Iness, J., and Mukherjee, B. (1998b). Optimizing amplifier placements in a multiwavelength optical LAN/MAN: The equally-powered-wavelengths case. *IEEE/OSA J. Lightwave Tech.*, 16(9): 1560–1569.

Ramamurthy, B., Iness, J., and Mukherjee, B. (1998c). Optimizing amplifier placements in a multiwavelength optical LAN/MAN: The unequally-powered-wavelengths case. *IEEE/ACM Trans on Networking*, 6(6):755–767.

Ramamurthy, S. (1998). *Optical Design of WDM Network Architectures*. PhD thesis, University of California, Davis, Department of Computer Science.

Ramamurthy, S. and Mukherjee, B. (1998). Fixed alternate routing and wavelength conversion in wavelength-routed optical networks. In *Proc. IEEE Globecom '98*, pages 2295–2302, Sydney, Australia.

Ramamurthy, S. and Mukherjee, B. (1999a). Survivable WDM mesh networks, Part I – protection. In *Proc. INFOCOM '99*, pages 744–751, New York.

Ramamurthy, S. and Mukherjee, B. (1999b). Survivable WDM mesh networks, Part II – restoration. In *Proc. IEEE International Conference on Communications (ICC '99)*, Vancouver, Canada.

Ramaswami, R. and Segall, A. (1997). Distributed network control for optical networks. *IEEE/ACM Trans on Networking*, 5(6):936–943.

Ramaswami, R. and Sivarajan, K. N. (1995). Optimal routing and wavelength assignment in all-optical networks. *IEEE/ACM Trans on Networking*, 3:489–500.

Sahasrabuddhe, L. H. and Mukherjee, B. (1999). Light-trees: Optical multicasting for improved performance in wavelength-routed networks. *IEEE Communications Magazine*, 37(2):67–73.

Wu, T. (1992). *Fiber Network Service Survivability*. Artech House, Norwood, MA.

Zang, H., Jue, J. P., and Mukherjee, B. (2000). A review of routing and wavelength assignment approaches for wavelength routed optical WDM networks. *Optical Networks Magazine*.

Zang, H., Sahasrabuddhe, L., Jue, J. P., Ramamurthy, S., and Mukherjee, B. (1999). Connection management for wavelength routed WDM networks. In *Proc. IEEE Globecom '99*, Rio de Janeiro, Brazil.

II

OPTICAL NETWORK COMPONENTS

Chapter 2

FIBER, LASERS, RECEIVERS, AND AMPLIFIERS

Byrav Ramamurthy
University of Nebraska-Lincoln
byrav@cse.unl.edu

Jason P. Jue
University of Texas at Dallas
jjue@utdallas.edu

Abstract   The success of optical WDM networks depends heavily on the available optical device technology. This chapter is intended as an introduction to some of the optical device issues in WDM networks. It discusses the basic principles of optical transmission in fiber, and reviews the current state of the art in optical device technology.

1.   INTRODUCTION

Research and development on optical (WDM) networks have matured considerably over the past few years, and a number of experimental prototypes have been deployed and are being tested in the U.S., Europe, and Japan. It is anticipated that the next generation of the Internet will employ WDM-based optical backbones.

Figure 2.1   Block diagram of a WDM transmission system.

The success of WDM networks relies heavily upon the available optical components. A block diagram of a WDM communication system is shown in Fig. 2.1. The network medium may be a simple fiber link, a passive star coupler (for a broadcast-and-select network), or a network of optical or electronic switches and fiber links. The transmitter block consists of one or more optical transmitters, which may be either fixed to a single wavelength, or may be tunable across a range of wavelengths. Each optical transmitter consists of a laser and a laser modulator and may also include an optical filter for tuning purposes. If multiple optical transmitters are used, then a multiplexer or coupler is needed to combine the signals from different laser transmitters onto a single fiber. The receiver block may consist of a tunable filter followed by a photodetector receiver, or a demultiplexer followed by an array of photodetectors. Examples of some WDM transmitters and receivers are shown in Fig. 2.2. Amplifiers

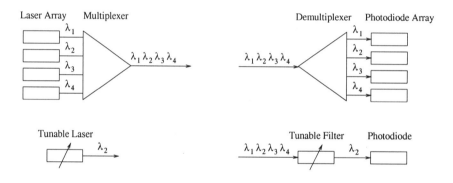

Figure 2.2   Transmitter and receiver structures.

may be required in various locations throughout the network to maintain the strength of optical signals.

This chapter serves as an introduction to WDM device issues. No background in optics or advanced physics is needed to follow the material presented in this chapter. For a more advanced and/or detailed discussion of optical/WDM devices, we refer the interested reader to [Green, 1993; Powers, 1996; Keiser, 1999; Agrawal, 1997; Henry, 1985; Brackett, 1990; Hecht, 1992; Hecht, 1993; Ramaswami, 1993; Borella et al., 1997].

This chapter presents an overview of optical fiber and devices such as couplers, optical transmitters, optical receivers and filters, and optical amplifiers. The chapter attempts to condense the physics behind the principles of optical transmission in fiber in order to provide some background for the reader.

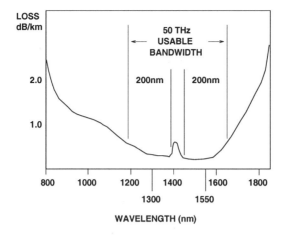

Figure 2.3   The low-attenuation regions of an optical fiber.

2.   OPTICAL FIBER

Fiber possesses many characteristics that make it an excellent physical medium for high-speed networking. Fig. 2.3 shows the two low-attenuation regions of optical fiber [Green, 1993]. Centered at approximately 1300 nm is a range of 200 nm in which attenuation is less than 0.5 dB per kilometer. The total bandwidth in this region is about 25 THz. Centered at 1550 nm is a region of similar size, with attenuation as low as 0.2 dB per kilometer. Combined, these two regions provide a theoretical upper bound of 50 THz of bandwidth[1]. The dominant loss mechanism in good fibers is Rayleigh scattering, while the peak in loss in the 1400 nm region is due to hydroxyl ion (OH^-) impurities in the fiber. Other sources of loss include material absorption and radiative loss.

By using these large low-attenuation areas for data transmission, the signal loss for a set of one or more wavelengths can be made very small, thus reducing the number of amplifiers and repeaters needed. In single-channel long-distance experiments, optical signals have been sent over hundreds of kilometers without amplification. Besides its enormous bandwidth and low attenuation, fiber also offers low error rates. Fiber optic systems typically operate at bit error rates (BERs) of less than 10^{-11}.

The small size and thickness of fiber allows more fiber to occupy the same physical space as copper, a property which is desirable when installing local networks in buildings. Fiber is flexible, difficult to break, reliable in corrosive environments, and deployable at short notice (which makes it particularly favorable for military communication systems). Also, fiber transmission is immune to electro-magnetic interference, and does not cause interference. Finally, fiber is made from one of the cheapest and most readily available substances on earth, viz. sand. This makes fiber environmentally sound, and unlike copper, its use will not deplete natural resources.

2.1.   OPTICAL TRANSMISSION IN FIBER

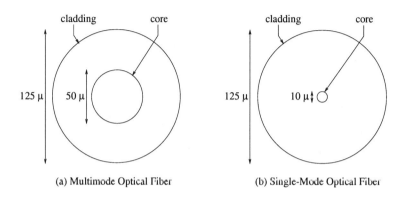

(a) Multimode Optical Fiber   (b) Single-Mode Optical Fiber

Figure 2.4   Multimode and single-mode optical fibers.

Before discussing optical components, it is essential to understand the characteristics of the optical fiber itself. Fiber is essentially a thin filament of glass which acts as a waveguide. A waveguide is a physical medium or a path which allows the propagation of electromagnetic waves, such as light. Due to the physical phenomenon of *total internal reflection*, light can propagate the length of a fiber with little loss. Fig. 2.4 shows the cross-section of the two types of fiber most commonly used: multimode and single mode. In order to understand the concept of a mode and to distinguish between these two types of fiber, a diversion into basic optics is needed.

Light travels through vacuum at a speed of $c_{vac} = 3 \times 10^8$ m/s. Light can also travel through any transparent material, but the speed of light will be slower in the material than in a vacuum. Let c_{mat} be the speed of light for a given material. The ratio of the speed of light in vacuum to that in a material is known as the material's *refractive index* (n), and is given by $n_{mat} = \frac{c_{vac}}{c_{mat}}$.

When light travels from material of a given refractive index to material of a different refractive index (i.e., when refraction occurs), the angle at which the light is transmitted in the second material depends on the refractive indices of the two materials as well as the angle at which light strikes the interface between the two materials. Due to Snell's Law, we have $n_a sin\theta_a = n_b sin\theta_b$, where n_a and n_b are the refractive indices of the first substance and the second substance, respectively; θ_a is the angle of incidence, or the angle with respect to normal that light hits the surface between the two materials; and θ_b is the angle of light in the second material. However, if $n_a > n_b$ and θ_a is greater than some critical value, the rays are reflected back into substance *a* from its boundary with substance *b*.

2.2. MULTIMODE VS. SINGLE-MODE FIBER

A mode in an optical fiber corresponds to one of possibly multiple ways in which a wave may propagate through the fiber. It can also be viewed as a standing wave in the transverse plane of the fiber. More formally, a mode corresponds to a solution of the wave equation which is derived from Maxwell's equations and subject to boundary conditions imposed by the optical fiber waveguide.

The advantage of multimode fiber is that its core diameter is relatively large; as a result, injection of light into the fiber with low coupling loss[2] can be accomplished by using inexpensive, large-area light sources, such as light-emitting diodes (LEDs).

The disadvantage of multimode fiber is that it introduces the phenomenon of *intermodal dispersion*. In multimode fiber, each mode propagates at a different velocity due to different angles of incidence at the core-cladding boundary. This effect causes different rays of light from the same source to arrive at the other end of the fiber at different times, resulting in a pulse which is spread out in the time domain. Intermodal dispersion increases with the distance of propagation. The effect of intermodal dispersion may be reduced through the use of *graded-index* fiber, in which the region between the cladding and the core of the fiber consists of a series of gradual changes in the index of refraction. However, even with graded-index multimode fiber, intermodal dispersion may still limit the bit rate of the transmitted signal and may limit the distance that the signal can travel.

One way to limit intermodal dispersion is to reduce the number of modes. This reduction in the number of modes can be accomplished by reducing the core diameter, by reducing the numerical aperture (given by the term $\sqrt{n_{core}^2 - n_{clad}^2}$), or by increasing the wavelength of the light.

Single-mode fiber eliminates intermodal dispersion, and can, hence, support transmission over much longer distances. However, it introduces the problem of concentrating enough power into a very small core. LEDs cannot couple enough light into a single-mode fiber to facilitate long distance communications. Such a high concentration of light energy may be provided by a semiconductor laser, which can generate a narrow beam of light.

2.3. ATTENUATION IN FIBER

Attenuation in optical fiber leads to a reduction of the signal power as the signal propagates over some distance. When determining the maximum distance that a signal can propagate for a given transmitter power and receiver sensitivity, one must consider attenuation. Let $P(L)$ be the power of the optical pulse at distance L km from the transmitter and A be the attenuation constant of the fiber (in dB/km). Attenuation is characterized by [Henry, 1985]

$$P(L) = 10^{-AL/10}P(0) \tag{2.1}$$

where $P(0)$ is the optical power at the transmitter. For a link length of L km, $P(L)$ must be greater than or equal to P_r, the receiver sensitivity. From Equation (2.1), we get

$$L_{max} = \frac{10}{A} \log_{10} \frac{P(0)}{P_r} \tag{2.2}$$

The maximum distance between the transmitter and the receiver (or the distance between amplifiers[3]) depends more heavily on the constant A than on the optical power launched by the transmitter. Referring back to Fig. 2.3, we note that the lowest attenuation occurs at approximately 1550 nm.

2.4.   DISPERSION IN FIBER

Dispersion is the widening of a pulse duration as it travels through a fiber. As a pulse widens, it can broaden enough to interfere with neighboring pulses (bits) on the fiber, leading to intersymbol interference. Dispersion thus limits the bit spacing and the maximum transmission rate on a fiber-optic channel.

As mentioned earlier, one form of dispersion is *intermodal dispersion*. This is caused when multiple modes of the same signal propagate at different velocities along the fiber. Intermodal dispersion does not occur in a single-mode fiber.

Another form of dispersion is *material* or *chromatic dispersion*. In a dispersive medium, the index of refraction is a function of the wavelength. Thus, if the transmitted signal consists of more than one wavelength, certain wavelengths will propagate faster than other wavelengths. Since no laser can create a signal consisting of an exact single wavelength, material dispersion will occur in most systems[4].

A third type of dispersion is *waveguide dispersion*. Waveguide dispersion is caused because the propagation of different wavelengths depends on waveguide characteristics such as the indices and shape of the fiber core and cladding.

At 1300 nm, material dispersion in a conventional single-mode fiber is near zero. Luckily, this is also a low-attenuation window (although loss is lower at 1550 nm). Through advanced techniques such as *dispersion shifting*, fibers with zero dispersion at a wavelength between 1300 nm and 1700 nm can be manufactured [Powers, 1996]. In a dispersion-shifted fiber, the core and cladding are designed such that the waveguide dispersion is negative with respect to the material dispersion, thus canceling the total dispersion. However, the dispersion will only be zero for a single wavelength.

2.5.   NONLINEARITIES IN FIBER

Nonlinear effects in fiber may potentially have a significant impact on the performance of WDM optical communication systems. Nonlinearities in fiber may lead to attenuation, distortion, and cross-channel interference. In a WDM system, these effects place constraints on the spacing between adjacent wavelength channels, limit the maximum power on any channel, and may also limit the maximum bit rate. Such effects include self-phase modulation (SPM), cross-phase modulation (XPM), stimulated Raman scattering (SRS), stimulated Brillouin scattering (SBS), and four-wave mixing (FWM).

It is shown that, in a WDM system using channels spaced 10 GHz apart and a transmitter power of 0.1 mW per channel, a maximum of about 100 channels can be obtained in the 1550 nm low-attenuation region [Chraplyvy, 1990].

The details of optical nonlinearities are very complex, and beyond the scope of this chapter. However, they are a major limiting factor in the available number channels in a WDM system, especially those operating over distances greater than 30 km [Chraplyvy, 1990]. The existence of these nonlinearities suggests that WDM protocols which limit the number of nodes to the number of channels do not scale well. For further details on fiber nonlinearities, the reader is referred to [Agrawal, 1989].

2.6.   COUPLERS

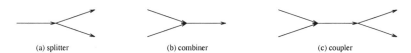

(a) splitter   (b) combiner   (c) coupler

Figure 2.5   **Splitter, combiner, and coupler.**

A coupler is a general term that covers all devices that combine light into or split light out of a fiber. A splitter is a coupler that divides the optical signal on one fiber to two or more fibers. The most common splitter is a 1×2 splitter, as shown in Fig. 2.5(a). The *splitting ratio*, α, is the fraction of input power that goes to each output. For a two-port splitter, the most common splitting ratio is 50:50, though splitters with any ratio can be manufactured [Powers, 1996]. Combiners (see Fig. 2.5(b)) are the reverse of splitters, and when turned around, a combiner can be used as a splitter. An input signal to the combiner suffers a power loss of about 3 dB. A 2×2 coupler (see Fig. 2.5(c)), in general, is a 2×1 combiner followed immediately by a 1×2 splitter, which has the effect of broadcasting the signals from two input fibers onto two output fibers. One implementation of a 2×2 coupler is the *fused biconical tapered coupler* which basically consists of two fibers fused together. In addition to the 50:50 power split incurred in a coupler, a signal also experiences *return loss*. If the signal enters an input of the coupler, roughly half of the signal's power goes to each output of the coupler. However, a small amount of power is reflected in the opposite direction and is directed back to the inputs of the coupler. Typically, the amount of power returned by a coupler is 40-50 dB below the input power. Another type of loss is *insertion loss*. One source of insertion loss is the loss incurred when directing the light from a fiber into the coupler device; ideally, the axes of the fiber core and the coupler input port must be perfectly aligned, but full perfection may not be achievable due to the very small dimensions.

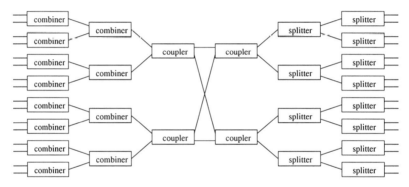

Figure 2.6   **A 16x16 passive-star coupler.**

The passive-star coupler (PSC) is a multiport device in which light coming into any input port is broadcast to every output port. The PSC is attractive because the optical power that each output receives P_{out} equals

$$P_{out} = \frac{P_{in}}{N} \qquad (2.3)$$

where P_{in} is the optical power introduced into the star by a single node and N is the number of output ports of the star. Note that this expression ignores the *excess loss*, caused by flaws introduced in the manufacturing process, that the signal experiences when passing through each coupling element. One way to implement the PSC is to use a combination of splitters, combiners, and couplers as shown in Fig. 2.6. Another implementation of the star coupler is the integrated-optics planar star coupler in which the star coupler and waveguides are fabricated on a semiconductor, glass (silica), or polymer substrate. A 19×19 star coupler on silicon has been demonstrated with excess loss of around 3.5 dB at a wavelength of 1300 nm [Dragone et al., 1989]. In [Okamoto et al.,

1991], an 8 × 8 star coupler with an excess loss of 1.6 dB at a wavelength of 1550 nm was demonstrated.

3.   OPTICAL TRANSMITTERS

In order to understand how a tunable optical transmitter works, we must first understand some of the fundamental principles of lasers and how they work. Good references on tunable laser technology include [Green, 1993] [Brackett, 1990] [Lee and Zah, 1989].

The word laser is an acronym for *Light Amplification by Stimulated Emission of Radiation.* The key words are stimulated emission, which is what allows a laser to produce intense high-powered beams of coherent light (light which contains one or more distinct frequencies).

3.1.   SEMICONDUCTOR DIODE LASERS

The most useful type of laser for optical networks is the semiconductor diode laser. The simplest implementation of a semiconductor laser is the bulk laser diode, which is a p-n junction with mirrored edges perpendicular to the junction (see Fig. 2.7). A detailed description of the operation of a semiconductor diode laser can be found in [Borella et al., 1997].

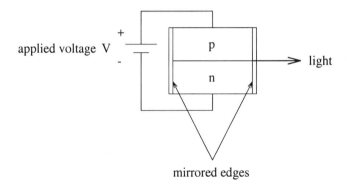

Figure 2.7   Structure of a semiconductor diode laser.

An improvement over the bulk laser diode is the multiple-quantum-well (MQW) laser. Quantum wells are thin alternating layers of semiconductor materials. The alternating layers create potential barriers in the semiconductors which confine the position of electrons and holes to a smaller number of energy states. The quantum wells are placed in the region of the p-n junction. By confining the possible states of the electrons and holes, it is possible to achieve higher-resolution, low-linewidth lasers (lasers which generate light with a very narrow frequency range).

3.2.   TUNABLE AND FIXED LASERS

While the previous section provided an overview of a generic model of a laser, the transmitters used in WDM networks often require the capability to tune to different wavelengths. This section briefly describes some of the more popular, tunable and fixed, single-frequency laser designs.

3.2.1   Laser Characteristics.   Some of the physical characteristics of lasers which may affect system performance are laser linewidth, frequency stability, and the number of longitudinal modes.

The laser linewidth is the spectral width of the light generated by the laser. The linewidth affects the spacing of channels and also affects the amount of dispersion that occurs when the light is propagating along a fiber.

Frequency instabilities in lasers are variations in the laser frequency. Three such instabilities are mode hopping, mode shifts, and wavelength chirp [Moore and Todd, 1993]. Mode hopping occurs primarily in injection-current lasers and is a sudden jump in the laser frequency caused by a change in the injection current above a given threshold. Mode shifts are changes in frequency due to temperature changes. Wavelength chirp is a variation in the frequency due to variations in injection current. In WDM systems, frequency instabilities may limit the placement and spacing of channels.

The number of longitudinal modes in a laser is the number of wavelengths that are amplified by the laser. The unwanted longitudinal modes produced by a laser may result in significant dispersion; therefore, it is desirable to implement lasers which produce only a single longitudinal mode.

Some primary characteristics of interest for tunable lasers are the *tuning range*, the *tuning time*, and whether the laser is continuously tunable (over its tuning range) or discretely tunable (only to selected wavelengths). The tuning range refers to the range of wavelengths over which the laser may be operated. The tuning time is the time required for the laser to tune from one wavelength to another.

3.2.2   Mechanically-Tuned Lasers.   Most mechanically-tuned lasers use a Fabry-Perot cavity that is adjacent to the lasing medium (i.e. an *external cavity*) to filter out unwanted wavelengths. Tuning is accomplished by physically adjusting the distance between two mirrors on either end of the cavity such that only the desired wavelength constructively interferes with its multiple reflections in the cavity. This approach to tuning results in a tuning range that encompasses the entire useful gain spectrum of the semiconductor laser [Brackett, 1990], but tuning time is limited to the order of milliseconds due to the mechanical nature of the tuning and the length of the cavity. The length of the cavity may also limit transmission rates unless an external modulator is used. External cavity lasers tend to have very good frequency stability.

3.2.3   Acoustooptically- and Electrooptically-Tuned lasers.   Other types of tunable lasers which use external tunable filters include acousto-optically- and electrooptically-tuned lasers. In an acoustooptic or electrooptic laser, the index of refraction in the external cavity is changed by using either sound waves or electrical current, respectively. The change in the index results in the transmission of light at different frequencies. In these types of tunable lasers, the tuning time is limited by the time required for light to build up in the cavity at the new frequency.

3.2.4   Injection-Current-Tuned Lasers.   Injection-current-tuned lasers form a family of transmitters which allow wavelength selection via a diffraction grating. The Distributed Feedback (DFB) laser uses a diffraction grating placed in the lasing medium. In general, the grating consists of a waveguide in which the index of refraction alternates periodically between two values. Only wavelengths which match the period and indices of the grating will be constructively reinforced. All other wavelengths will destructively interfere, and will not propagate through the waveguide. The condition for propagation is given by:

$$D = \lambda/2n$$

where D is the period of the grating [Moore and Todd, 1993]. The laser is tuned by injecting a current which changes the index of the grating region.

If the grating is moved to the outside of the lasing medium, the laser is called a Distributed Bragg Reflector (DBR) laser. The tuning in a DBR laser is discrete rather than continuous, and tuning times of less than 10 ns have been measured [Brackett, 1990].

3.2.5   Laser Arrays.   An alternative to tunable lasers is the laser array, which contains a set of fixed-tuned lasers and its advantages and applications are explained below. A laser array consists of a number of lasers which are integrated into a single component, with each laser operating at a different wavelength. The advantage of using a laser array is that, if each of the wavelengths in the array is modulated independently, then multiple transmissions may take place simultaneously. The drawback is that the number of available wavelengths in a laser array is fixed and is currently limited to about 20 wavelengths today. Laser arrays with up to 21 wavelengths have been demonstrated in the laboratory [Zah et al., 1992], while a laser array with four wavelengths has actually been deployed in a network prototype [Lee et al., 1996].

Table 2.1   Tunable optical transmitters and their associated tuning ranges and times.

Tunable Transmitter	Approx. Tuning Range (nm)	Tuning Time
Mechanical (external cavity)	500	1-10 ms
Acoustooptic	83	10 μs
Electrooptic	7	1-10 ns (estimated)
Injection-Current (DFB and DBR)	10	1-10 ns

3.3.   SUMMARY

Table 2.1 summarizes the characteristics of the different types of tunable transmitters. We observe that there is a tradeoff between the tuning range of a transmitter and its tuning time.

4.   OPTICAL RECEIVERS AND FILTERS

Tunable optical filter technology is a key in making WDM networks realizable. Good sources of information on these devices include [Green, 1993] [Brackett, 1990] [Kobrinski and Cheung, 1989].

4.1.   PHOTODETECTORS

In receivers employing *direct detection*, a photodetector converts the incoming photonic stream into a stream of electrons. The electron stream is then amplified and passed through a threshold device. Whether a bit is a logical 0 or 1 depends on whether the stream is above or below a certain threshold for a bit duration. In other words, the decision is made based on whether or not light is present during the bit duration.

The basic detection devices for direct detection optical networks are the PN photodiode (a p-n junction) and the PIN photodiode (an intrinsic material is placed between "p" and "n" type material). In its simplest form, the photodiode is basically a reverse-biased p-n junction. Through the photoelectric effect, light incident on the junction will create electron-hole pairs in both the "n" and the "p" regions of the photodiode. The electrons released in the "p" region will cross over to the "n" region, and the holes created in the "n" region will cross over to the "p" region, thereby resulting in a current flow.

The alternative to direct detection is *coherent detection* in which phase information is used in the encoding and detection of signals. Coherent-detection-based receivers use a local monochromatic laser as an oscillator. The incoming optical stream is added to the signal from the oscillator, and the resulting signal is detected by a photodiode. The photodiode output is integrated over the symbol duration and a detection threshold is used to obtain the bit stream. While coherent detection is more elaborate than direct detection, it allows the reception of weak signals in a noisy background. However, in optical systems, it is difficult to maintain the phase information required for coherent detection. Since semiconductor lasers have non-zero linewidths, the transmitted signal consists of a number of frequencies with varying phases and amplitudes. The effect is that the phase of the transmitted signal experiences random but significant fluctuations around the desired phase. These phase fluctuations make it difficult to recover the original phase information from the transmitted signal, thus limiting the performance of coherent detection systems.

4.2.   TUNABLE OPTICAL FILTERS

This section discusses several types of tunable optical filters and the properties of each type, while Section 4.3 examines fixed-tuned optical filters. The feasibility of many local WDM networks is dependent upon the speed and range of tunable filters. Overviews of tunable filter technology can be found in [Green, 1993] and [Brackett, 1990].

4.2.1   Filter Characteristics.   Tunable optical filters are characterized primarily by their tuning range and tuning time. The tuning range specifies the range of wavelengths which can be accessed by a filter. A wide tuning range allows systems to utilize a greater number of channels. The tuning time of a filter specifies the time required to tune from one wavelength to another. Fast tunable filters are required for many WDM local area networks (LANs) based on broadcast-and-select architectures.

4.2.2   The Etalon.   The etalon consists of a single cavity formed by two parallel mirrors. Light from an input fiber enters the cavity and reflects a number of times between the mirrors. By adjusting the distance between the mirrors, a single wavelength can be chosen to propagate through the cavity, while the remaining wavelengths destructively interfere. The distance between the mirrors may be adjusted mechanically by physically moving the mirrors, or may be adjusted by changing the index of the material within the cavity. Many modifications (e.g., multi-cavity and multi-pass) to the etalon can be made to improve the number of resolvable channels [Brackett, 1990]. In a multi-pass filter, the light passes through the same cavity multiple times, while in a multi-cavity filter, multiple etalons are cascaded to effectively increase the width of the filter's transfer function (or finesse). An example of a mechanically-tuned etalon is the Fabry-Perot filter. In [Humblet and Hamdy, 1990], it was found that the maximum number of channels for a single-cavity Fabry-Perot filter is 0.65F, where F is the finesse. A two-pass filter was found to have a maximum of 1.4F channels, and two-cavity filters up to $0.44F^2$ channels.

While the Fabry-Perot etalon can be made to access virtually the entire low-attenuation region of the fiber and can resolve very narrow passbands, it has a tuning time on the order of tens of milliseconds, due to its mechanical tuning. This makes it unsuitable for many packet-switched applications in which the packet duration is much smaller than the tuning time. The Fabry-Perot etalon is used as a tunable receiver in the RAINBOW optical WDM local area network prototype [Dono et al., 1990].

4.2.3   The Mach-Zehnder Chain.   In a Mach-Zehnder (MZ) interferometer, a splitter splits the incoming wave into two waveguides, and a combiner recombines the signals at the outputs of the waveguides (see Fig. 2.8). An adjustable delay element controls the optical path length in one of the waveguides, resulting in a phase difference between the two signals when they are recombined. Wavelengths for which the phase difference is 180° are filtered out. By constructing a chain of these elements, a single desired optical wavelength can be selected.

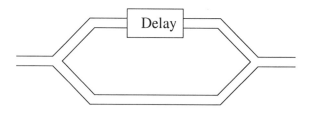

Figure 2.8   Structure of a Mach-Zehnder interferometer.

4.2.4   Acoustooptic Filters.   A fast tuning time is obtained when acoustooptic filters are used. Radio frequency (RF) waves are passed through a transducer. The transducer is a piezoelectric crystal that converts sound waves to mechanical movement. The sound waves change the crystal's index of refraction, which enables the crystal to act as a grating. Light incident upon the transducer will diffract at an angle that depends on the angle of incidence and the wavelength of the light [Green, 1993]. By changing the RF waves, a single optical wavelength can be chosen to pass through the material while the rest of the wavelengths destructively interfere.

The tuning time of the acoustooptic filter is limited by the flight time of the surface acoustic wave to about 10 μs [Brackett, 1990]. However, the tuning range for acoustooptic filters covers the entire 1300 nm to 1560 nm

Table 2.2 Tunable optical filters and their associated tuning ranges and times.

Tunable Receiver	Approx. Tuning Range (nm)	Tuning Time
Fabry-Perot	500	1-10 ms
Acoustooptic	250	10 μs
Electrooptic	16	1-10 ns
LC Fabry-Perot	30	0.5 - 10 μs

spectrum [Brackett, 1990]. This tuning range may potentially allow about 100 channels.

One drawback of acoustooptic filters is that, because of their wide transfer function, they are unable to filter out crosstalk from adjacent channels if the channels are closely spaced. Therefore, the use of acoustooptic filters in a multiwavelength system places a constraint on the channel spacing, thus limiting the allowable number of channels.

4.2.5   Electrooptic Filters.   Since the tuning time of the acoustooptic filter is limited by the speed of sound, crystals whose indices of refraction can be changed by electrical currents can be used. Electrodes, which rest in the crystal, are used to supply current to the crystal. The current changes the crystal's index of refraction, which allows some wavelengths to pass through while others destructively interfere [Green, 1993]. Since the tuning time is limited only by the speed of electronics, tuning time can be on the order of several nanoseconds, but the tuning range (and thus, the number of resolvable channels) remains quite small at 16 nm (on the order of 10 channels) [Brackett, 1990].

4.2.6   Liquid-Crystal Fabry-Perot Filters.   Liquid-crystal filters appear to be a promising new filter technology. The design of a liquid-crystal filter is similar to the design of a Fabry-Perot filter, but the cavity consists of a liquid crystal (LC). The refractive index of the LC is modulated by an electrical current to filter out a desired wavelength, as in an electrooptic filter. Tuning time is on the order of microseconds (sub-microsecond times are expected to be achievable), and tuning range is 30-40 nm [Sneh and Johnson, 1994]. These filters have low power requirements and are inexpensive to fabricate. The filter speed of LC filter technology promises to be high enough to handle high-speed packet switching in broadcast-and-select WDM networks.

4.2.7   Summary.   Table 2.2 summarizes the state of the art in tunable receivers.   As has been stated earlier, tuning range and tuning time seem inversely proportional, except for LC Fabry-Perot filters.

4.3.   FIXED FILTERS

An alternative to tunable filters is to use fixed filters or grating devices. Grating devices typically filter out one or more different wavelength signals from a single fiber. Such devices may be used to implement optical multiplexers and demultiplexers or receiver arrays.

4.3.1   Grating Filters.   One implementation of a fixed filter is the diffraction grating. The diffraction grating is essentially a flat layer of transparent material (e.g., glass or plastic) with a row of parallel grooves cut into it [Hecht, 1992]. The grating separates light into its component wavelengths by reflecting light incident with the grooves at all angles. At certain angles, only one wavelength adds constructively; all others destructively interfere. This allows us to select the wavelength(s) we want by placing a filter tuned to the proper wavelength at the proper angle. Alternatively, some gratings are transmissive rather than reflective and are used in tunable lasers (see DFB lasers in Section 3.2.4).

An alternative implementation of a demultiplexer is the *waveguide grating router* (WGR) in which only one input is utilized. WGRs will be discussed in the next chapter.

4.3.2   Fiber Bragg Gratings.   In a fiber Bragg grating, a periodical variation of the index of refraction is directly photo-induced in the core of an optical fiber. A Bragg grating will reflect a given wavelength of light back to the source while passing the other wavelengths. Two primary characteristics of a Bragg grating are the reflectivity and the spectral bandwidth. Typical spectral bandwidths are on the order of 0.1 nm, while a reflectivity in excess of 99% is achievable [Inoue et al., 1995]. While inducing a grating directly into the core of a fiber leads to low insertion loss, a drawback of Bragg gratings is that the refractive index in the grating varies with temperature, with increases in temperature resulting in longer wavelengths being reflected. An approach for compensating for temperature variations is presented in [Arya et al., 1996]. Fiber Bragg gratings may be used in the implementation of multiplexers, demultiplexers, and tunable filters.

4.3.3   Thin-Film Interference Filters.   Thin-film interference filters offer another approach for filtering out one or more wavelengths from a number of wavelengths. These filters are similar to fiber Bragg grating devices with the exception that they are fabricated by depositing alternating layers of low index and high index materials onto a substrate layer. Thin-film filter technology suffers from poor thermal stability, high insertion loss, and poor spectral profile. However, advances have been made which address some of these issues [Scobey and Spock, 1996].

5.   OPTICAL AMPLIFIERS

Although an optical signal can propagate a long distance before it needs amplification, both long-haul and local lightwave networks can benefit from optical amplifiers. All-optical amplification may differ from opto-electronic amplification in that it may act only to boost the power of a signal, not to restore the shape or timing of the signal. This type of amplification is known as 1R (regeneration), and it provides total data transparency (the amplification process is independent of the signal's modulation format). 1R-amplification is emerging as the choice for transparent all-optical networks of tomorrow. However, in today's digital networks (e.g., Synchronous Optical Network (SONET) and Synchronous Digital Hierarchy (SDH)), which use the optical fiber only as a transmission medium, the optical signals are amplified by first converting the information stream into an electronic data signal, and then retransmitting the signal optically. Such amplification is referred to as 3R (regeneration, reshaping, and reclocking). The reshaping of the signal reproduces the original pulse shape of each bit, eliminating much of the noise. Reshaping applies primarily to digitally-modulated signals, but in some cases may also be applied to analog signals. The reclocking of the signal synchronizes the signal to its original bit timing pattern and bit rate. Reclocking applies only to digitally-modulated signals. Another approach to amplification is 2R (regeneration and reshaping), in which the optical signal is converted to an electronic signal which is then used to directly modulate a laser. 3R and 2R techniques provide less transparency than the 1R technique; and in future optical networks, the aggregate bit rate of even just a few channels might make 3R and 2R techniques less practical.

Also, in a WDM system, each wavelength would need to be separated before being amplified electronically, and then recombined before being retransmitted. Thus, in order to eliminate the need for optical multiplexers and demultiplexers in amplifiers, optical amplifiers must boost the strength of optical signals without first converting them to electrical signals. A drawback is that noise, as well as the signal, will be amplified.

Optical amplification uses the principle of stimulated emission, similar to the approach used in a laser. The two basic type of optical amplifiers are semiconductor laser amplifiers and rare-earth-doped-fiber amplifiers, which will be discussed in the following sections. A general overview of optical amplifiers can be found in [O'Mahony, 1993].

5.1.   AMPLIFIER CHARACTERISTICS

Some basic parameters of interest in an optical amplifier are gain, gain bandwidth, gain saturation, polarization sensitivity, and amplifier noise.

Gain measures the ratio of the output power of a signal to its input power. Amplifiers are sometimes also characterized by *gain efficiency* which measures the gain as a function of input power in dB/mW.

The *gain bandwidth* of an amplifier refers to the range of frequencies or wavelengths over which the amplifier is effective. In a network, the gain bandwidth limits the number of wavelengths available for a given channel spacing.

The *gain saturation* point of an amplifier is the value of output power at which the output power no longer increases with an increase in the input power. When the input power is increased beyond a certain value, the carriers (electrons) in the amplifier are unable to output any additional light energy. The saturation power is typically defined as the output power at which there is a 3-dB reduction in the amplifier gain from the small-signal gain value.

Polarization sensitivity refers to the dependence of the gain on the polarization of the signal. The sensitivity is measured in dB and refers to the gain difference between the transverse-electric (TE) and transverse-magnetic (TM) polarizations.

In optical amplifiers, the dominant source of *noise* is *amplified spontaneous emission* (ASE), which arises from the spontaneous emission of photons in the active region of the amplifier (see Fig. 2.9). The amount of noise generated by the amplifier depends on factors such as the amplifier gain spectrum, the noise bandwidth, and the *population inversion parameter* which specifies the degree of population inversion that has been achieved between two energy levels. Amplifier noise is especially a problem when multiple amplifiers are *cascaded*. Each subsequent amplifier in the cascade amplifies the noise generated by previous amplifiers.

5.2.   SEMICONDUCTOR-LASER AMPLIFIER

A semiconductor laser amplifier (see Fig. 2.9) consists of a modified semiconductor laser. A weak signal is sent through the active region of the semiconductor, which, via stimulated emission, results in a stronger signal being emitted from the semiconductor.

The two basic types of semiconductor laser amplifiers are the Fabry-Perot amplifier, which is basically a semiconductor laser, and the traveling-wave amplifier (TWA). The primary difference between the two is in the reflectivity of the end mirrors. Fabry-Perot amplifiers have a reflectivity of around 30%, while TWAs have a reflectivity of around 0.01% [O'Mahony, 1993]. In order to prevent lasing in the Fabry-Perot amplifier, the bias current

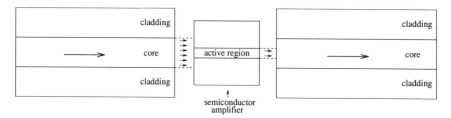

Figure 2.9   A semiconductor optical amplifier.

is operated below the lasing threshold current. The higher reflections in the Fabry-Perot amplifier cause Fabry-Perot resonances in the amplifier, resulting in narrow passbands of around 5 GHz. This phenomenon is not very desirable for WDM systems; therefore, by reducing the reflectivity, the amplification is performed in a single pass and no resonances occur. Thus, TWAs are more appropriate than Fabry-Perot amplifiers for WDM networks.

Today's semiconductor amplifiers can achieve gains of 25 dB with a gain saturation of 10 dBm, polarization sensitivity of 1 dB, and bandwidth range of 40 nm [O'Mahony, 1993].

Semiconductor amplifiers based on multiple quantum wells (MQW) are currently being studied. These amplifiers have higher bandwidth and higher gain saturation than bulk devices. They also provide faster on-off switching times. The disadvantage is a higher polarization sensitivity.

An advantage of semiconductor amplifiers is the ability to integrate them with other components. For example, they can be used as gate elements in switches. By turning a drive current on and off, the amplifier basically acts like a gate, either blocking or amplifying the signal.

5.3.   DOPED-FIBER AMPLIFIER

Optical doped-fiber amplifiers are lengths of fiber doped with an element (rare earth) which can amplify light (see Fig. 2.10). The most common doping element is erbium, which provides gain for wavelengths between 1525 nm and 1560 nm. At the end of the length of fiber, a laser transmits a strong signal at a lower wavelength (referred to as the *pump wavelength*) back up the fiber. This pump signal excites the dopant atoms into a higher energy level. This allows the data signal to stimulate the excited atoms to release photons. Most erbium-doped fiber amplifiers (EDFAs) are pumped by lasers with a wavelength of either 980 nm or 1480 nm. The 980 nm pump wavelength has shown gain efficiencies of around 10 dB/mW, while the 1480 nm pump wavelength provides efficiencies of around 5 dB/mW. Typical gains are on the order of 25 dB. Experimentally, EDFAs have been shown to achieve gains of up to 51 dB with the maximum gain limited by internal Rayleigh backscattering in which some of the light energy of the signal is scattered in the fiber and directed back towards the signal source [Hansen et al., 1992]. The 3-dB gain bandwidth for the EDFA is around 35 nm (see Fig. 2.11), and the gain saturation power is around 10 dBm [O'Mahony, 1993].

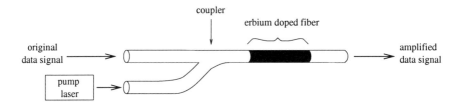

Figure 2.10   Erbium-doped fiber amplifier.

For the 1300 nm region, the praseodymium-doped fluoride fiber amplifier (PDFFA) has recently been receiving attention. These amplifiers have low cross-talk and noise characteristics, while attaining high gains. They are able to operate over a range of around 50 nm in the 1280 nm to 1330 nm range. In [Yamada et al., 1995], a PDFFA was developed which had a 40.6 dB gain. Recent developments on PDFFAs are presented in [Whitley, 1995].

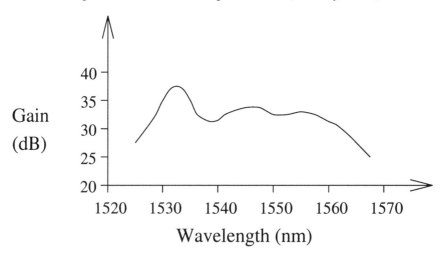

Figure 2.11   The gain spectrum of an erbium-doped fiber amplifier with input power = -40 dBm.

A limitation to optical amplification is the unequal gain spectrum of optical amplifiers. The EDFA gain spectrum is shown in Fig. 2.11 (from [Sam, 1997]). While an optical amplifier may provide gain across a range of wavelengths, it will not necessarily amplify all wavelengths equally. This characteristic, accompanied by the fact that optical amplifiers amplify noise as well as signal, and the fact that the active region of the amplifier can spontaneously emit

Table 2.3 Amplifier characteristics.

Amplifier Type	Gain Region	Gain Bandwidth	Gain
Semiconductor	Any	40 nm	25 dB
EDFA	1525-1560 nm	35 nm	25-51 dB
PDFFA	1280-1330 nm	50 nm	20-40 dB

photons which also cause noise, limits the performance of optical amplifiers. Thus, a multiwavelength optical signal passing through a series of amplifiers will eventually result in the power of the wavelengths being uneven.

A number of approaches to equalizing the gain of an EDFA have been studied. In [Tachibana et al., 1991], a notch filter (a filter which attenuates the signal at a selected frequency) centered at around 1530 nm is used to suppress the peak in the EDFA gain (see Fig. 2.11). However, when multiple EDFAs are cascaded, another peak appears around the 1560 nm wavelength. In [Wilner and Hwang, 1993], a notch filter centered at 1560 nm is used to equalize the gain for a cascade of EDFAs. Another approach to flattening the gain is to adjust the input transmitter power such that the powers on all received wavelengths at the destination are equal [Chraplyvy et al., 1992]. A third approach to gain equalization is to demultiplex the individual wavelengths and then attenuate selected wavelengths such that all wavelengths have equal power. In [Elrefaie et al., 1993], this approach is applied to a WDM interoffice ring network.

6.   CONCLUSION

Recent advances in the field of optics have paved the way for the practical implementation of WDM networks. In this chapter, we have provided a brief overview of some of the optical WDM devices currently available or under development, as well as some insight into the underlying technology.

Acknowledgments

The authors thank Dr. Michael S. Borella, Dr. Dhritiman Banerjee, and Prof. Biswanath Mukherjee for their contributions to the material presented in this chapter.

Notes

1. However, usable bandwidth is limited by fiber nonlinearities (see Section 2.5).

2. Coupling loss measures the power loss experienced when attempting to direct light into a fiber.

3. The amplifier sensitivity is usually equal to the receiver sensitivity, while the amplifier output is usually equal to optical power at a transmitter.

4. Even if an unmodulated source consisted of a single wavelength, the process of modulation would cause a spread of wavelengths.

References

(1997). Single forward pumping EDFA,. EDFA Datasheet, Samsung Electronics Co. Ltd.

Agrawal, G. P. (1989). *Nonlinear Fiber Optics*. Academic Press, Inc.

Agrawal, G. P. (1997). *Fiber-Optic Communication Systems*. New York: Wiley, 2nd edition.

Arya, V., Sherrer, D. W., Wang, A., Claus, R. O., and Jones, M. (1996). Temperature compensation scheme for refractive index grating-based optical fiber devices. In *Proceedings of the SPIE*, volume 2594, pages 52–59.

Borella, M. S., Jue, J. P., Banerjee, D., Ramamurthy, B., and Mukherjee, B. (1997). Optical components for WDM lightwave networks. *Proceedings of the IEEE*, 85(8):1274–1307.

Brackett, C. A. (1990). Dense wavelength division multiplexing networks: Principle and applications. *IEEE Journal on Selected Areas in Communications*, 8(6):948–964.

Chraplyvy, A. R. (1990). Limits on lightwave communications imposed by optical-fiber nonlinearities. *IEEE/OSA Journal of Lightwave Technology*, 8(10):1548–1557.

Chraplyvy, A. R., Nagel, J. A., and Tkach, R. W. (1992). Equalization in amplified WDM lightwave transmission systems. *IEEE Photonics Technology Letters*, 4(8):920–922.

Dono, N. R., Green, P. E., Liu, K., Ramaswami, R., and Tong, F. F.-K. (1990). A wavelength division multiple access network for computer communication. *IEEE Journal on Selected Areas in Communications*, 8(6):983–993.

Dragone, C., Henry, C. H., Kaminow, I. P., and Kistler, R. C. (1989). Efficient multichannel integrated optics star coupler on silicon. *IEEE Pohtonics Technology Letters*, 1(8):241–243.

Elrefaie, A. F., Goldstein, E. L., Zaidi, S., and Jackman, N. (1993). Fiber-amplifier cascades with gain equalization in multiwavelength unidirectional inter-office ring network. *IEEE Photonics Technology Letters*, 5(9):1026–1031.

Green, P. E. (1993). *Fiber Optic Networks*. Prentice Hall.

Hansen, S. L., Dybdal, K., and Larsen, L. C. (1992). Gain limit in erbium-doped fiber amplifiers due to internal Rayleigh backscattering. *IEEE Photonics Technology Letters*, 4(6):559–561.

Hecht, J. (1992). *Understanding Lasers: An Entry-Level Guide*. IEEE Press, New York.

Hecht, J. (1993). *Understanding Fiber Optics*, volume 2. Sams Publishing, Indianapolis, IN.

Henry, P. S. (1985). Lightwave primer. *IEEE Journal of Quantum Electronics*, QE-21(12):1862–1879.

Humblet, P. A. and Hamdy, W. M. (1990). Crosstalk analysis and filter optimization of single- and double-cavity Fabry-Perot filters. *IEEE Journal on Selected Areas in Communications*, 8(6):1095–1107.

Inoue, A., Shigehara, M., Ito, M., Inai, M., Hattori, Y., and Mizunami, T. (1995). Fabrication and application of fiber Bragg grating – A review. *Optoelectronics - Devices and Technologies*, 10(1):119–130.

Keiser, G. (1999). *Optical Fiber Communications*. McGraw-Hill, New York, 3rd edition.

Kobrinski, H. and Cheung, K.-W. (1989). Wavelength-tunable optical filters: Applications and technologies. *IEEE Comm. Magazine*, 27(10):53 – 63.

Lee, T.-P. et al. (1996). Multiwavelength DFB laser array transmitters for ONTC reconfigurable optical network testbed. *IEEE/OSA Journal of Lightwave Technology*, 14(6):967–976.

Lee, T.-P. and Zah, C.-E. (1989). Wavelength-tunable and single-frequency lasers for photonic communications networks. *IEEE Communications Magazine*, 27(10):42–52.

Moore, J. B. and Todd, D. E. (1993). Recent developments in distributed feedback and distributed Bragg reflector lasers for wide-band long-haul fiber-optic communication systems. In *Proceedings IEEE Southeastcon '93*, page 9, Charlotte, NC.

Okamoto, K., Takahashi, H., Suzuki, S., Sugita, A., et al. (1991). Design and fabrication of integrated-optic 8×8 star coupler. *Electronic Letters*, 27(9):774–775.

O'Mahony, M. J. (1993). Optical amplifiers. In Midwinter, J. E., editor, *Photonics in Switching*, volume 1, pages 147–167. Academic Press, San Diego, CA.

Powers, J. P. (1996). *An Introduction to Fiber Optic Systems*. Irwin, Chicago, IL, 2nd edition.

Ramaswami, R. (1993). Multiwavelength lightwave networks for computer communication. *IEEE Communications Magazine*, 31(2):78–88.

Scobey, M. A. and Spock, D. E. (1996). Passive DWDM components using MicroPlasma optical interference filters. In *OFC '96 Technical Digest*, pages 242–243, San Jose, CA.

Sneh, A. and Johnson, K. M. (1994). High-speed tunable liquid crystal optical filter for WDM systems. In *Proceedings, IEEE/LEOS '94 Summer Topical Meetings on Optical Networks and their Enabling Technologies*, pages 59–60, Lake Tahoe, NV.

Tachibana, M., Laming, R. I., Morkel, P. R., and Payne, D. N. (1991). Erbium-doped fiber amplifier with flattened gain spectrum. *IEEE Photonics Technology Letters*, 3(2):118–120.

Whitley, T. J. (1995). A review of recent system demonstrations incorporating 1.3-μm praseodymium-doped fluoride fiber amplifiers. *IEEE/OSA Journal of Lightwave Technology*, 13(5):744–760.

Wilner, A. E. and Hwang, S. M. (1993). Passive equalization of nonuniform EDFA gain by optical filtering for megameter transmission of 20 WDMA channels through a cascade of EDFA's. *IEEE Photonics Technology Letters*, 5(9):1023–1026.

Yamada, M., Shimizu, M., Kanamori, T., Ohishi, Y., Terunuma, Y., Oikawa, K., Yoshinaga, H., Kikushima, K., Miyamoto, Y., and Sudo, S. (1995). Low-noise and high-power Pr^{3+}-doped fluoride fiber amplifier. *IEEE Photonics Technology Letters*, 7(8):869–871.

Zah, C.-E., Favire, F. J., Pathak, B., Bhat, R., et al. (1992). Monolithic integration of a multiwavelength compressive-strained multiquantum-well distributed – feedback laser array with star coupler and optical amplifiers. *Electronic Letters*, 28(25):2361–2362.

Chapter 3

SWITCHES, WAVELENGTH ROUTERS, AND WAVELENGTH CONVERTERS

Byrav Ramamurthy

University of Nebraska-Lincoln

byrav@cse.unl.edu

Abstract   This chapter introduces some of the basic components in WDM networks, discusses various implementations of these components, and provides insights into their capabilities and limitations. The components discussed are fixed and reconfigurable wavelength routers, optical packet switches, and wavelength converters.

1.   INTRODUCTION

While optical transmission systems have been deployed widely for several years, a new generation of optical networks is emerging now which requires components for *optical switching*. Components such as optical switches, wavelength routers, and wavelength converters which enable optical circuit-switching and/or optical packet-switching are described in this chapter. We refer the interested reader to [Mestdagh, 1995; Borella et al., 1997; Mukherjee, 1997; Ramaswami and Sivarajan, 1998; Stern and Bala, 1999] for additional details and studies of the performance of such devices in optical networks.

2.   SWITCHING ELEMENTS

Most current networks employ electronic processing and use the optical fiber only as a transmission medium. Switching and processing of data are performed by converting an optical signal back to its "native" electronic form. Such a network relies on electronic switches. These switches provide a high degree of flexibility in terms of switching and routing functions; however, the speed of electronics is unable to match the high bandwidth of an optical fiber. Also, an electro-optic conversion at an intermediate node in the network introduces extra delay. These factors have motivated a push towards the development of all-optical networks in which optical switching components are able to switch high bandwidth optical data streams without electro-optic conversion. In a class of switching devices currently being developed, the control of the switching function is performed electronically with the optical stream being transparently routed from a given input of the switch to a given output. Such transparent switching allows for the switch to be independent of the data rate and format of the optical signals. For WDM systems, switches which are wavelength dependent are also being developed.

Switches can be divided into two classes. A switch in the first of these classes, called *relational devices*, establishes a relation between the inputs and the outputs. The relation is a function of the control signals applied to the device and is independent of the contents of the signal or data inputs. A property of this device is that the information entering and flowing through it cannot change or influence the current relation between the inputs and the outputs. An example of this type of device is the directional coupler as it is used in switching applications. Thus, the strength of a relational device, which allows signals at high bit rates to pass through it, is that it cannot sense the presence of individual bits that are flowing through itself. This characteristic is also known as *data transparency*. The same feature may sometimes also be a weakness since it causes loss of flexibility (i.e., individual portions of a data stream cannot be switched independently).

The second class of devices is referred to as *logic devices*. In such a device, the data, or the information-carrying signal that is incident on the device, controls the state of the device in such a way that some Boolean function, or combination of Boolean functions, is performed on the inputs. For this class of devices, at least some of the components within an entire system must be able to change states or *switch* as fast as or faster than the signal bit rate [Hinton, 1990]. This ability gives the device some added flexibility but limits the maximum bit rate that can be accommodated.

Thus, *relational devices* are needed for circuit-switching, and *logic devices* are needed for packet switching.

In the following sections, we review a number of different optical switch elements and architectures.

2.1.   FIBER CROSSCONNECT ELEMENTS

A fiber crossconnect element switches optical signals from input ports to output ports. These type of elements are usually considered to be wavelength insensitive, i.e., incapable of demultiplexing different wavelength signals on a given input fiber.

A basic crossconnect element is the 2×2 crosspoint element. A 2×2 crosspoint element routes optical signals from two input ports to two output

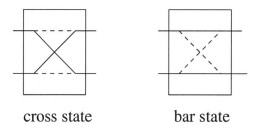

cross state   bar state

Figure 3.1   2 × 2 crossconnect elements in the cross state and bar state.

ports and has two states: cross state and bar state (see Fig. 3.1). In the cross state, the signal from the upper input port is routed to the lower output port, and the signal from the lower input port is routed to the upper output port. In the bar state, the signal from the upper input port is routed to the upper output port, and the signal from the lower input port is routed to the lower output port.

Optical crosspoint elements have been demonstrated using two types of technologies: (a) the generic directive switch [Alferness, 1988], in which light via some structure is physically directed to one of two different outputs, and (b) the gate switch, in which optical amplifier gates are used to select and filter input signals to specific output ports.

2.1.1   Directive Switches.   The directional coupler (see Fig. 3.2(a) [Alferness, 1988]) consists of a pair of optical channel waveguides that are parallel and in close proximity over some finite interaction length. Light input to one of the waveguides couples to the second waveguide via evanescent[1] coupling. The coupling strength corresponds to the interwaveguide separation and the waveguide mode size which in turn depends upon the optical wavelength and the confinement factor[2] of the waveguide. If the two waveguides are identical, complete coupling between the two waveguides occurs over a characteristic length which depends upon the coupling strength. However, by placing electrodes over the waveguides, the difference in the propagation constants in the two waveguides can be sufficiently increased via the linear electrooptic effect so that no light couples between the two waveguides. Therefore, the cross state corresponds to zero applied voltage, and the bar state corresponds to a non-zero switching voltage. Unfortunately, the interaction length needs to be very accurate for good isolation, and these couplers are wavelength specific.

Switch fabrication tolerances, as well as the ability to achieve good switching for a relatively wide range of wavelengths, can be overcome by using the so-called reversed delta-beta coupler (see Fig. 3.2(b)). In this device, the electrode is split into at least two sections. The cross state is achieved by applying equal and opposite voltages to the two electrodes. This approach has been shown to be very successful [Alferness, 1988].

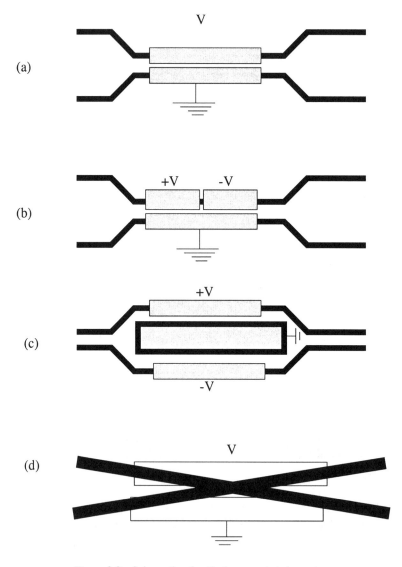

Figure 3.2   Schematic of optical crosspoint elements.

The balanced bridge interferometric switch (see Fig. 3.2(c)) consists of an input 3-dB coupler, two waveguides sufficiently separated so that they do not couple, electrodes to allow changing the effective path length over the two arms, and a final 3-dB coupler. Light incident on the upper waveguide is split in half by the first coupler. With no voltage applied to the electrodes, the optical path length of the two arms enters the second coupler in phase. The second coupler acts like the continuation of the first, and all the light is crossed over to the second waveguide to provide the cross state. To achieve the bar state, voltage is applied to an electrode, placed over one of the interferometer arms to electrooptically produce a 180° phase difference between the two arms. In this case, the two inputs from the arms of the interferometer combine at the second 3-dB coupler out of phase, with the result that light remains in the upper waveguide.

The intersecting waveguide switch is shown in Fig. 3.2(d). This device can be viewed as a directional coupler (see Fig. 3.2(a)) with no gap between the waveguides in the interaction region. When properly fabricated, both cross and bar states can be electrooptically achieved with good crosstalk performance.

Other types of switches include the mechanical fiber-optic switch and the thermo-optic switch. These devices offer slow switching (about milliseconds) and may be employed in circuit-switched networks. One mechanical switch, for example, consists of two ferrules, each with polished end faces that can rotate to switch the light appropriately [Anderson, 1995]. Thermo-optic waveguide switches, on the other hand, are fabricated on a glass substrate and are operated by the use of the thermo-optic effect. One such device uses a zero-gap directional-coupler configuration with a heater electrode to increase the waveguide index of refraction [Lee and Su, 1994].

2.1.2   Gate Switches.   In the N × N gate switch, each input signal first passes through a 1 × N splitter. The signals then pass through an array of N^2 gate elements, and are then recombined in N × 1 combiners and sent to the N outputs. The gate elements can be implemented using optical amplifiers which can either be turned on or off to pass only selected signals to the outputs. The amplifier gains can compensate for coupling losses and losses incurred at the splitters and combiners. A 2 × 2 amplifier gate switch is illustrated in Fig. 3.3. A disadvantage of the gate switch is that the splitting and combining losses limit the size of the switch.

Amplifier gate switches of size 8 × 8 are commercially available. In [opt, 1995] an 8 × 8 switch is described which uses semiconductor optical amplifiers to provide lossless switching. It operates around the 1300 nm region, has an optical bandwidth of 40 nm, has low polarization dependence (1 dB), and has a fairly low crosstalk (below -40 dB). The disadvantages are that the switch is bulky (weight of 50 lbs.) and expensive.

2.2.   NON-RECONFIGURABLE WAVELENGTH ROUTER

A wavelength-routing device can route signals arriving at different input fibers (ports) of the device to different output fibers (ports) based on the wavelengths of the signals. Wavelength routing is accomplished by demultiplex-

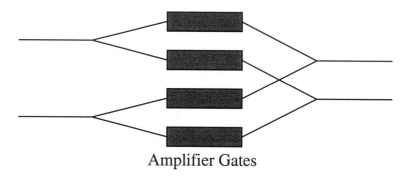

Amplifier Gates

Figure 3.3   A 2 × 2 amplifier gate switch.

ing the different wavelengths from each input port, optionally switching each wavelength separately, and then multiplexing signals at each output port. The device can be either *non-reconfigurable*, in which case there is no switching stage between the demultiplexers and the multiplexers, and the routes for different signals arriving at any input port are fixed (these devices are referred to as routers rather than switches), or *reconfigurable*, in which case the routing function of the switch can be controlled electronically. In this section we will discuss wavelength routers, while Section 2.3 will cover reconfigurable wavelength switches.

A non-reconfigurable wavelength router can be constructed with a stage of demultiplexers which separate each of the wavelengths on an incoming fiber, followed by a stage of multiplexers which recombine wavelengths from various inputs to a single output. The outputs of the demultiplexers are hardwired to the inputs of the multiplexers. Let this router have P incoming fibers, and P outgoing fibers. On each incoming fiber, there are M wavelength channels. A 4 × 4 non-reconfigurable wavelength router with M = 4 is illustrated in Fig. 3.4. The router is non-reconfigurable because the path of a given wavelength channel, after it enters the router on a particular input fiber, is fixed. The wavelengths on each incoming fiber are separated using a grating demultiplexer. And finally, information from multiple WDM channels are multiplexed before launching them back onto an output fiber. In between the demultiplexers and multiplexers, there are direct connections from each demultiplexer output to each multiplexer input. Which wavelength on which input port gets routed to which output port depends on a "routing matrix" characterizing the router; this matrix is determined by the internal "connections" between the demultiplexers and multiplexers.

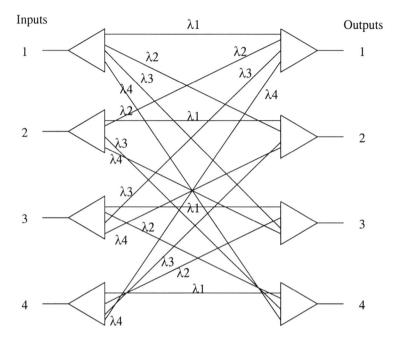

Figure 3.4   A 4 × 4 non-reconfigurable wavelength-router.

2.2.1   Waveguide Grating Routers.

One implementation of a wavelength router is the waveguide grating router (WGR), which is also referred to as an arrayed waveguide grating (AWG) multiplexer. A WGR provides a fixed routing of an optical signal from a given input port to a given output port based on the wavelength of the signal. Signals of different wavelengths coming into an input port will each be routed to a different output port. Also, different signals using the same wavelength can be input simultaneously to different input ports, and still not interfere with each other at the output ports. Compared to a passive-star coupler in which a given wavelength may only be used on a single input port, the WGR with N input and N output ports is capable of routing a maximum of N^2 connections, as opposed to a maximum of N connections in the passive-star coupler. Also, because the WGR is an integrated device, it can easily be fabricated at low cost. The disadvantage of the WGR is that it is a device with a fixed routing matrix which cannot be reconfigured.

The WGR, shown in Fig. 3.5, can be used as a non-reconfigurable wavelength router, or it can be used to build a tunable optical transmitter or a tunable optical receiver. It consists of two passive-star couplers connected by a grating array. The first star coupler has N inputs and N' outputs (where $N \ll N'$), while the second one has N' inputs and N outputs. The inputs to the first star are separated by an angular distance of α, and their outputs are separated by

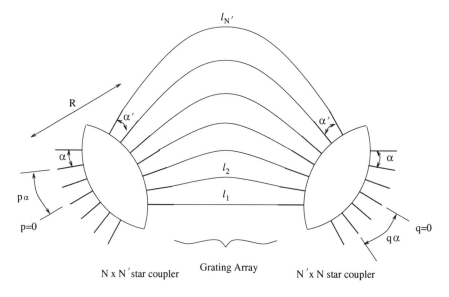

Figure 3.5   The waveguide grating router (WGR).

angular distance α'. The grating array consists of N' waveguides, with lengths $l_1, l_2, \cdots, l_{N'}$ where $l_1 < l_2 < \cdots < l_{N'}$. The difference in length between any two adjacent waveguides is a constant Δl.

In the first star coupler, a signal on a given wavelength entering from any of the input ports is split and transmitted to its N' outputs which are also the N' inputs of the grating array. The signal travels through the grating array, experiencing a different phase shift in each waveguide depending on the length of the waveguides and the wavelength of the signal. The constant difference in the lengths of the waveguides creates a phase difference of $\beta \cdot \Delta l$ in adjacent waveguides, where $\beta = \frac{2\pi n_{eff}}{\lambda}$ is the propagation constant in the waveguide, n_{eff} is the effective refractive index of the waveguide, and λ is the wavelength of the light. At the input of the second star coupler, the phase difference in the signal will be such that the signal will constructively recombine only at a single output port.

Two signals of the same wavelength coming from two different input ports will not interfere with each other in the grating because there is an additional phase difference created by the distance between the two input ports. The two signals will be combined in the grating, but will be separated out again in the second star coupler and directed to different outputs. This phase difference is given by $kR(p - q)\alpha\alpha'$, where k is a propagation constant which does not depend on wavelength, R is the constant distance between the two foci of the optical star, p is the input port number of the router, and q is the output port number of the router. The total phase difference is:

$$\phi = \frac{2\pi \cdot \Delta l}{\lambda} + kR(p-q)\alpha\alpha' \tag{3.1}$$

The transmission power from a particular input port p to a particular output port q is maximized when ϕ is an integer multiple of 2π. Thus, only wavelengths λ for which ϕ is a multiple of 2π will be transmitted from input port p to output port q. Alternatively, for a given input port and a given wavelength, the signal will only be transmitted to the output port which causes ϕ to be a multiple of 2π.

2.3.   RECONFIGURABLE WAVELENGTH-ROUTING SWITCH

A reconfigurable wavelength-routing switch (WRS), also referred to as a wavelength-selective crossconnect (WSXC), uses photonic switches inside the routing element. The functionality of the reconfigurable WRS, illustrated in Fig. 3.6, is as follows. The WRS has P incoming fibers and P outgoing fibers. On each incoming fiber, there are M wavelength channels. Similar to the non-reconfigurable router, the wavelengths on each incoming fiber are separated using a grating demultiplexer.

The outputs of the demultiplexers are directed to an array of M $P \times P$ optical switches between the demultiplexer and the multiplexer stages. All signals on a given wavelength are directed to the same switch. The switched signals are then directed to multiplexers which are associated with the output ports. Finally, information from multiple WDM channels are multiplexed before launching them back onto an output fiber.

Space-division optical-routing switches may be built from 2×2 optical crosspoint elements [Schmidt and Alferness, 1990] arranged in a banyan-based fabric. The space division switches (which may be one per wavelength [Sharony et al., 1993]) can route a signal from any input to any output on a given wavelength. Such switches based on relational devices [Hinton, 1990] are capable of switching very high-capacity signals. The 2×2 crosspoint elements that are used to build the space-division switches may be slowly tunable and they may be reconfigured to adapt to changing traffic requirements. Switches of this type can be constructed from off-the-shelf components available today.

Networks built from such switches are more flexible than passive, non-reconfigurable, wavelength-routed networks, because they provide additional control in setting up connections. The routing is a function of both the wavelength chosen at the source node, as well as the configuration of the switches in the network nodes.

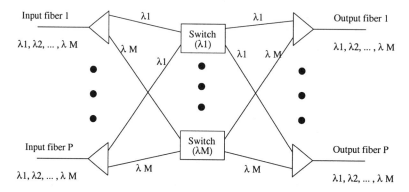

Figure 3.6   A P × P reconfigurable wavelength-routing switch with M wavelengths.

2.4.   PHOTONIC PACKET SWITCHES

Most of the switches discussed above are *relational devices*, i.e., they are useful in a circuit-switched environment where a connection may be set up over long periods of time. Here, we review optical packet switches that have been proposed in the literature. These switches are composed of *logic devices*, instead of *relational devices* used before, so that the switch configuration is a function of the data on the input signal.

In a packet-switched system, there exists the problem of resource contention when multiple packets contend for a common resource in the switch. In an electronic system, contention may be resolved through the use of buffering; however, in the optical domain, contention resolution is a more complex issue, since it is difficult to implement components which can store optical data. A number of switch architectures which use delay lines to implement optical buffering have been proposed. A delay line is simply a long fiber which introduces propagation delays that are on the order of packet transmission times.

2.4.1   The Staggering Switch.   The *staggering switch*, which is an "almost-all-optical" packet switch has been proposed in [Haas, 1993]. In an "almost-all-optical" network, the data path is fully optical, but the control of the switching operation is performed electronically. One of the advantages of such switching over its electronic counterpart is that it is transparent, i.e., except for the control information, the payload may be encoded in an arbitrary format or at an arbitrary bit rate. The main problem in the implementation of packet-switched optical networks is the lack of random-access optical memory.

The staggering switch architecture is based on an output-collision-resolution scheme that is controlled by a set of delay lines with unequal delays. The architecture is based on two rearrangeably nonblocking stages interconnected by optical delay lines with different amounts of delay. The work in [Haas, 1993] investigates the probability of packet loss and the switch latency as a function of link utilization and switch size. In general, with proper setting of the number of delay lines, the switch can achieve an arbitrarily low probability of packet loss. Fig. 3.7 gives a simple overview of the switch architecture.

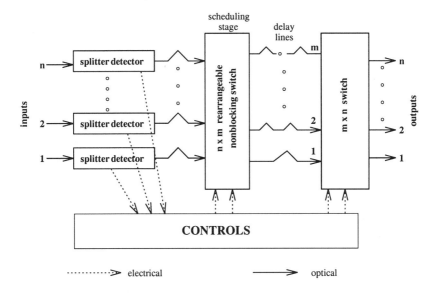

Figure 3.7   The Staggering Switch architecture.

2.4.2   Contention Resolution by Delay Lines (CORD).   Another architecture which deals with contention in a packet-switched optical network is the Contention Resolution by Delay Lines (CORD) architecture [Chlamtac et al., 1996]. The CORD architecture consists of a number of 2×2 crossconnect elements and delay lines (see Fig. 3.8). Each delay line functions as a buffer for a single packet. If two packets contend for the same output port, one packet may be switched to a delay line while the other packet is switched to the proper output. The packet which was delayed can then be switched to the same output after the first packet has been transmitted.

3.   WAVELENGTH CONVERSION

Consider the network in Fig. 3.9. It shows a wavelength-routed network containing two WDM cross-connects (S1 and S2) and five access stations (A through E). Three lightpaths have been set up (C to A on wavelength λ_1, C to B on λ_2, and D to E on λ_1). To establish a lightpath, we require that the *same* wavelength be allocated on all the links in the path. This requirement is known

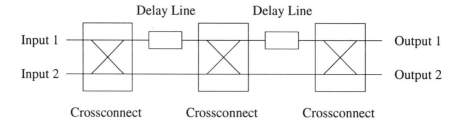

Figure 3.8   The CORD architecture.

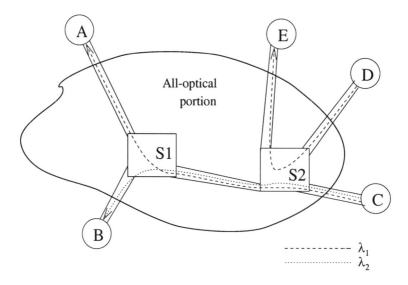

Figure 3.9   An all-optical wavelength-routed network.

as the *wavelength-continuity constraint* (e.g., see [Banerjee and Mukherjee, 1996]). This constraint distinguishes the wavelength-routed network from a circuit-switched network which blocks calls only when there is no capacity along any of the links in the path assigned to the call. Consider the example in Fig. 3.10(a). Two lightpaths have been established in the network: (i) between Node 1 and Node 2 on wavelength λ_1, and (ii) between Node 2 and Node 3 on wavelength λ_2. Now suppose a lightpath between Node 1 and Node 3 needs to be set up. Establishing such a lightpath is impossible even though there is a free wavelength on each of the links along the path from Node 1 to Node 3. This is because the available wavelengths on the two links are *different*. Thus, a wavelength-routed network may suffer from higher blocking as compared to a circuit-switched network.

It is easy to eliminate the wavelength-continuity constraint, if we were able to *convert* the data arriving on one wavelength along a link into another

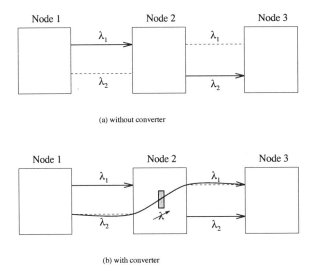

(a) without converter

(b) with converter

Figure 3.10   Wavelength-continuity constraint in a wavelength-routed network.

wavelength at an intermediate node and forward it along the next link. Such a technique is actually feasible and is referred to as *wavelength conversion*. In Fig. 3.10(b), a wavelength converter at Node 2 is employed to convert data from wavelength λ_2 to λ_1. The new lightpath between Node 1 and Node 3 can now be established by using the wavelength λ_2 on the link from Node 1 to Node 2, and then by using the wavelength λ_1 to reach Node 3 from Node 2. Notice that a single lightpath in such a *wavelength-convertible* network can use a different wavelength along each of the links in its path. Thus, wavelength conversion may improve the efficiency in the network by resolving the wavelength conflicts of the lightpaths.

The function of a wavelength converter is to convert data on an input wavelength onto a possibly different output wavelength among the N wavelengths in the system (see Fig. 3.11). In this figure and throughout this section, λ_s denotes the input signal wavelength; λ_c, the converted wavelength; λ_p, the pump wavelength; f_s, the input frequency; f_c, the converted frequency; f_p, the pump frequency; and CW, the continuous wave (unmodulated) generated as the pump signal.

An ideal wavelength converter should possess the following characteristics [Durhuus et al., 1996]:

- transparency to bit rates and signal formats,

- fast setup time of output wavelength,

- conversion to both shorter and longer wavelengths,

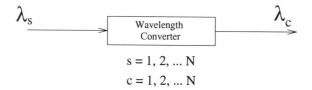

λ_s Wavelength λ_c
                          Converter

$$s = 1, 2, ... N$$
$$c = 1, 2, ... N$$

Figure 3.11   Functionality of a wavelength converter.

- moderate input power levels,

- possibility for same input and output wavelengths (i.e., no conversion),

- insensitivity to input signal polarization,

- low-chirp output signal with high extinction ratio[3] and large signal-to-noise ratio, and

- simple implementation.

3.1. WAVELENGTH CONVERSION TECHNOLOGIES

Several researchers have attempted to classify and compare the several techniques available for wavelength conversion [Durhuus et al., 1996; Mikkelsen et al., 1996; Sabella and Iannone, 1996; Wiesenfeld, 1996; Yoo, 1996]. The classification of these techniques presented in this section follows that in [Wiesenfeld, 1996]. Wavelength conversion techniques can be broadly classified into two types: *opto-electronic wavelength conversion*, in which the optical signal must first be converted into an electronic signal; and *all-optical wavelength conversion*, in which the signal remains in the optical domain. All-optical conversion techniques may be sub-divided into techniques which employ *coherent effects* and techniques which use *cross modulation*.

3.1.1   Opto-Electronic Wavelength Conversion.   In opto-electronic wavelength conversion [Fujiwara et al., 1988], the optical signal to be converted is first translated into the electronic domain using a photodetector (labeled R in Fig. 3.12, from [Mestdagh, 1995]). The electronic bit stream is stored in the buffer (labeled FIFO for the First-In-First-Out queue mechanism). The electronic signal is then used to drive the input of a tunable laser (labeled T) tuned to the desired wavelength of the output (see Fig. 3.12). This method has been demonstrated for bit rates up to 10 Gbps [Yoo, 1996]. However, this method is much more complex and consumes a lot more power than the other methods described below [Durhuus et al., 1996]. Moreover, the process of opto-electronic (O/E) conversion adversely affects the transparency of the signal, requiring the optical data to be in a specified modulation format and at a specific bit rate. All information in the form of phase, frequency, and analog amplitude of the optical signal is lost during the conversion process. One form of transparency is *digital transparency*, in which digital signals of any bit rate up to a certain limit can be accommodated by the system [Yoo, 1996].

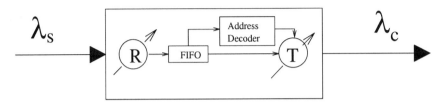

Figure 3.12   An opto-electronic wavelength converter.

3.1.2   Wavelength Conversion Using Coherent Effects.
Wavelength conversion methods using coherent effects are typically based on wave-mixing effects (see Fig. 3.13). Wave-mixing arises from a nonlinear optical response of a medium when more than one wave is present. It results in the generation of another wave whose intensity is proportional to the product of the interacting wave intensities. Wave-mixing preserves both phase and amplitude information, offering strict transparency. It is also the only approach that allows simultaneous conversion of a set of multiple input wavelengths to another set of multiple output wavelengths and could potentially accommodate signals with bit rates exceeding 100 Gbps [Yoo, 1996]. In Fig. 3.13, the value $n = 3$ corresponds to Four-Wave Mixing (FWM) and $n = 2$ corresponds to Difference Frequency Generation (DFG). These techniques are described below.

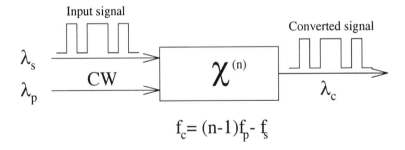

Figure 3.13   A wavelength converter based on nonlinear wave-mixing effects.

Four-Wave Mixing (FWM).   FWM (also referred to as four-photon mixing) is a third-order nonlinearity in silica fibers, which causes three optical waves of frequencies f_i, f_j, and f_k ($k \neq i, j$) to interact in a multichannel WDM system [Tkach et al., 1995] to generate a fourth wave of frequency given by

$$f_{ijk} = f_i + f_j - f_k$$

FWM is also achievable in other passive waveguides such as semiconductor waveguides and in an active medium such as a semiconductor optical amplifier (SOA). This technique provides modulation-format independence [Schnabel et al., 1994] and high bit rate capabilities [Ludwig and Raybon, 1994]. However, the conversion efficiency from pump energy to signal energy of this technique is not very high, and it decreases swiftly with increasing conversion span (shift between pump and output signal wavelengths) [Zhou et al., 1994].

Difference Frequency Generation (DFG).   DFG is a consequence of a second-order nonlinear interaction of a medium with two optical waves: a pump wave and a signal wave [Yoo, 1996]. DFG is free from satellite signals which appear in FWM-based techniques. This technique offers a full range of transparency without adding excess noise to the signal [Yoo et al., 1995]. It is also bi-directional and fast, but it suffers from low efficiency and high polarization sensitivity. The main difficulties in implementing this technique lies in the phase-matching of interacting waves [Yoo et al., 1996] and in fabricating a low-loss waveguide for high conversion efficiency [Yoo, 1996].

3.1.3   Wavelength Conversion using Cross Modulation.   Cross-modulation wavelength conversion techniques utilize active semiconductor optical devices such as semiconductor optical amplifiers (SOA) and lasers. These techniques belong to a class known as optical-gating wavelength conversion [Yoo, 1996].

Semiconductor Optical Amplifiers (SOA) in XGM and XPM mode.   The principle behind using an SOA in the cross-gain modulation (XGM) mode is shown in Fig. 3.14 (from [Durhuus et al., 1996]). The intensity-modulated input signal modulates the gain in the SOA due to gain saturation. A continuous-wave (CW) signal at the desired output wavelength (λ_c) is modulated by the gain variation so that it carries the same information as the original input signal. The CW signal can either be launched into the SOA in the same direction as the input signal (co-directional), or launched into the SOA in the opposite direction as the input signal (counter-directional). The XGM scheme gives a wavelength-converted signal that is inverted compared to the input signal. While the XGM scheme is simple to realize and offers penalty-free conversion at 10 Gbps [Durhuus et al., 1996], it suffers from the drawbacks due to inversion of the converted bit stream and extinction ratio degradation for the converted signal.

Figure 3.14 A wavelength converter using co-propagation based on XGM in a SOA.

The operation of a wavelength converter using SOA in cross-phase modulation (XPM) mode is based on the fact that the refractive index of the SOA is dependent on the carrier density in its active region. An incoming signal that depletes the carrier density will modulate the refractive index and thereby result in phase modulation of a CW signal (wavelength λ_c) coupled into the converter [Durhuus et al., 1996; Lacey et al., 1996]. The SOA can be integrated into an interferometer so that an intensity-modulated signal format results at the output of the converter. Techniques involving SOAs in XPM mode have been proposed using nonlinear optical loop mirrors (NOLMs) [Eiselt et al., 1993], Mach-Zender interferometers (MZI) [Durhuus et al., 1994] and Michelson interferometers (MI) [Mikkelsen et al., 1994]. Fig. 3.15 shows an asymmetric MZI wavelength converter based on SOA in XPM mode (from [Durhuus et al., 1996]). With the XPM scheme, the converted output signal can be either inverted or non-inverted, unlike in the XGM scheme where the output is always inverted. The XPM scheme is also very power efficient compared to the XGM scheme [Durhuus et al., 1996].

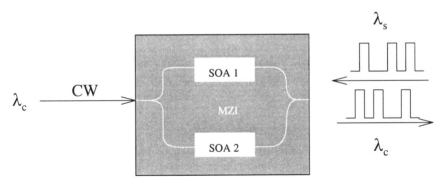

Figure 3.15 An interferometric wavelength converter based on XPM in SOAs.

Semiconductor Lasers. Using single-mode semiconductor lasers, lasing mode intensity is modulated by input signal light through lasing mode gain saturation. The obtained output (converted) signal is inverted compared to the input signal. This gain suppression mechanism has been employed in a

Distributed Bragg Reflector (DBR) laser to convert signals at 10 Gbps [Yasaka et al., 1996].

3.1.4   Summary.   In this subsection, we reviewed the various techniques and technologies used in the design of a wavelength converter. The actual choice of the technology to be employed for wavelength conversion in a network depends on the requirements of the particular system. However, it is clear that opto-electronic converters offer only limited digital transparency. Moreover, deploying multiple opto-electronic converters in a WDM cross-connect, e.g., in a WRS, requires sophisticated packaging to avoid crosstalk among channels. This leads to increased cost per converter, further making this technology less attractive than all-optical converters [Yoo, 1996]. Other disadvantages of opto-electronic converters include complexity and large power consumption [Durhuus et al., 1996]. Among all-optical converters, converters based on SOAs using the XGM and the XPM conversion scheme presently seem well suited for system use. Converters based on FWM, though transparent to different modulation formats, perform inefficiently [Durhuus et al., 1996]. However, wave-mixing converters are the only category of wavelength converters that offer the full range of transparency while also allowing simultaneous conversion of a set of input wavelengths to another set of output wavelengths. In this respect, DFG-based methods offer great promise. Further details on comparison of various wavelength-conversion techniques can be found in [Mikkelsen et al., 1996; Wiesenfeld, 1996; Yoo, 1996].

In the next subsection, we examine various switch architectures that may be employed in a wavelength-convertible network.

3.2.   WAVELENGTH CONVERSION IN SWITCHES

As wavelength converters become readily available, a vital question comes to mind: Where do we place them in the network? An obvious location is in the switches (cross-connects) in the network. A possible architecture of such a wavelength-convertible switching node is the dedicated wavelength-convertible switch (see Fig. 3.16, from [Lee and Li, 1993]). In this architecture, referred to as a wavelength-interchanging cross-connect (WIXC), each wavelength along each output link in a switch has a *dedicated* wavelength converter (shown as boxes labeled WC in Fig. 3.16), i.e., an $M \times M$ switch in an N-wavelength system requires MN converters. The incoming optical signal from a link at the switch is first wavelength demultiplexed into separate wavelengths. Each wavelength is switched to the desired output port by the nonblocking optical switch. The output signal may have its wavelength changed by its wavelength converter. Finally, various wavelengths combine to form an aggregate signal coupled to an outbound fiber. The switch architecture shown in Fig. 3.16 is

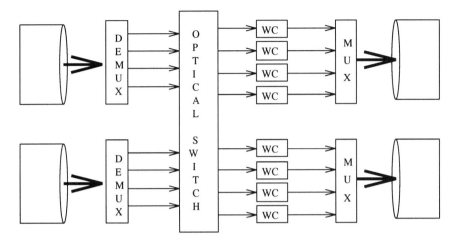

Figure 3.16 A switch which has dedicated converters at each output port for each wavelength.

similar to that of the reconfigurable WRS (also WSXC) shown in Fig. 3.6 with additional wavelength converters added after the switching elements.

However, the dedicated wavelength-convertible switch is not very cost efficient since all the wavelength converters may not be required all the time [Lee and Li, 1993; Subramaniam et al., 1996; Iness, 1997]. An effective method to cut costs is to share the converters. Two architectures have been proposed for switches sharing converters [Lee and Li, 1993]. In the share-per-node structure (see Fig. 3.17(a)), all the converters at the switching node are collected in a converter bank. A converter bank is a collection of a few wavelength converters (e.g., two in each of the boxes labeled WC in Fig. 3.17), each of which is assumed to have identical characteristics and can convert any input wavelength to any output wavelength. This bank can be accessed by any wavelength on any incoming fiber by appropriately configuring the larger optical switch in Fig. 3.17(a). In this architecture, only the wavelengths which require conversion are directed to the converter bank. The converted wavelengths are then switched to the appropriate outbound fiber link by the second optical switch. In the share-per-link structure (see Fig. 3.17(b)), each outgoing fiber link is provided with a dedicated converter bank which can be accessed only by those lightpaths traveling on that particular outbound link. The optical switch can be configured appropriately to direct wavelengths towards a particular link, either with conversion or without conversion.

When opto-electronic wavelength conversion is used, the functionality of the wavelength converter can also be performed at the access stations instead of at the switches. The share-with-local switch architecture proposed in [Lee and Li, 1993] (see Fig. 3.18) and the simplified network access station architecture proposed in [Kovačević and Acampora, 1996b] (see Fig. 3.19) fall under this

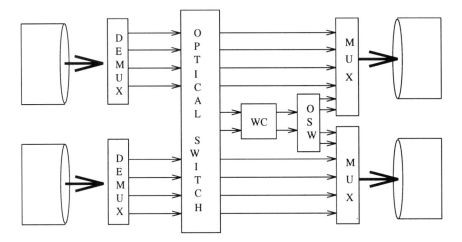

(a) Share-per-node wavelength-convertible switch architecture

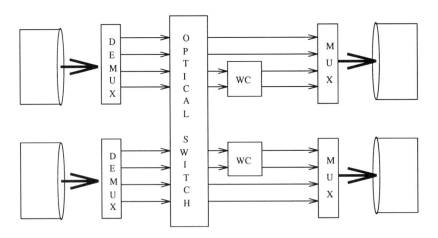

(b) Share-per-link wavelength-convertible switch architecture

Figure 3.17   Switches which allow sharing of converters.

category. In the share-with-local switch architecture, selected incoming optical signals are converted to electrical signals by a receiver bank. A signal can then be either dropped locally or retransmitted on a different wavelength by a transmitter bank. In Fig. 3.19, an optical signal on wavelength W1 can be switched to a network access station where it is converted to an electronic signal. The signal can then be retransmitted by the network access station on a new wavelength W2.

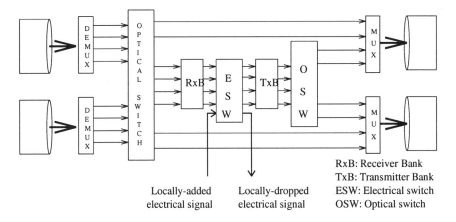

Figure 3.18   The share-with-local wavelength-convertible switch architecture.

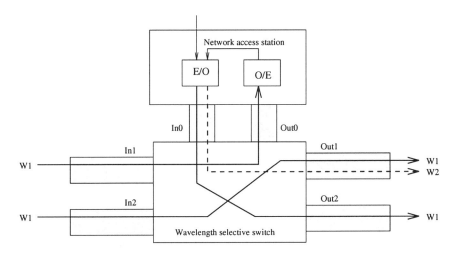

Figure 3.19   Architecture which supports electronic wavelength conversion.

3.3.   SUMMARY

In the above section, we examined the various facets of the wavelength-conversion technology from its *realization* using current opto-electronic devices to its *incorporation* in a wavelength-routed network design. Several studies have investigated the benefits of wavelength conversion in an optical network [Kovačević and Acampora, 1996a; Subramaniam et al., 1996; Birman, 1996; Barry and Humblet, 1996]. An overview of such studies as well as the impact of wavelength conversion technology on network control and management algorithms can be found in [Ramamurthy and Mukherjee, 1998].

4.   CONCLUSION

Novel devices such as those presented in this chapter have revolutionized the design of next-generation optical networks. As optical device technology continues to improve, network designers must be ready to take advantage of new device capabilities, while keeping in mind the limitations of such devices.

Acknowledgments

The author thanks Prof. Jason P. Jue, Prof. Biswanath Mukherjee, Dr. Michael S. Borella, and Dr. Dhritiman Banerjee for their contributions to the material presented in this chapter.

Notes

1. An evanescent wave is the part of a propagating wave which travels along or outside of the waveguide boundary.

2. The confinement factor determines the fraction of power that travels within the core of the waveguide.

3. The *extinction ratio* is defined as the ratio of the optical power transmitted for a bit "0" to the power transmitted for a bit "1".

References

(1995). Optivision, Inc. home page. http://www.optivision.com/.

Alferness, R. C. (1988). Titanium-diffused lithium niobate waveguide devices. In Tamir, T., editor, *Guided-Wave Optoelectronics*, chapter 4. Springer-Verlag, New York, NY.

Anderson, D. (1995). Low-cost mechanical fiber-optic switch. In *OFC '95 Technical Digest*, volume 8, pages 185–186, San Diego, CA, USA.

Banerjee, D. and Mukherjee, B. (1996). Practical approaches for routing and wavelength assignment in large all-optical wavelength-routed networks. *IEEE Journal on Selected Areas in Communications*, 14(5):903–908.

Barry, R. A. and Humblet, P. A. (1996). Models of blocking probability in all-optical networks with and without wavelength changers. *IEEE Journal on Selected Areas in Communications*, 14(5):858–867.

Birman, A. (1996). Computing approximate blocking probabilities for a class of all-optical networks. *IEEE Journal on Selected Areas in Communications*, 14(5):852–857.

Borella, M. S., Jue, J. P., Banerjee, D., Ramamurthy, B., and Mukherjee, B. (1997). Optical components for WDM lightwave networks. *Proceedings of the IEEE*, 85(8):1274–1307.

Chlamtac, I., Fumagalli, A., Kazovsky, L. G., et al. (1996). CORD: Contention resolution by delay lines. *IEEE Journal on Selected Areas in Communications*, 14(5):1014–1029.

Durhuus, T. et al. (1994). All optical wavelength conversion by SOA's in a Mach-Zender configuration. *IEEE Photonic Technology Letters*, 6:53–55.

Durhuus, T. et al. (1996). All-optical wavelength conversion by semiconductor optical amplifiers. *IEEE/OSA Journal of Lightwave Technology*, 14(6):942–954.

Eiselt, M., Pieper, W., and Weber, H. G. (1993). Decision gate for all-optical retiming using a semiconductor laser amplifier in a loop mirror configuration. *Electronic Letters*, 29:107–109.

Fujiwara, M. et al. (1988). A coherent photonic wavelength-division switching system for broadband networks. *Proc., European Conf. Communication (ECOC '88)*, pages 139–142.

Haas, Z. (1993). The 'staggering switch': An electronically controlled optical packet switch. *IEEE/OSA Journal of Lightwave Technology*, 11(5/6):925–936.

Hinton, H. S. (1990). Photonic switching fabrics. *IEEE Communications Magazine*, 28(4):71–89.

Iness, J. (1997). *Efficient Use of Optical Components in WDM-based Optical Networks*. PhD thesis, Dept. of Computer Science, University of California, Davis, CA.

Kovačević, M. and Acampora, A. S. (1996a). Benefits of wavelength translation in all-optical clear-channel networks. *IEEE Journal on Selected Areas in Communications*, 14(5):868–880.

Kovačević, M. and Acampora, A. S. (1996b). Electronic wavelength translation in optical networks. *IEEE/OSA Journal of Lightwave Technology*, 14(6):1161–1169.

Lacey, J. P. R., Pendock, G. J., and Tucker, R. S. (1996). Gigabit-per-second all-optical 1300-nm to 1550-nm wavelength conversion using cross-phase modulation in a semiconductor optical amplifier. *Proc., Optical Fiber Communication (OFC '96)*, 2:125–126.

Lee, C. and Su, T. (1994). 2*2 single mode zero-gap directional coupler thermo-optic waveguide switch on glass. *Applied Optics*, 33(30):7016–7022.

Lee, K.-C. and Li, V. O. K. (1993). A wavelength-convertible optical network. *IEEE/OSA Journal of Lightwave Technology*, 11(5/6):962–970.

Ludwig, R. and Raybon, G. (1994). BER measurements of frequency converted signals using four-wave mixing in a semiconductor laser amplifier at 1, 2.5, 5, and 10 Gbit/s. *Electronic Letters*, 30:338–339.

Mestdagh, D. J. G. (1995). *Fundamentals of Multiaccess Optical Fiber Networks*. The Artech House Optoelectronics Library. Artech House.

Mikkelsen, B. et al. (1994). Polarization insensitive wavelength conversion of 10 Gbit/s signals with SOAs in a Michelson interferometer. *Electronic Letters*, 30(3):260–261.

Mikkelsen, B. et al. (1996). Wavelength conversion devices. *Proc., Optical Fiber Communication (OFC '96)*, 2:121–122.

Mukherjee, B. (1997). *Optical Communication Networks*. McGraw-Hill, New York, NY.

Ramamurthy, B. and Mukherjee, B. (1998). Wavelength conversion in WDM networking. *IEEE Journal on Selected Areas in Communications*, 16(7):1061 – 1073.

Ramaswami, R. and Sivarajan, K. N. (1998). *Optical Networks: A Practical Perspective*. Morgan Kaufmann Publishers, Inc., San Francisco, CA.

Sabella, R. and Iannone, E. (1996). Wavelength conversion in optical transport networks. *Fiber and Integrated Optics*, 15(3):167–191.

Schmidt, R. V. and Alferness, R. C. (1990). Directional coupler switches, modulators, and filters using alternating $\delta\beta$ techniques. In Hinton, H. S. and Midwinter, J. E., editors, *Photonic Switching*, pages 71–80. IEEE Press.

Schnabel, R. et al. (1994). Polarization insensitive frequency conversion of a 10-channel OFDM signal using four-wave mixing in a semiconductor laser amplifier. *IEEE Photonic Technology Letters*, 6(1):56–58.

Sharony, J., Cheung, K., and Stern, T. E. (1993). The wavelength dilation concept in lightwave networks - Implementation and system considerations. *IEEE/OSA Journal of Lightwave Technology*, 11(5/6): 900–907.

Stern, T. E. and Bala, K. (1999). *Multiwavelength Optical Networks – A Layered Approach*. Addison-Wesley, Reading, MA.

Subramaniam, S., Azizoğlu, M., and Somani, A. K. (1996). All-optical networks with sparse wavelength conversion. *IEEE/ACM Transactions on Networking*, 4(4):544–557.

Tkach, R. W. et al. (1995). Four-photon mixing and high-speed WDM systems. *IEEE/OSA Journal of Lightwave Technology*, 13(5):841–849.

Wiesenfeld, J. M. (1996). Wavelength conversion techniques. *Proc., Optical Fiber Communication (OFC '96)*, Tutorial TuP 1:71–72.

Yasaka, H. et al. (1996). Finely tunable 10-Gb/s signal wavelength conversion from 1530 to 1560-nm region using a super structure grating distributed Bragg reflector laser. *IEEE Photonic Technology Letters*, 8(6):764–766.

Yoo, S. J. B. (1996). Wavelength conversion technologies for WDM network applications. *IEEE/OSA Journal of Lightwave Technology*, 14(6):955–966.

Yoo, S. J. B., Caneau, C., Bhat, R., and Koza, M. A. (1995). Wavelength conversion by quasi-phase-matched difference frequency generation in AlGaAs waveguides. *Proc., Optical Fiber Communication (OFC '95)*, 8:377–380.

Yoo, S. J. B., Caneau, C., Bhat, R., and Koza, M. A. (1996). Transparent wavelength conversion by difference frequency generation in AlGaAs waveguides. *Proc., Optical Fiber Communication (OFC '96)*, 2:129–131.

Zhou, J. et al. (1994). Four-wave mixing wavelength conversion efficiency in semiconductor traveling-wave amplifiers measured to 65 nm of wavelength shift. *IEEE Photonic Technology Letters*, 6(8):984–987.

III

WAVELENGTH ROUTING NETWORKS

Chapter 4

DESIGN OF LOGICAL TOPOLOGIES FOR WAVELENGTH ROUTED NETWORKS

Rudra Dutta
Department of Computer Science
North Carolina State University
Raleigh, NC 27695-7534
rdutta@csc.ncsu.edu

George N. Rouskas
Department of Computer Science
North Carolina State University
Raleigh, NC 27695-7534
rouskas@csc.ncsu.edu

Abstract   A *virtual topology* over a wavelength routed WAN consists of clear optical channels between nodes called *lightpaths*. These carry traffic end-to-end without electronic switching, creating an optical layer of the topology. Virtual topology design aims at combining the best of optical switching and electronic routing abilities. Designing a virtual topology on a physical network consists of deciding the lightpaths to be set up in terms of their source and destination nodes and wavelength assignment.

In this chapter we provide a complete formulation of the problem and survey the literature on the topic. We restrict ourselves to transport networks rather than local area networks, and static topology design as opposed to topologies in which individual lightpaths are set up and torn down in response to traffic demands.

1.   INTRODUCTION

Wide area "All Optical Networks" with *wavelength division multiplexing* (WDM), using *wavelength routing*, are considered to be candidates for future wide area backbone networks. The ability to tap into attractive properties of optics, including the very high bandwidth potential of optical fiber, makes these networks attractive for backbone transport networks. At the same time, the WDM technique can be used to bridge the mismatch between user and fiber equipment. A fuller discussion of wide area optical networks can be found in [Green, 1992; Mukherjee, 1997; Ramaswami and Sivarajan, 1998; Green, 1996].

In recent times, there has been growing interest in virtual topology design problems on these networks. *Virtual topology* design over a WDM WAN is intended to combine the best features of optics and electronics. The architecture uses clear channels between nodes, called *lightpaths*, so named because they traverse several physical links but information traveling on a lightpath is carried optically from end-to-end. Usually a lightpath is implemented by choosing a path of physical links and reserving a particular wavelength on each of these links for the lightpath. This is known as the *wavelength continuity constraint*, indicating that a lightpath consists of a single wavelength over a sequence of physical links. Because of limitations on the number of wavelengths that can be used, and hardware constraints at the network nodes, it is not possible to set up a clear channel between every pair of source and destination nodes. The particular set of lightpaths we decide to establish on a physical network constitutes the virtual (otherwise called the logical) topology.

The tradeoff involved here is between bandwidth and electronic processing overhead. Forming lightpaths locks up bandwidth in the corresponding links on the assigned wavelength, but the traffic on the lightpath does not have to undergo opto-electronic conversion at intermediate nodes. A good virtual topology trades some of the ample bandwidth inherent in the fiber to obtain a solution that is the best of both worlds.

Optical fiber can be used simply as a point-to-point link carrying only one channel, using one wavelength. The use of WDM increases the bandwidth available and the use of virtual topologies effects reduction of delay, allowing more efficient use of bandwidth by appropriate routing. Fig. 4.1 shows a simple physical network in which lightpaths, indicated by dotted lines, have been set up to allow communication by a clear channel between nodes which are not directly connected by a fiber link. An attractive feature of the process of stepping up from point-to-point fibers to WDM and then virtual topologies is that it can be undertaken in an incremental manner with current networks [Mukherjee et al., 1996]. The virtual topology provides a certain measure of independence from the physical topology, because different virtual topologies can be set up on the same physical topology, and allows us to choose a topology which will result in greater network performance, given network conditions such as average traffic between network nodes.

In general, virtual topology design problems can be formulated as optimization problems aimed at maximizing network throughput or other performance

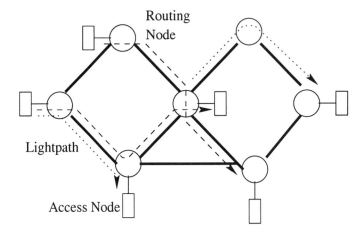

Figure 4.1   A WDM network. The routing nodes are interconnected by point-to-point fiber links and may have access nodes connected to them. The dashed lines and dotted lines show lightpaths.

measures of interest. Typically, the exact solution can be shown to be NP-hard, and heuristic approaches are needed to find realistic good solutions. For this purpose, the problem can be decomposed into four subproblems, which we discuss in detail in Section 3.2.

The above discussion focuses on the issue of throughput or delay optimization, which are related to network performance. There are at least two other important related issues. The first is related to the cost required to set up and operate the network, which is an important practical consideration. Thus, a particular virtual topology may result in lower delay and higher throughput than another, but if the latter virtual topology involves the use of fewer expensive network components such as optical switches or converters, resulting in a lower overall implementation cost, then in practice it may well be chosen over the "better" one. This issue is discussed in Section 2.

The second issue relates to the reconfigurability of optical networks using virtual topology. Reconfigurability is seen as one of the strengths of optical networks in general and the virtual topology approach in particular. A virtual topology is designed on the basis of traffic patterns and a physical topology. Being able to implement a new virtual topology provides adaptability (when traffic patterns change), self-healing capability (when the physical topology changes due to failure of network components) and upgradability (when the physical topology changes due to the addition or upgrading of network components). Thus being able to redesign a virtual topology and configure the network to the new one from the old one is of interest to the virtual topology problem in general, and we have considered it within the scope of this survey.

Scope   A similar virtual topology design problem exists for broadcast optical networks, used as LANs, also known as multihop networks. Virtual topology design for these networks is a different problem. One of the reasons is that with a broadcast medium, the physical topology does not constrain the virtual topologies that can be implemented. Another reason is that since each lightpath in the network needs a unique wavelength, there is no possibility of wavelength reuse as with WDM WANs. A survey of these problems for multihop networks can be found in [Labourdette, 1998].

The virtual topology design problem outlined has been formulated in terms of static traffic demands. That is, the bandwidth demand from one node to another is considered to be known when designing the virtual topology. This is distinct from topology design problems for networks in which we are interested in estimating and obtaining optimum blocking probabilities under dynamic traffic demands, that is, calls which are established and terminated on demand [Chlamtac et al., 1993]. In the present case, if the traffic pattern changes significantly, it would act as the input data for a new virtual topology design. The old virtual topology would be reconfigured to the new virtual topology, a topic we discuss in Section 5.

The advantages of optical technology lie in switching and transmission, not processing or storage. Thus, electronic switching and transmission are more suitable in access networks where the bandwidth requirements are low and processing requirements (as in routing or consolidating) are relatively high. The virtual topology design problem is accordingly defined on transport networks only, not access networks.

The rest of this chapter is organized as follows. In Section 2, the architecture of wavelength routed WANs is described and notations pertaining to these are introduced. Section 3 describes approaches related to network performance optimization, including mathematical formulations and algorithms. Some particular approaches not conforming to any of these categories are described in Section 4. Section 5 addresses the reconfiguration issue. Section 6 concludes the chapter.

2.   ARCHITECTURE AND NOTATIONS

Wavelength Division Multiplexing (WDM) refers to the use of distinct wavelengths over an optical fiber to implement separate channels. An optical fiber can carry several channels in parallel, each on a particular wavelength. An *add/drop multiplexer* (ADM) is an optical system that is used to modify the flow of traffic through a fiber at a routing node [Gerstel et al., 1998]. An ADM passes traffic on certain wavelengths through without interruption or opto-electronic conversions, while other wavelengths are added or dropped, carrying traffic originating or terminating at the node. A *Wavelength Router*

(WR) is a more powerful system than an ADM. It takes in a signal at each of the wavelengths at an input port, and routes it to a particular output port, independent of the other wavelengths [Chlamtac et al., 1993; Ramaswami and Sivarajan, 1996]. A WR with N input and N output ports capable of handling W wavelengths can be thought of as W independent $N \times N$ switches. These switches have to be preceded by a wavelength demultiplexer and followed by a wavelength multiplexer to implement a WR. They are sometimes also called Wavelength Routing Switches (WRS) or wavelength crossconnects. A *Wavelength Converter* is an optical device that can be used in an optical router, to convert the wavelength a channel is being carried on without intermediate opto-electronic conversion [Ramaswami and Sasaki, 1997]. Wavelength conversion allows a clear optical channel to be carried on different wavelengths on different physical links. Different levels of wavelength conversion capability are possible. *Full wavelength conversion* capability implies that any input wavelength may be converted to any other wavelength. *Limited wavelength conversion* denotes that each input wavelength may be converted to any of a specific set of wavelengths, which is not the set of all wavelengths for at least one input wavelength. If a node has limited or full wavelength conversion capability, then the conversion to be effected can be configured as part of the virtual topology design.

The advantage of wavelength conversion is that the virtual topology that can be implemented is less constrained, since the wavelength continuity constraint is removed. Thus wavelength use is more efficient. However, the use of converters increases cost, as well as the complexity of the problem. The cost can be decreased by using limited conversion rather than full conversion, and assuming a small number of converters rather than conversion capability in every node. But these assumptions introduce the problems of specifying the nature of the limited conversion and placement of converters in the network, which greatly increase the difficulty of topology design.

The virtual topology designed and implemented on a physical network not only determines the performance of the network in terms of metrics like throughput, but also carries a cost associated with the virtual topology, determined by how many and what network components are used to implement that virtual topology. Attempting to model the network cost is a topic related to the virtual topology design problem. The primary goal of such studies is to provide a measure of the relative impact of various system components on system cost, and hence provide guidelines for economically efficient virtual topology design, rather than actually determine the cost of implementing a virtual topology. Comparatively few studies have been undertaken in this area. Guidelines that result from such studies may relate to choosing some initial parameters for the virtual topology, as suggested in [Banerjee and Mukherjee, 1997], or may be integrated into the optimization procedure to find the virtual topology. The latter approach is taken in [Chen and Banerjee, 1995], where a heuristic is designed for the topology design problem with a goal of maximizing wavelength utilization in the Wavelength Routers, which would certainly have an impact on the cost of the virtual topology.

2.1.   NOTATIONS

In this section, we define some terminology and notations and introduce some concepts which will be used in the following sections, and which are common to most formulations of the virtual topology problem.

Physical Topology   A graph $G_p(V, E_p)$ in which each node in the network is a vertex, and each fiber optic link between two nodes is an arc. Each fiber link is also called a **physical link**, or sometimes just a link. The graph is usually assumed to be undirected, because each fiber link is assumed to be bidirectional. There is a weight associated with each of the arcs which is usually the fiber distance or propagation delay over the corresponding fiber.

Lightpath   A lightpath is a clear optical channel between two nodes. That is, traffic on a lightpath does not get converted into electronic forms at any intermediate nodes, but remains and is routed as an optical signal throughout. With the usual wavelength continuity constraint, the lightpath becomes a sequence of physical links forming a path from source to destination, along with a single wavelength which is set aside on each of these links for this lightpath.

Virtual Topology   A graph $G_v(V, E_v)$ in which the set of nodes is the same as that of the physical topology graph, and each lightpath is an arc. It is also called the **logical topology**, and the lightpaths are also called **logical links**. This graph is assumed to be directed, since a lightpath may exist from node A to node B while there is none from node B to node A. This graph is also weighted, with the *lightpath distance* of each lightpath (see below) acting as the weight of the corresponding arc.

Link Indicator   Whether a physical link exists in the physical topology from a node l to another node m, denoted by p_{lm} which is 1 if such a link exists in the physical topology and 0 if not.

Lightpath Indicator   Whether a lightpath exists from a node i to another node j, denoted by b_{ij} which is 1 if such a lightpath exists in the virtual topology and 0 if not.

Lightpath Distance   The propagation delay over a lightpath, denoted by d_{ij} for the lightpath from node i to node j. It is the sum of the propagation delays over the physical links which make up the lightpath in the virtual topology.

Physical Degree   The physical degree of a node is the number of physical links that directly connect that node to other nodes.

Virtual Degree   The virtual (or logical) degree of a node is the number of lightpaths connecting that node to other nodes. The number of lightpaths originating and terminating at a node may be different, and we denote them by *virtual out-degree* and *virtual in-degree* respectively. We speak simply of the virtual degree if these are assumed to be equal, as they often are. If this degree is assumed to be same for all nodes of the network, then this is called the virtual degree of the network. The virtual degree is determined in part by the physical degree, but is also affected by the consideration of what volume of electronic switching can be done at a node [Ramaswami and Sivarajan, 1996].

Physical Hops   The number of physical links that make up a lightpath is called the physical hop length of that lightpath.

Logical Hops   The number of lightpaths a given traffic packet has to traverse, in order to reach from source to destination node over a particular virtual topology, is called the virtual or logical hop length of the path from that source to that destination in that virtual topology.

Traffic Matrix   A matrix which specifies the average traffic between every pair of nodes in the physical topology. If there are N nodes in the network, the traffic matrix is an $N \times N$ matrix $\Lambda = [\lambda^{(sd)}]$, where $\lambda^{(sd)}$ is the average traffic from node s to node d in some suitable units, such as arriving packets per second, or a quantized bandwidth requirement. This matrix provides in numerical terms the nature of how the total network traffic is distributed between different source-destination node pairs, that is, the pattern of the network traffic.

Virtual Traffic Load   When a virtual topology is established on a physical topology, the traffic from each source node to destination node must be routed over some lightpath. The aggregate traffic resulting over a lightpath is the load offered to that logical link. If a lightpath exists from node i to node j, the load offered to that lightpath is denoted by λ_{ij}. The component of this load due to traffic from source node s to destination node d is denoted by $\lambda_{ij}^{(sd)}$. The maximum of the logical loads is called the **congestion**, and denoted by $\lambda_{\max} = \max_{i,j} \lambda_{ij}$.

2.2.   ARCHITECTURE

In this section we characterize in more detail the WDM wavelength routed network we have been describing above, and which Fig. 4.1 illustrates. The network consists of several routing nodes which are connected to each other by point-to-point optical fibers, specified by the physical topology. Each of the routing nodes may have access nodes connected to it. For the purposes of virtual topology design, however, only the aggregate traffic between routing nodes is important. Thus we can assume that each routing node has exactly one access node connected to it. We concentrate on the routing nodes and refer to them simply as nodes. The traffic matrix specifies the aggregate traffic from every node to each of the other nodes.

The fiber links connecting the nodes each support a specific number of wavelengths, say W. Each of the nodes is equipped with a WR capable of routing these W wavelengths. In general, no wavelength conversion capability is assumed to exist at any of the nodes.

Lightpaths are set up on the physical topology, creating the virtual topology. A lightpath is set up by configuring the source and destination nodes to originate and terminate a specific wavelength, then choosing a path from the source to destination node and configuring the WR at each intermediate node on that path to forward that wavelength optically to the next node. Two lightpaths that share a physical link must be assigned different wavelengths. The total number of wavelengths used on all links must be W or less. It is usually assumed that the numbers of lightpaths terminating and originating at each node are equal, and this number is same for each node. Thus the network is usually assumed to have a unique logical degree.

Traffic is routed from each source to destination node over a single lightpath if one exists for that source and destination, or a sequence of more than one lightpaths or logical hops. It is usually assumed to simplify the optimization problem that traffic for a single source-destination pair may be bifurcated over different virtual routes. The aim of creating the virtual topology is to ensure that more traffic can be carried with fewer opto-electronic conversions along the way. The extreme case of this would be if a lightpath could be set up from each source to each destination; however, the number of wavelengths available is usually too limited to allow this. At the other extreme is a virtual topology which is identical to the physical topology, so that opto-electronic conversion occurs at every intermediate node. With reasonable and achievable virtual topologies, the number of opto-electronic conversions should not be very large. Together with the fact that in high speed wide area networks the propagation delay dominates over the queueing delay (as long as links are not loaded close to capacity), queueing delays are typically neglected in the problem formulation [Ramaswami and Sivarajan, 1996].

The goal of the virtual topology design process is usually to optimize some network performance metric. Thus, a particular formulation of the problem may seek to minimize network congestion, or minimize average packet delay. In the optimization, usually the number of wavelengths available is taken as a constraint. If both minimizations are desired, then one of them is usually expressed as a constraint by relating it to a known physical network characteristic. In general both are important because too little emphasis placed on the congestion aspect usually results in a virtual topology very similar to the physical topology, and too little emphasis placed on the delay aspect can result in virtual topologies which bear little relation to the physical topology, with long lightpaths that increase delay [Ramaswami and Sivarajan, 1996].

3.   PERFORMANCE OPTIMIZATION

In this section we provide an exact formulation of the virtual topology design problem using the packet traffic approach, and discuss specific techniques and heuristics used to solve it.

3.1.   FORMULATION

The exact formulation of the virtual topology problem is usually given as a Mixed Integer Linear Program. The formulation provided here follows closely that in [Krishnaswamy and Sivarajan, 1998], and also those in [Ramaswami and Sivarajan, 1996; Mukherjee et al., 1994; Mukherjee et al., 1996]. The symbols and terminology are as defined in Section 2.1. New terminology is defined as necessary.

Additional Definitions   Let $H = [h_{ij}]$ be the **allowed physical hop matrix**, where h_{ij} denotes the maximum number of physical hops a lightpath from node i to node j is allowed to take. This hop matrix is one of the ways to characterize the bounds which lightpaths in the virtual topology must be within. Let $c_{ij}^{(k)}$ be the **lightpath wavelength indicator**, *i.e.*, $c_{ij}^{(k)}$ is 1 if a lightpath from node i to node j uses the wavelength k, 0 otherwise. Let $c_{ij}^{(k)}(l, m)$ be the **link-lightpath wavelength indicator**, to indicate whether the lightpath from node i to node j uses the wavelength k and passes through the physical link from node l to node m. Let Δ_l denote the logical degree of the virtual topology.

Objective: Subject to the constraints below, minimize the congestion of the network, that is,

$$\min \lambda_{\max} \tag{4.1}$$

Degree Constraints

$$\sum_j b_{ij} \leq \Delta_l, \quad \forall i \tag{4.2}$$

$$\sum_j b_{ji} \leq \Delta_l, \quad \forall i \tag{4.3}$$

Traffic Constraints

$$\lambda_{ij} \leq \lambda_{\max}, \quad \forall(i,j) \tag{4.4}$$

$$\lambda_{ij} = \sum_{sd} \lambda_{ij}^{(sd)}, \quad \forall(i,j) \tag{4.5}$$

$$\lambda_{ij}^{(sd)} \leq b_{ij}\lambda^{(sd)}, \quad \forall(i,j),(s,d) \tag{4.6}$$

$$\sum_{j} \lambda_{ij}^{(sd)} - \sum_{j} \lambda_{ji}^{(sd)} = \left\{ \begin{array}{ll} \lambda^{(sd)}, & s = i \\ -\lambda^{(sd)}, & d = i \\ 0, & s \neq i, d \neq i \end{array} \right\} \forall(s,d) \tag{4.7}$$

Wavelength Constraints

$$\sum_{k=0}^{W-1} c_{ij}^{(k)} = b_{ij}, \quad \forall(i,j) \tag{4.8}$$

$$c_{ij}^{(k)}(l,m) \leq c_{ij}^{(k)}, \quad \forall(i,j),(l,m),k \tag{4.9}$$

$$\sum_{ij} c_{ij}^{(k)}(l,m) \leq 1, \quad \forall(l,m),k \tag{4.10}$$

$$\sum_{k=0}^{W-1}\sum_{l} c_{ij}^{(k)}(l,m)p_{lm} - \sum_{k=0}^{W-1}\sum_{l} c_{ij}^{(k)}(m,l)p_{ml}$$
$$= \left\{ \begin{array}{ll} b_{ij}, & m = j \\ -b_{ij}, & m = i \\ 0, & m \neq i, m \neq j \end{array} \right\} \forall(i,j),m \tag{4.11}$$

Hop Constraints

$$\sum_{lm} c_{ij}^{(k)}(l,m) \leq h_{ij}, \quad \forall(i,j),k \tag{4.12}$$

Discussion   Most of the above constraints are self-explanatory. Many of them enforce the consistency between the various parameters and variables of the formulation. Constraint (4.7) asserts the conservation of traffic at lightpath endpoints. Expression (4.11) asserts the conservation of every wavelength at every physical node for each lightpath.

The parameters, or inputs, to the formulation are the traffic matrix Λ, the hop bound matrix H, the number of wavelengths in a fiber W, the desired logical degree Δ_l, and the details of the physical topology graph. The variables, whose values at optimum are the "output" of the MILP, relate to the virtual topology graph, wavelength assignment in the virtual topology, and the traffic routing over the virtual topology. The lightpath indicators b_{ij} provide the virtual topology graph. The lightpath wavelength and link-lightpath wavelength indicators provide the wavelength assignments to the lightpaths in the virtual topology and also the physical links implementing each lightpath. Lastly, the virtual traffic load variables λ_{ij} and $\lambda_{ij}^{(sd)}$ provide the routing of the traffic between each source and destination on the virtual topology. This formulation allows for no more than one lightpath from one node to another.

Formulations of this problem are possible that address only some and not all of these aspects. In Section 3.2 we discuss such approaches. Even when all these aspects are addressed, or the same aspect is addressed, different formulations of the problem are possible. Specific formulations can be found in the literature but are not discussed here.

This formulation gets quickly intractable with the size of the network. One of the ways it can be made more tractable is to aggregate traffic from a given source node to all destination nodes, that is, not formulate the problem in terms of the traffic components between each source-destination pair $\lambda^{(sd)}$, but traffic components for each source node $\lambda^{(s)}$ only. This results in a more tractable formulation because the number of variables and constraints is smaller, otherwise the formulation is similar. Of course, a solution to such an aggregation does not provide a complete solution, moreover there may be no corresponding complete solution. However, the aggregate problem, being less constrained than the original one, helps set achievability bounds on the full problem, such as lower bounds on the achievable congestion [Ramaswami and Sivarajan, 1996; Krishnaswamy and Sivarajan, 1998]. Bounds which can be calculated with significantly lower computational costs than solving the full problem are useful in evaluating heuristics, as discussed in Section 3.2.

Usually, such an aggregate formulation is used after relaxing the MILP above into an LP, that is, allowing the lightpath, lightpath wavelength and link-lightpath wavelength indicator variables to take up values from the continuous interval [0, 1] rather than constraining them to be binary variables. The relaxation, like the aggregate formulation, results in a less constrained formulation, When the MILP is relaxed, an extra "cutting plane" constraint is introduced [Ramaswami and Sivarajan, 1996; Krishnaswamy and Sivarajan, 1998], to ensure that the definition of congestion remains consistent with the MILP formulation when traffic components may be weighted with the "fractional lightpaths" that the relaxation introduces.

3.2.   HEURISTICS

The problem, whose exact formulation is given in Section 3.1, and some of its subproblems are known to be NP-hard [Chlamtac et al., 1992; Mukherjee et al., 1994; Krishnaswamy and Sivarajan, 1998; Banerjee and Mukherjee, 1996]. Thus for networks of moderately large sizes it is not practical to attempt to solve this problem exactly. Heuristics to obtain good approximations are needed. In the rest of this section we discuss heuristic approaches to the virtual topology design problem or to related subproblems.

Subproblems   The full virtual topology design problem can be approximately decomposed into four subproblems. The decomposition is approximate or inexact. Solving the subproblems in sequence and combining the solutions may not result in the optimal solution for the fully integrated problem. It is also possible that some later subproblem may have no solution given the solution obtained for an earlier subproblem, so no solution at all to the original problem may be obtained. Although this decomposition follows [Mukherjee et al., 1996], it is also consistent with the decompositions of [Ramaswami and Sivarajan, 1996; Mukherjee et al., 1994; Krishnaswamy and Sivarajan, 1998; Banerjee and Mukherjee, 1996]. The subproblems are as follows.

1 **Topology Subproblem:** Determine the virtual topology to be imposed on the physical topology, that is determine the lightpaths in terms of their source and destination nodes.

2 **Lightpath Routing Subproblem:**   Determine the physical links which each lightpath consists of, that is route the lightpaths over the physical topology.

3 **Wavelength Assignment Subproblem:** Determine the wavelength each lightpath uses, that is assign a wavelength to each lightpath in the virtual topology so that wavelength restrictions are obeyed for each physical link.

4 **Traffic Routing Subproblem:** Route packet traffic between source and destination nodes over the virtual topology obtained.

In terms of the formulation provided in Section 3.1, the topology subproblem consists of determining the values of the lightpath indicator variables b_{ij}, the lightpath routing subproblem consists of determining the values of the variables $c_{ij}^{(k)}(l, m)$, the wavelength assignment subproblem consists of determining the values of the variables $c_{ij}^{(k)}$, and the traffic routing subproblem consists of determining the values of the variables $\lambda_{ij}^{(sd)}$. It may be noted that the above description of the lightpath routing subproblem is approximate, since determining $c_{ij}^{(k)}$ would be redundant after determining $c_{ij}^{(k)}(l, m)$.

The traffic routing subproblem may appear to be not essential to the virtual topology design issue. Indeed, once the virtual topology is fixed by solving the first three subproblems, the traffic routing subproblem is the known one of routing traffic over a given topology, for which many algorithms exist.

However, it is included in the list of subproblems since in the exact formulation it is an integral part of the problem to determine how traffic flows over the virtual topology being designed, as it should be to optimize network performance metrics.

As we remarked above, the decomposition into subproblems is inexact. Exact solution of all the subproblems is also not possible since some of the subproblems are NP-hard as well. Heuristics must be employed to obtain good solutions to the subproblems. This also leads to the possibility of obtaining no solution to the full problem. Some constraints are usually relaxed so that at least some solution is obtained from the heuristics, which can be then tested for near optimality using achievability bounds as we discuss in the following section. One of the constraints which is commonly relaxed is that of the maximum number of wavelengths that can be carried by a fiber. Sanity checks must be performed at the end to verify that the solution obtained is feasible.

The virtual topology problem can be decomposed into different subproblems than the ones we list above. Such different decompositions are used in many of the studies we survey. However, we consider the above decomposition to be reasonable and fairly consistent with any others proposed in the literature we survey, and we shall refer only to this decomposition while discussing such studies.

Bounds To evaluate an approximate solution produced by a heuristic, we would like to know how close the obtained solution is to the optimal one. Since we are using the heuristic because of the very reason that the optimal solution cannot be obtained in the first place, we must resort to comparing the solution obtained with known bounds on the optimal solutions derived from theoretical considerations. These are the achievability bounds we have mentioned before (they are bounds on what can be achieved in principle) and we discuss them below.

Lower Bounds on Congestion The goal of virtual topology design is often to minimize network congestion, as in our formulation in Section 3.1. A lower bound on the congestion obtained from theoretical considerations allows us to know that an even smaller value of congestion cannot be achieved by any solution, and helps us evaluate the solution produced by some heuristic. We discuss several lower bounds on congestion below. Our discussion follows closely that of [Ramaswami and Sivarajan, 1996], and also that of [Krishnaswamy and Sivarajan, 1998], as well as literature on virtual topology problems in broadcast LAN scenarios as referred to in [Ramaswami and Sivarajan, 1996; Labourdette, 1998]. More details can be found in these sources.

Physical topology independent bound: This bound utilizes the fact that the load on each logical link would be the same, and this would be the congestion, if the total traffic in the network were equally distributed among all the lightpaths. The value of this congestion would then act as a lower bound on any virtual topology that could be designed for the network under the given traffic conditions. This bound takes into account the total traffic demand, but not the distribution of total traffic among the different source-destination pairs (that is, the traffic pattern). As such, it assumes that traffic for any source-destination pair can be assigned to any lightpath in the virtual topology, and hence, it ignores the physical topology.

Let $\overline{H}$ be the traffic weighted average number of logical hops in the virtual topology. If E_l denotes the number of lightpaths in the virtual topology and r denotes the total arrival rate of packets to the network, then it is easy to see that $\lambda_{\max} \geq r\overline{H}/E_l$, and setting a lower bound on $\overline{H}$ results in a lower bound on the congestion. For the traffic weighted number of hops to be minimum, source-destination pairs with the largest amount of traffic must be connected by a small number of logical hops. Since there can be a maximum of $N\Delta_l$ node pairs connected by a single logical hop, we assume these are exactly the node pairs with the largest traffic between them, and similarly for two, three, and larger number of logical hops. The traffic weighted average number of logical hops in this case is a lower bound. This is given by the expression $\overline{H} \geq \sum_k kS_k$ where S_k is the sum of the traffic fractions (with respect to total network traffic r) which are assumed to be carried in k hops. These traffic fractions can be determined as follows: arrange the traffic fractions in descending order of magnitude, and divide them in blocks, the i-th block being made up of $N\Delta_l^i$ successive fractions in that list. Thus the first block consists of the first $N\Delta_l$ traffic fractions, the second block consists of the next $N\Delta_l^2$ traffic fractions, and so on. Then the sum S_k is the sum of the traffic fractions which form the k-th block.

Minimum flow tree bound: This bound is derived from similar considerations as above, but on a per node basis. In this bound, we take into account the restriction that each source node can only source Δ_l lightpaths altogether, in addition to the considerations above. Thus this is a stronger bound. The traffic weighted average number of logical hops $\overline{H}$ is bounded by assuming that each node is connected by one logical hop to the Δ_l nodes to which it has the largest amounts of traffic, by two hops to the Δ_l^2 nodes to which it has the next largest amounts of traffic, and so on. We omit the derivation and exact expression of this bound, which can be found in [Ramaswami and Sivarajan, 1996].

Iterative bound: This type of bound is developed in [Ramaswami and Sivarajan, 1996; Krishnaswamy and Sivarajan, 1998] by aggregating and then relaxing the MILP formulation and solving it as mentioned in Section 3.1. The additional constraint imposed on the relaxed aggregate formulation is that the congestion be higher than a lower bound on the congestion known *a priori*, such as the minimum flow tree bound discussed above. To improve the tightness, the value obtained for the congestion by solving the relaxed aggregate LP can be used as a new value of the *a priori* bound and the LP solved again to yield a further tightened bound on the congestion. This iterative process can be carried out repeatedly to improve the tightness of the bound.

Independent topologies bound: This bound is proposed in [Banerjee et al., 1997] as another method of taking into account the physical topology in computing a bound on the congestion. Successive topologies are derived to maximize one-hop, two-hop, three-hop traffic etc., and these topologies are not allowed to constrain each other. Finally the resulting congestion is read out as a bound. The authors note that this bound is only a little tighter than the flow tree based bound if traffic is uniform, but becomes much tighter for highly nonuniform traffic.

Lower Bounds on the Number of Wavelengths It is usually necessary in virtual topology design to complete the design using as few distinct wavelengths as possible, since in practice there is a limit on the number of wavelengths a fiber can carry. This limit may be known and introduced in the exact formulation as in the formulation of Section 3.1, but such a limit is often not included in heuristic approaches. A lower bound on the number of wavelength needed for a particular problem is then useful in evaluating the solution provided by the heuristic. Also, in the presence of practical limitations, an easy-to-compute lower bound on the number of wavelengths can provide a quick negative answer to the question of whether a virtual topology design problem is at all feasible or not. Two such bounds can be found in [Ramaswami and Sivarajan, 1996]. The first bound is derived from the simple consideration that the node with the minimum physical degree δ_p must source Δ_l lightpaths. Then the number of wavelengths required is bounded from below by $W \geq \lceil \Delta_l/\delta_p \rceil$. The second bound is derived by assuming that each node sources lightpaths to exactly those nodes it can reach with the minimum number of physical hops. We omit the derivation and exact expression of this bound, which can be found in [Ramaswami and Sivarajan, 1996].

Bounds on the number of wavelengths required can also be found under specific assumptions regarding the solution to the different subproblems. Two such bounds are demonstrated in [Chlamtac et al., 1993], one based on the assumption that the virtual topology being implemented is a hypercube and a specific node mapping algorithm is used, the other relating the number of wavelengths needed for topologies with and without the wavelength continuity constraint.

Heuristic Approaches and Techniques   In the design of heuristics or approximate solutions to the virtual topology problem, emphasis is placed on different aspects of the problems by different authors. In the majority of the literature, heuristics are designed for only some and not all the subproblems. Some assumption regarding the nature of the virtual topology to be implemented is often a starting point for heuristic methods. Below we discuss heuristics found in the literature surveyed under three different categories. In the first, it is assumed that the virtual topology to be implemented is a well-known regular topology, such as a hypercube or a shufflenet. In the second, the lightpaths of the virtual topology are assumed to be already known in terms of sources and destinations for each instance of the problem, and the lightpath routing and wavelength assignment subproblems are addressed. No particular assumption is made regarding the virtual topology in the last category. Some of the interest in the study of regular topologies in the context of virtual topologies for WANs came from the assumption that to some extent the virtual topology could dictate the physical topology, that is, fibers could be laid to supplement a physical topology before implementing a virtual topology. As more and more fiber has been laid in practice and has become part of single wavelength optical networks utilizing the fibers as point-to-point links, the concern has shifted to extracting more utilization out of these fibers using WDM and virtual topologies, rather than having to lay more fibers. Thus studies relating to arbitrary physical topologies have attracted more interest in recent times.

Regular Topologies   Regular topologies such as hypercubes or shufflenets have several advantages as virtual topologies. They are well understood, and results regarding bounds and averages are comparatively easier to derive. Routing of traffic on a regular topology is usually also simpler and results are available in the literature, so the traffic routing subproblem usually becomes trivial. Also, regular topologies possess inherent load balancing characteristics.

Once a regular topology is decided upon as the one to be implemented as a virtual topology, it remains to decide which physical node will realize each given node in the regular topology (this will be referred to as the node mapping subproblem) and which sequence of physical links between two physical nodes will be used to realize each given edge in the regular topology, that is, lightpath (this will be called the path mapping subproblem). This procedure is also called embedding a regular topology in the physical topology. In terms of the subproblems introduced in Section 3.2, the choice of the regular topology together with the node mapping problem make up the virtual topology subproblem, and the path mapping problem corresponds to the lightpath routing subproblem. Obviously, the number of nodes in the (regular) virtual topology may not be chosen with complete freedom, instead it must obey the constraints of the regular topology. In case the physical topology has a few nodes less than the regular topology, this can usually be circumvented by adding fictitious nodes to it before embedding [Mukherjee et al., 1994]. If a few more nodes are present in the physical topology, then some of the ones with less traffic may be combined for the purpose of embedding, though this introduces further approximations to the virtual topology solutions. In general, the node mapping and path mapping problems leave out of consideration the network traffic pattern, and utilize metrics such as fiber distance and wavelength reuse to route lightpaths over the physical topology. Thus there is a tacit assumption of a uniform or close to uniform traffic pattern in the use of regular topologies as virtual topologies. The mappings must also be free of wavelength clashes and must obey any predefined limit on the number of wavelengths. These problems are themselves known or conjectured to be NP-hard [Chlamtac et al., 1993; Mukherjee et al., 1994], hence heuristics are needed for them.

We omit a detailed discussion of the studies available in the literature on this topic and only mention a few of them. In [Chlamtac et al., 1993], the physical topology is first mapped into an "equivalent string" preserving some characteristics of the topology, and then the selected regular topology is embedded into this string. Three different regular topologies are compared for their suitability as virtual topologies in [Marsan et al., 1993]. The comparison is based on the number of logical hops between nodes and the number of wavelengths required. In [Mukherjee et al., 1994], heuristics are developed for the node mapping problem for regular topologies (specifically, hypercubes), and the solution is carried through to path mapping as well as wavelength assignment. Thus, the node mapping part of the virtual topology subproblem, the lightpath routing, and the lightpath wavelength assignment subproblems are addressed. A greedy algorithm and a simulated annealing algorithm are specified for obtaining an initial mapping and then refining the mapping based on considerations of minimizing overall network message delay. In [Mukherjee et al., 1996], a very similar simulated annealing heuristic is presented, with the difference that the traffic routing subproblem is assumed to be solved using the flow deviation method. The flow deviation method is a good heuristic alternative to an exactly optimal linear programming routing flow solution. The literature in which it was developed is referred to in [Mukherjee et al., 1996] and also [Labourdette, 1998]. This method starts from an initial flow assignment, and iteratively deviates flows over alternate paths, avoiding links carrying the largest amounts of traffic.

Pre-specified Topologies   We now discuss studies which focus on the light-path routing subproblem, and possibly the wavelength assignment and traffic routing subproblems. In other words, the virtual topology in terms of a list of lightpaths with their source and destination nodes is supposed to be given for each instance of the problem.

The traffic pattern in the network would certainly have been taken into account when the lightpaths were decided upon. Because of this, in some approaches we discuss below, the traffic pattern is not taken into account, though it is also possible to utilize the traffic pattern information in routing lightpaths. The lightpath routing and wavelength assignment subproblems can then be viewed as having goals defined purely in terms of the lightpaths, such as minimization of the number of distinct wavelengths needed.

We mention a few of the studies in the literature, omitting detailed discussion as before. In [Chlamtac et al., 1992], not only the source and destination, but also the routing of the lightpaths is assumed to be given. That is, the lightpath wavelength assignment subproblem is addressed, and it is called the Static Lightpath Establishment (SLE) problem. It is proved that SLE as stated is equivalent to the n-graph-colorability problem, and hence NP-complete. A greedy heuristic algorithm to assign wavelengths to a given set of lightpaths with the aim of using as few wavelengths as possible is presented. The study presented in [Chen and Banerjee, 1995] assumes that the virtual topology subproblem has been solved and the set of lightpaths to be established is available in terms of the source and destination nodes of the lightpaths. The objective for the routing and wavelength assignment problem presented is to maximize wavelength utilization at the switches. This objective is presented in terms of the utilization of the Wavelength Routers at each network node. To formally define the problem, the concept of a "Latin Square" is introduced. Lightpath routing and wavelength assignment is posed in terms of completion of partial Latin Squares. Two heuristic algorithms are presented to complete partial Latin Squares at individual network nodes, and then a scheme is specified to use these in combination to solve the lightpath routing and wavelength assignment problem at the network level. In [Banerjee and Mukherjee, 1996], the virtual topology is assumed to be given in terms of a list of lightpaths with their source and destination nodes for each instance of the problem. The lightpath routing problem is formulated in terms of lightpath traffic as a multi-commodity flow problem which is known to be NP-complete. It is suggested that the problem size can be reduced considerably by customizing the formulation for each instance of the problem, by pruning the search tree for lightpath routes and relaxing integer constraints. The study employs known heuristic methods, including randomized rounding and graph coloring, with provably good characteristics, to address the lightpath routing and wavelength assignment problems.

Arbitrary Topologies There are various studies proposing heuristic methods for arbitrary virtual topologies. These studies address the virtual topology subproblem itself, as well as some or all of the subsequent subproblems of virtual topology design. Most of these methods take into account the effect of the network traffic pattern, since arbitrary virtual topologies are usually called for in response to non-uniform traffic patterns and irregular physical topologies. Some of the heuristics proposed are similar to each other.

In [Zhang and Acampora, 1995], the problem is looked upon as the establishment of an optical connection graph over a WAN based on the average traffic demand, and then using demand based routing on this connection graph, that is, dynamic virtual circuits, which allocate whole lightpaths at a time. The connection graph subproblem presented is therefore identical to the first three subproblems of the virtual topology problem as presented in Section 3.2. The problem is formulated as a nonlinear integer programming problem, and an approximate decomposition is presented. The heuristic algorithm is then presented, which is based on a greedy approach, and attempts to utilize the number of wavelengths used for a maximum number of lightpaths.

Several different heuristics are presented in [Ramaswami and Sivarajan, 1996], including one which also attempts to create lightpaths between nodes in order of decreasing traffic demands. Lightpaths are established between the nodes that have the maximum amount of traffic between them. If all traffic is accounted for but each node does not have the required degree, the rest of the lightpaths are placed at random obeying the constraints. A modified version of this heuristic is also presented in which a pair of lightpaths in opposite directions is initially set up for each physical edge, then the original algorithm is exactly followed. This ensures that traffic can always be routed on the shortest physical path between any two nodes and hence can satisfy any physically realizable delay constraints. Another heuristic depends on the iterative bound developed in this study by relaxing the MILP formulation as described in Section 3.2, and rounding off the lightpath indicators. Finally, a heuristic is presented that does not take into account the traffic pattern at all, but concentrates on creating lightpaths that use only a few physical edges, since this should conserve wavelengths.

A similar heuristic maximizing one logical hop traffic is briefly described in [Banerjee and Mukherjee, 1997], but a heuristic with the opposite objective is also suggested. This heuristic aims at maximizing multihop traffic, since concentrating only on single hop traffic can lead to congestion due to multihop traffic. Some results are provided in which the two approaches appear to perform very similarly to each other. Details of wavelength assignment are not discussed.

The study in [Banerjee et al., 1997] also suggests that attempts to maximize one logical hop traffic concentrate on the comparatively larger traffic components, and may cause the smaller traffic components to be routed unreasonably and cause congestion on some physical links. A scheme involving mapping the network to a bipartite graph is specified to avoid unbalanced loading. A known graph algorithm is then specified as a heuristic for wavelength assignment.

In [Krishnaswamy and Sivarajan, 1998], a heuristic algorithm following the LP relaxation heuristic from [Ramaswami and Sivarajan, 1996], but more complete, is presented. The lightpath wavelength indicator variables are also rounded and then a least resistance algorithm followed to choose a single routing for each lightpath. A final phase of the design eliminates wavelength clashes.

4.   RELATED APPROACHES

In this section we discuss some techniques and algorithms that are different from those described in Section 3.2, but which are related to the problem of virtual topology design for wavelength routed networks.

Incremental Benefit Analysis: In [Mukherjee et al., 1996], a study of the incremental benefits of introducing a virtual topology over optical WANs is undertaken. A realistic traffic pattern is obtained from the T1 NSFNET backbone data of January 1992. Three schemes are applied on this data to scale up the traffic pattern. The first scheme merely used efficient routing to establish a baseline and the other two used WDM and virtual topology with WDM respectively. The most dramatic result was in the increase of the scale-up factor, from 49 and 57 in the first two schemes to 106 in the third, and the link utilization which went from 32% and 23% in the minimum loaded link to 71% in the last one (the maximum link load remained 99% in all three schemes). Thus this analysis provides demonstration of the benefits of implementing a virtual topology, as well as the incremental nature in which it may be undertaken.

Limited Conversion: The motivation for the study presented in [Ramaswami and Sasaki, 1997] is the lower cost associated with limited conversion of wavelengths at Wavelength Routers as opposed to full conversion, as we remarked in Section 2. The wavelength assignment subproblem is the focus of this study. Several results are obtained in theoretical terms about ring networks with specific wavelength conversion capabilities. These results are followed by constructive proofs rather than simply existence proofs, so that a blueprint is provided for the actual construction of such ring networks. Some results are derived for more general physical network topologies.

Traffic Grooming: As we have remarked in Section 1, each lightpath has a high bandwidth and it may not be possible for single users to utilize this bandwidth. Lightpaths must be viewed as transport channels in the backbone network, in which traffic from multiple user applications is multiplexed in by access networks. In a sense, this is the justification for including the traffic routing subproblem in the virtual topology design problem, since traffic for individual applications must be routed onto the virtual topology provided, so that lightpaths carry traffic obtained by aggregating lower speed traffic streams. The pattern of multiplexing traffic onto lightpaths affects the efficiency of optical forwarding of information through Wavelength Routers, since all information in an entire lightpath will need to undergo opto-electronic conversion and electronic routing at an intermediate node if even one lower speed traffic stream from that lightpath has to be terminated at the intermediate node. This also reflects in the cost (in numbers and capabilities) of network components needed. [Gerstel et al., 1998] addresses some of these issues. Different ring architectures are specified and compared on the basis of results derived regarding the average number of transceivers at the nodes, number of wavelengths, average number of physical hops and characterization of traffic patterns on which they perform best. Similar issues arising in arbitrary physical topologies and extension to actual grooming methods would appear to be areas worth further investigation.

Generalized Lightpaths: In [Sahasrabuddhe and Mukherjee, 1999], the concept of a lightpath is generalized into that of a *light tree*, which, like a lightpath, is a clear channel implemented with a single wavelength with a given source node. But unlike the lightpath, a light tree has multiple destination nodes, thus a light tree is a point-to-multipoint channel. It is emphasized that topologies using light trees would be more optimal for any given situation since the light tree is a more general construct, and there may be possibilities of optical multicast.

As we already know, optimal solutions are not practically obtainable, and with a more general construct and hence a much larger search space this is going to be even more true. Heuristic solutions will have to be designed to obtain good solutions, and must be tailored to suit the larger search space. With unicast traffic problems, the light-tree approach trades off more bandwidth to further improve delay, congestion, and physical hop characteristics than the lightpath approach. This is the tradeoff we mentioned in Section 1. The challenge in this case will be to design heuristics that can cope with the increased complexity of the problem and yet produce solutions in which a good tradeoff is achieved.

5.   RECONFIGURATION ISSUES

As we have already remarked, the problem of reconfiguring a network from one virtual topology to another is a related problem to virtual topology design. Two possible approaches to this problem are discussed in this section.

Cost Approach   In this approach, it is assumed that the current virtual topology as well as the new virtual topology that the network must be reconfigured to are known, together with the physical topology details. The concern is to minimize the cost of the reconfiguration. The cost can be expressed in terms of the number of Wavelength Routers that need to have their optical switching reprogrammed, or the total number of optical switchings that need to be changed to implement the new lightpaths and eliminate old ones. These metrics are appropriate since they reflect the amount of time the network must be taken off line to make the changes, as well as the reprogramming effort for the reconfiguration. Other similar metrics may also be applicable. It may be the case that the network cannot be taken off line at all, but that a succession of intermediate virtual topologies have to be designed to eliminate single, or groups of, routers which can be reconfigured and put back in operation. Much more complicated metrics reflecting total time taken to reconfigure as well as the effort to redesign the intermediate topologies need to be developed in this case.

We have not found any study of these reconfiguration problems in the literature for wavelength routed WANs, though studies involving the reconfiguration of virtual topologies for broadcast LANs exist, as detailed in the survey of related literature carried out in [Labourdette, 1998]. These studies involve link-exchange and branch-exchange techniques to minimize the cost of converting one virtual topology into another, and similar methods may be possible for wavelength routed network which are the topic of this survey.

Optimization Approach   Another approach is to assume that only the current virtual topology is given, together with the changed traffic pattern and/or physical topology that makes reconfiguration necessary. This is the approach taken in [Banerjee and Mukherjee, 1997]. The reconfiguration algorithm proposed involves solving the new virtual topology problem on its own without reference to the current virtual topology to obtain a new optimal solution, with a new optimal value for the objective function which is noted. The virtual topology design problem is then reformulated with an additional constraint that constrains the old objective function to this noted value, and a new objective function that involves minimizing the number of lightpaths that must be either added or removed.

While this method is guaranteed to find a solution that results in a virtual topology that is optimal for the new conditions, it does not achieve a balance between finding an optimal new virtual topology and one that involves as little change from the old one as possible. It is possible that a very costly reconfiguration will be undertaken for only a slight gain in network performance. More balanced formulations of this problem may be possible, and heuristics designed on such formulations are likely to perform better in practice.

6.   CONCLUDING REMARKS

The problem of virtual topology design for wide area wavelength routed optical networks covers a considerable area, and many approaches to this and related problems have been taken in the literature. The lightpaths of a virtual topology are set up to trade off the ample bandwidth available in the fiber with the opto-electronic conversion and electronic processing at intermediate nodes. Exact formulations of the problem are known to be computationally intractable, so heuristics for determining and implementing a virtual topology have been proposed. Most heuristics attempt to address parts of the problem rather than the whole, by decomposing the problem approximately into subproblems, or address special cases of network topology.

Virtual topology design is a growing research area. New areas of investigation include extending results obtained for special cases to broader context, and extending the freedom allowed in formulating the problem to take advantage of improving equipment capabilities.

References

Banerjee, D. and Mukherjee, B. (1996). A practical approach for routing and wavelength assignment in large wavelength-routed optical networks. *IEEE JSAC*, 14(5):903–908.

Banerjee, D. and Mukherjee, B. (1997). Wavelength-routed optical networks: Linear formulation, resource budgeting tradeoffs, and a reconfiguration study. *Proc. IEEE INFOCOM*, pages 269–276.

Banerjee, S., Yoo, J., and Chen, C. (1997). Design of wavelength-routed optical networks for packet switched traffic. *Journal of Lightwave Technology*, 15(9):1636–1646.

Chen, C. and Banerjee, S. (1995). Optical switch configuration and lightpath assignment in wavelength routing multihop lightwave networks. *Proc. IEEE INFOCOM*, pages 1300–1307.

Chlamtac, I., Ganz, A., and Karmi, G. (1992). Lightpath communications: An approach to high bandwidth optical wans. *IEEE Trans. Communications*, 40(7):1171–1182.

Chlamtac, I., Ganz, A., and Karmi, G. (1993). Lightnets: Topologies for high-speed optical networks. *Journal of Lightwave Technology*, 11(5/6):951–961.

Gerstel, O., Ramaswami, R., and Sasaki, G. (1998). Cost effective traffic grooming in WDM rings. *Proc. IEEE INFOCOM*, pages 69-77.

Green, P. (1992). *Fiber optic network*. Englewood Cliffs, NJ Prentice Hall.

Green, P. (1996). Optical networking update. *IEEE JSAC*, 14(5):764–779.

Krishnaswamy, R. and Sivarajan, K. (1998). Design of logical topologies: a linear formulation for wavelength routed optical networks with no wavelength changers. *Proc. IEEE INFOCOM*, pages 919–927.

Labourdette, J.-F. P. (1998). Traffic optimization and reconfiguration management of multiwavelength multihop broadcast lightwave networks. *Computer Networks and ISDN Systems*, 30(9-10):981–998.

Marsan, M., Bianco, A., Leonardi, E., and Neri, F. (1993). Topologies for wavelength-routing all-optical networks. *IEEE/ACM Trans. Networking*, 1(5):534–546.

Mukherjee, B. (1997). *Optical communication networks*. McGraw-Hill.

Mukherjee, B., Ramamurthy, S., Banerjee, D., and Mukherjee, A. (1994). Some principles for designing a wide-area optical network. *Proc. IEEE INFOCOM*, pages 110–119.

Mukherjee, B., Ramamurthy, S., Banerjee, D., and Mukherjee, A. (1996). Some principles for designing a wide-area optical network. *IEEE/ACM Trans. Networking*, 4(5):684–696.

Ramaswami, R. and Sasaki, G. (1997). Multiwavelength optical networks with limited wavelength conversion. *Proc. IEEE INFOCOM*, pages 489–498.

Ramaswami, R. and Sivarajan, K. (1996). Design of logical topologies for wavelength-routed optical networks. *IEEE JSAC*, 14(5):840–851.

Ramaswami, R. and Sivarajan, K. (1998). *Optical networks: a practical perspective*. Morgan Kaufmann Publishers.

Sahasrabuddhe, L. and Mukherjee, B. (1999). Light-trees: Optical multicasting for improved performance in wavelength-routed networks. *IEEE Communications Magazine*, pages 67–73.

Zhang, Z. and Acampora, A. (1995). A heuristic wavelength assignment algorithm for multihop WDM networks with wavelength routing and wavelength re-use. *IEEE/ACM Trans. Networking*, 3(3):281–288.

Chapter 5

A TAXONOMY OF SWITCHING TECHNIQUES

Chunming Qiao
Department of CSE
SUNY at Buffalo, Buffalo, NY 14260
qiao@computer.org

Myungsik Yoo
Department of EE
SUNY at Buffalo, Buffalo, NY 14260

Abstract   Many switching techniques, ranging from the well-known *circuit-* and *packet-switching* to new switching techniques proposed for optical (WDM) networks such as *wavelength-routing* and *optical burst switching* (OBS) will be described and compared in terms of their suitable applications and implementation technologies. This chapter's emphasis will be on the unique characteristics of optical networks and issues related to optical switching techniques. In particular, we will discuss how OBS can be effective in building the next generation Optical Internet.

1.   INTRODUCTION

In this chapter, we will concentrate on *switched* networks consisting of reconfigurable switches. Here, switches broadly refer to devices that may also be called routers, cross-connects and add-drop multiplexers (ADMs). In particular, an *optical* switch is the one that can switch an optical *data* signal without converting it from the optical domain to the electronic domain, and then back to the optical domain (i.e. O/E/O conversions), although the switch may still be controlled by electronic signals. Switches are common in a backbone (or core) network, as opposed to an access network. In a typical (switched)

---

[1]This research is supported in part by National Science Foundation under grants MIP-9409864 and ANIR-9801778.

network, not all the nodes (where a node consists of a switch and its controller) are linked to each other directly, and hence a path from one to another will span multiple links (hops) and go through intermediate nodes.

Recently, the increasing demand for a transparent networking infrastructure that can provide integrated voice and data services has inspired research on optical networks. As optical communication technologies, and in particular, wavelength division multiplexing (WDM) technologies for both transmission and switching mature, research focus has been gradually shifting from local area networks (LANs) based mainly on WDM star-couplers to metropolitan and wide area networks (MANs and WANs) based mainly on rings and meshes using optical (WDM) switches. The WDM links and switches also form a so-called WDM layer, with respect to the above electronic layer that consists of Internet Protocol (IP) routers, Asynchronous Transfer Mode (ATM) switches, and/or Synchronous Optical NETworks/Synchronous Digital Hierarchy (SONET/SDH)ADMs [Bertsekas and Gallager, 1992; McDysan and Spohn, 1994; Spragins et al., 1994]. In addition, although many switching techniques ranging from *circuit-switching* to *packet-switching* have been studied for both voice and data communications for over a hundred years (ever since telephone switching offices were established), new switching techniques for the optical layer such as *wavelength-routing* and *optical burst switching* (OBS) are still being developed.

In this chapter, we will describe switching techniques that prevail in today's voice and data networks, as well as those showing promises for optical (WDM) networks. Note that it is not our intention, nor it seems possible, to include all relevant materials in this chapter. In addition, some terms (and acronyms) may have different meanings to different people, but when describing them, we will provide a common interpretation (if any), or our own interpretation (especially when describing terms/acronyms we have coined by ourselves, e.g. OBS [Qiao, 1997; Qiao and Yoo, 1999]). Finally, in order to facilitate the presentation, we will introduce a few new terms and acronyms (such as *reserve-a-fixed-duration* or RFD).

The rest of the chapter is organized as follows. In Section 2, we will first review *circuit-switching* and *packet-switching* based on either datagrams (e.g. IP packet routing) or virtual circuit (e.g. ATM cell switching), including Multiprotocol label switching (or MPLS) [Callon et al., 1997]. In Section 3, we will describe *burst-switching*, list its variations and explain why it differs from its better-known counterparts, namely, circuit- and packet- switching. In Section 4, we will establish a mapping between these switching techniques and those proposed for the optical layer, with a focus on how the unique characteristics of optical networks affect the choice of suitable switching techniques. Finally, we will discuss how optical burst switching (OBS) can be effectively applied to the next generation Optical Internet in Section 5, and summarize the chapter in Section 6.

2.   CIRCUIT- AND PACKET-SWITCHING

There are two basic switching paradigms, one is circuit-switching and the other is packet-switching. The former is mainly for voice communications while the latter is mainly for data communication. Hereafter, the term "data" will be used to refer to payload in general which includes voice as well as ordinary data unless there is clearly a need to distinguish the two. In contrast, all overhead (i.e. non-payload) will be referred to as "control" (i.e. "signaling") information which may include network addresses for routing purposes, and error checking/correction codes. In addition, we will use the term "connection" to refer to a communication session at the application layer, e.g., a phone call for which a circuit is established as in today's telephone networks, or a *Telnet* session during which no circuit is set up, and data is transferred in packets (as to be discussed in more details next).

2.1.   CIRCUIT-SWITCHING

In circuit-switching, there are three distinct phases: circuit set-up, data transfer and circuit tear-down. In the first phase, only control information (e.g. a set-up request, an acknowledgment) is exchanged to set up an end-to-end circuit between a source and a destination[1], which uses a dedicated channel of a fixed bandwidth, e.g. a time-slot[2] or a frequency, on each link along a path from the source to the destination. The intermediate switches are also configured to "latch" the channels to form a circuit. Afterwards, data (and only data) is transmitted in the second phase which lasts for the duration of the connection. Finally, after data transfer is complete, the circuit is released in the third phase.

Circuit-switching is suitable for an application requiring data transmissions at a constant bit-rate that matches the channel bandwidth, and the connection duration is long relative to the circuit set-up time. Since no processing is needed at any intermediate node once a circuit is set up, circuit-switching does not require the use of fast switches (although fast switches may help reduce circuit set-up time) and any buffer at the intermediate nodes (except for the delaying mechanism needed for interchanging time slots [Hui, 1990]).

Given that there is usually more than one path between any given source and destination pair, *routing*, which basically refers to the determination of the path to be taken, is required. In circuit-switching, routing is a part of circuit set-up, which may be done under centralized control or distributed control.

A variation of circuit-switching used in TASI (time assignment speech interpolation) systems is called *fast circuit-switching*, where the first phase involves only routing, that is, it does not set up a circuit. Circuit set-up (or tear down) takes place when the beginning (or end) of a burst is detected by sending a special control signal, and is *fast* since routing has already been done. In addition, circuit set-up can be a two-way process as in circuit-switching or a one-way process where a burst is sent after the special control signal without waiting for the acknowledgment that the circuit has been set up.

2.2.   PACKET-SWITCHING

Packet-switching typically uses distributed routing control. One major difference between circuit-switching and packet-switching (even if distributed routing control is used in both cases) is that in the latter, data can be sent without setting up a circuit. More specifically, in packet-switching, data (e.g. a message) is transmitted in packets, each of which contains a header with some control information, and is sent to its destination via intermediate nodes in a "store-and-forward" fashion. That is, when a packet arrives at a node, it is stored first and then after the packet header is processed, forwarded to the next node. This implies that in packet-switching, a switch is configured only after the data (i.e. a packet) arrives.

A packet can have either a fixed length (as in digitized voice packet-switching), or a variable length with a limited maximum size. A similar technique is *message-switching* in which an entire message (of a large size) may be sent along with a header, thus reducing (the percentage of) control overhead. However, due to the store-and-forward nature, this requires a larger buffer at each node than breaking the message into smaller packets. In addition, it may take longer for the message to arrive at its destination because the message has to be completely assembled (and received) at the source (and each intermediate node, respectively) before its transmission can start.

Packet-switching is more suitable for bursty traffic since it allows statistical sharing of the channel bandwidth among packets for different source and destination pairs. Packet-switching requires buffering at the intermediate nodes, and fast switches to keep the percentage of the control overhead down.

Datagram and Virtual-Circuit (VC)

There are two basic variations, one based on datagrams and the other based on virtual circuits (VCs). In datagram based packet-switching, a header is similar to a circuit set-up request (under distributed control) in that they both contain similar control information and are processed in a similar way at each intermediate node. However, a packet's header and its payload are normally sent on the same channel without an *offset* time (i.e. a time gap or an idle period) in-between, while in circuit-switching, a set-up request and data are sent on two separate channels with an offset time equal to the circuit set-up time (which is at least as long as the round-trip delay). Note that Internet Protocol (IP) uses datagram-based packet-switching.

In VC-based packet-switching, there are two phases, one for setting up a VC (or routing) just as in fast circuit-switching, and the other for sending packets over the VC (or switching). A third phase for tearing down the VC may also be needed. Note that setting up a VC is different from setting up a circuit in that the former does not require any dedicated bandwidth (channel) on each link - it just creates an entry in the switching table at each intermediate node along a selected path. Such an entry basically maps an incoming *label* (e.g. an VC identification number) to an output port. In the second phase of the VC-based packet-switching, each packet contains a label, and when an exact matching between the label and one already in the switching table is found, the packet is forwarded to the appropriate output port (and possibly assigned with a new label).

Finding the exact label matching is easier (faster) than making a routing decision[3]. ATM, which at one time was called fast packet-switching, uses a similar technique called cell-switching, where a cell is just a small packet of a fixed length (53 bytes). In fact, ATM switches in general have a lower cost (in dollars) to performance (in throughput) ratio than IP routers.

Multiprotocol Label Switching (MPLS)

The idea of Multiprotocol Label Switching (MPLS) [Callon et al., 1997], which is currently being standardized by the Internet Engineering Task Force (IETF), is similar to that of VC-based packet-switching. More specifically, instead of making a routing decision for each packet independently as in datagram-based packet-switching, a label-switched path (LSP), which is similar to a VC, will be established so that the routing decision for each packet is made only once at its source by assigning a label to each packet. This allows MPLS to simplify packet-forwarding and support explicit routing without requiring each packet to carry an explicit route. In addition, since packets having the same pair of source and destination addresses may be assigned different labels, while those having different pairs may be assigned the same label, and each label can be associated with a routing policy or a class of services, MPLS facilitates traffic engineering.

The establishment of LSPs can be control-driven, i.e., performed by a network according to its topology and connectivity as in Cisco's Tag-switching [Rekhter et al., 1997]. But it may also be data-driven, unlike in VC-based packet-switching. For example, in IP-switching where IP runs over ATM switches, instead of establishing a ATM VC between a source and a destination as in classical IP over ATM or Multiprotocol over ATM (MPOA), a LSP will be set up after a few IP packets of a *flow* has been sent [Newman et al., 1998] (here, a flow could refer to all the IP packets from a source to a destination.

At a finer granularity level, it could refer to all the IP packets from a source host to a destination host, or just those belonging to the same TCP connection). More specifically, the first few IP packets will be routed by IP at each and every intermediate nodes. However, as soon as the destination recognizes the flow (e.g. when the number of IP packets it received from the source exceeds a given threshold), it triggers the establishment of a LSP in a backward direction, that is, each downstream node assigns a label for use (as an output label) by the upstream node via a label distribution protocol (LDP). After the LSP is established, the source will break each subsequent IP packet of the flow into cells carrying an appropriate label. These cells will then be switched at all the intermediate nodes, thus eliminating the need for IP routing (of the rest of the flow) at any intermediate node.

Of course, MPLS is not just for running IP over ATM. In fact, it can support multiple network layer (i.e. layer 3 or L3) protocols over a number of link layer (i.e. layer 2 or L2) protocol.

3.   BURST-SWITCHING

While running IP over ATM switches as described above allows one to tap into a large number of ATM switches invested without having to deal with the complex ATM signaling (which may be considered redundant given the existence of IP routing software), the approach results in a relatively high cell-tax (of about 10 %) since 5 of the 53 bytes in every ATM cell are overhead[4].

To reduce the high percentage of the control overhead due to cells or small packets, a *burst*, which can be either a digitized talk spurt or a data message, may be switched as a unit. While the concept of *burst-switching* may lack universal understanding, we do not consider all switching techniques which allocate network resources (e.g. bandwidth and buffer space) on a burst by burst basis to be *burst-switching*. For example, one such technique is *message switching* mentioned earlier. Nevertheless, when applied to voice communications, burst-switching is similar to fast circuit-switching in that both will have a phase for routing a call first and send a burst (a talk spurt) using a one-way reservation process and in addition, the burst will *cut through* switches.

3.1.   ONE-WAY RESERVATION

We restrict burst-switching to three basic variations, which not only allocate network resources on a burst-by-burst basis, but also *integrate* the elements from both circuit- and packet-switching. These variations all set up a circuit for the duration of a burst based on *one-way reservation* of the channel (or bandwidth). In one-way reservation, a source sends a set-up request, which is followed by a burst before receiving an acknowledgment back. This reduces the pre-transmission delay of the burst, which is important as the burst transmission time can be relatively short, especially in high-speed (e.g. Gbps) networks.

The three basic variations of burst-switching differ in the way the bandwidth is released. More specifically, in *tell-and-go* (or TAG), as soon as the burst is transmitted, the sender sends an explicit release signal to tear down the circuit, just as is done in phase 3 of circuit-switching. In what we call *reserve-a-fixed-duration* (or RFD), each set-up request specifies the duration for which the circuit is to be set up. Finally, in what we call *in-band-terminator* or IBT, a burst contains a header (just as a packet header) and a *terminator* to indicate the end of the burst, as in [Haselton, 1983; Amstutz, 1983]. Thus, as far as what triggers the bandwidth release is concerned, TAG and IBT are similar to circuit- and packet-switching, respectively, while RFD is unique.

3.2.   OFFSET TIME AND SWITCH CUT-THROUGH

As far as how bandwidth is allocated, both TAG and RFD are similar to circuit-switching and the third variation is similar to message-switching, but burst-switching is not the same as either circuit or packet-switching.

First, in circuit-switching, setting up a circuit is a *two-way* process involving sending a set-up request and receiving an acknowledgment by a source node before it can send any data. In one-way reservation, there is no clearly distinguishable phases 1 and 2 due to the absence of an acknowledgment prior to data transmission. More specifically, in both TAG and RFD, the offset time between a set-up request and its corresponding data, denoted by T, is shorter than that in circuit-switching. In fact, the data can be sent even before the entire circuit has been set up (i.e. the last few channels are latched together). As long as this offset time T is so large that by the time the data arrives at a switch, the switch has already been set and the bandwidth on the outgoing link reserved by the set-up request, the data needs not to be buffered at any intermediate node as in circuit-switching.

Note that if T made too small (e.g. $T = 0$) in TAG and RFD or if IBT is used, the data needs to be buffered (or more precisely *delayed*) at an intermediate node, say X, where the partially established circuit ends while waiting for the processing of the set-up request to complete. However, as soon as an outgoing channel is latched, the burst can continue its journey towards the next node, even though the last few bits of the burst may still be arriving at node X. This technique is called virtual cut-through [Kermani and Kleinrock, 1979], which is different from store-and-forward in packet-switching (or message-switching). Of course, in the worst case where the processing delay of the set-up request (or header) is long and/or congestion occurs such that no output channels is available for a while, the entire burst may have to be either buffered at an intermediate node or, if there is not enough buffer for the burst, dropped or deflected (i.e. routed to an alternate output port).

We also note that although variations of the so-called fast reservation protocol (FRP) (see [ITU-T Rec. I.371, 1995; Boyer and Tranchier, 1992; Shimonishi et al., 1996; Varvarigos and Lang, 1996; Varvarigos and Sharma, 1997] for example) consider a number of cells/packets as a burst for the purpose of transmission capacity (and/or buffer space) allocation, each cell/packet in the burst is still switched individually based on its label/header in a store-and-forward fashion. Hence, they may be considered as a result of combining VC-based packet-switching and the burst-level capacity allocation, but are different from burst-switching.

4.   SWITCHING IN OPTICAL NETWORKS

In this section, we describe the corresponding switching techniques proposed for optical networks (particularly WDM networks) with a focus on their unique features.

4.1.   WAVELENGTH ROUTING

Wavelength routing is a form of circuit-switching. In wavelength routed networks, a lightpath which is an all-optical data path along which data does not need to go through any O/E/O conversion, is established before data can be sent [Chlamtac et al., 1992]. Such lightpaths are called "wavelength-routed" because each uses a dedicated wavelength channel on every link along a physical path, and hence, once data is transmitted on a specific wavelength by its source, how the data will be routed (or switched) at the intermediate nodes will be determined by the ("color" of the) wavelength only. Wavelength-routing is based mainly on the following two premises. First, it was expected that the main functionality of a WDM layer is to provide lightpaths between two selected electronic devices (e.g. SONET ADMs) that are not physically adjacent, or in other words, are separated by multiple (fiber) links interconnected with optical switches. Such lightpaths can *not only* provide a high-speed, high-bandwidth pipe that is transparent to bit rate and coding format, *but also* reduce the number of expensive electronic equipment such as SONET ADMs with proper traffic grooming and wavelength assignment algorithms. Secondly, the optical switches (wavelength routers) based on opto-mechanical, acousto-optic or thermo-optic technologies are currently too slow for efficient packet-switching.

A unique property of wavelength-routed networks (and optical TDM networks) is that because all-optical wavelength conversion (or time-slot interchanging) technologies is not mature, a lightpath may have to use the same wavelength (or time-slot) on different links. This approach is called *path-*

multiplexing (or PM), as opposed to link-multiplexing (or LM) where different wavelengths (or time-slots) can be used on different links [Qiao and Mei, 1999]. This property gives rise to many related research issues such as the design of "virtual topology" (formed by long-lasting lightpaths similar to leased lines used by some companies to form their private networks), the benefits of wavelength-conversion and optimal placement of (limited number of) wavelength converters. A survey of most of the research done can be found in [Mukherjee, 1997; Ramaswami and Sivarajan, 1998; Stern and Bala, 1999].

Note that, since more and more bandwidth intensive applications (e.g. high definition television distribution) will appear, and some high-end users will require one or more lightpaths only for a relatively short period (e.g. from minutes to weeks), a WDM layer may need to establish and tear down wavelength-routed lightpaths dynamically, similar to today's circuit-switched telephone networks. Although most of the research on lightpath establishments assumes centralized control, a few papers on distributed reservation protocols have been published recently [Qiao and Mei, 1996; Mei and Qiao, 1997; Ramaswami and Segall, 1996; Yuan et al., 1996; Sengupta et al., 1997].

As in electronic networks, distributed control can improve the reliability as well as scalability in wavelength-routed networks. However, as far as how a circuit is established under distributed control, two major differences between an electronic network and a WDM network are worth noting. The first is that as mentioned earlier, a circuit in WDM network may have to be established in PM. As a result, reserving multiple available wavelengths on each link at a time (as in [Qiao and Mei, 1996]) may yield a higher success probability than reserving just one wavelength at a time (e.g. as in [Qiao and Mei, 1996; Ramaswami and Segall, 1996]).

The second major difference is that in an electronic network, it is possible to send the status information on each link to all other nodes so that each node has the global knowledge of the network, as in the Open Shortest-Path First (OSPF) protocol [Moy, 1998] used in the Internet. However, this may not be feasible in a WDM network, especially the one using PM, since each wavelength needs to be managed (and allocated/deallocated) individually as a unit, and a link may have multiple fibers, each carrying up to tens of wavelengths. As a result, distributed reservation protocols based on local knowledge, i.e. the wavelength usage information on the outgoing links of a node, (as in [Qiao and Mei, 1996]) may be needed in addition to those based on global knowledge (as in [Ramaswami and Segall, 1996]).

The performance of distributed wavelength reservation protocols using PM have been reported in [Qiao and Mei, 1996; Mei and Qiao, 1997; Yuan et al., 1996; Sengupta et al., 1997]. It is worth noting that the performance advantage of LM over PM under distributed control, reported in [Qiao and Mei, 1996; Mei and Qiao, 1997], can be much more significant than under centralized control.

This suggests that using the costly wavelength-converters may be justifiable in WDM networks using distributed control.

Dynamically establishing lightpaths may also be performed in the context of MPLS. More specifically, after an IP flow is recognized, a lightpath can be established for all future IP packets of the flow [Bannister et al., 1999].

One of the limitations of wavelength routing is that it is inefficient for Internet traffic, which is self-similar (or bursty at all time scales). This is because for bursty traffic, the bandwidth utilization of a lightpath is poor. In addition, given a limited number of wavelengths, only a limited number of lightpaths can be established at the same time, and hence the connectivity of the virtual topology at the WDM layer is likely to be weak. Even if lightpaths are established dynamically (e.g. under distributed control as described above), the set-up time of a lightpath based on two-way reservation may be too long for a burst containing a few megabit (Mb) of data (e.g. a small file) given the high transmission rate (E.g. 2.5Gbps or OC-48).

4.2.   OPTICAL PACKET/CELL/LABEL SWITCHING

Given the forecast that the amount of the data traffic, which is bursty in nature, will soon surpass voice traffic, and in particular, the phenomenal success of the Internet (and WWW), it is reasonable to expect that a WDM layer will also employ switching techniques similar to packet-switching and burst-switching when fast switches based on technologies such as Lithium Niobate and semiconductor optical amplifiers (SOAs) become available.

Optical packet switching is similar to traditional electronic packet switching, except that packet payload (i.e. data) will remain in the optics, while its header may be processed electronically or optically [Blumenthal et al., 1994]. However, only limited optical processing is possible due to the existing primitive optical logic, making VC-based optical packet-switching or optical cell switching more attractive than datagram-based optical packet-switching.

In fact, even with electronic processing (i.e. photonic switching), the header processing time should be kept minimum. In addition, each packet should have a fixed size (padding is used in case of insufficient data), and also be very small (e.g. as small as an ATM cell). These two requirements are due to the fact that there is no optical equivalent of the random access memory (RAM), and accordingly, (1) an optical data signal can only be delayed for a limited amount of time via the use of fiber-optic delay lines (FDLs) [Masetti et al., 1993] before the header processing has to complete, and (2) the length of each packet, in terms of the product of its transmission time and the speed of light, cannot exceed that of the available FDL in order for the optical packet to be "stored". These are just some of the reasons why optical packet-switching uses fixed-length packets almost exclusively.

When the packets are of the same size, it is natural to use TDM where the time axis is divided into time slots. Even without using TDM, synchronization is a big issue in optical packet switching since each node needs to recognize the header and the end of a packet, and align the packets on different input/output ports before and/or after they are switched. In addition, it is difficult to re-align a modified/replaced header (or label) with its payload at each node.

Note that, traditionally, packet switching assumes that each packet is to be transmitted at the full bandwidth of the link, implying a high bit rate (e.g. 40 Gbps on a TDM fiber-optic link) for both its payload and header. Optical packet-switching has also been adapted to using WDM technologies by transmitting packets at the bandwidth of a wavelength. In particular, in order to facilitate implementation, headers can be transmitted on a separate wavelength or a subcarrier channel. Such use of "out-of-band" control is normally a feature found in burst-switching as described earlier. In fact, recently proposed optical label switching [Chang, 1998], which uses SCM to carry labels for IP packets (of variable lengths), and forwards an IP packet as soon as the label is recognized (and rewritten when needed), blurs the distinction between optical packet-switching and optical burst-switching.

4.3. OPTICAL BURST SWITCHING (OBS)

Optical burst switching (OBS) is a way to achieve a balance between the coarse-grained wavelength routing and the fine-grained optical packet/cell switching [Qiao, 1997]. Since it is difficult to optically recognize the end of each burst, optical burst switching (OBS) will likely be based on RFD and TAG (instead of IBT). While burst-switching based on TAG and IBT has been studied for electronic networks, RFD-based burst-switching has not. Here, we first describe a new OBS protocol based on RFD called Just-Enough-Time (JET), which is proposed in [Yoo and Qiao, 1997; Yoo et al., 1997], and then compare it with TAG-based OBS protocols.

Just-Enough-Time (JET)

JET works as shown in Fig. 5.1. More specifically, a source sends out a control packet (i.e. set-up request), which is followed by a burst after an offset time, $T \geq \sum_{h=1}^{H} \delta(h)$, where $\delta(h)$ is the (expected) control delay (e.g. the processing time incurred by the control packet) at hop $1 \leq h \leq H$ (see Fig. 5.1 (a), where $H = 3$ and $\delta(h) = \delta$). Because the burst is buffered at the source (in the electronic domain), no FDLs are necessary at each intermediate node to delay the burst while the control packet is being processed.

A unique feature of JET which distinguishes it from other RFD protocols is that JET uses delayed reservation (DR) as shown in Fig. 5.1 (b), whereby the bandwidth on the output link at node i (e.g. $i = 1, 2$) is reserved from the burst arrival time, t, instead of from the time at which the processing of the control packet finishes, t' (since the offset time remaining after i hops is $T(i) = T - \sum_{h=1}^{i} \cdot \delta(h)$, we have $t = t' + T(i)$). In addition, the bandwidth will be reserved until the burst departure time, $t + l$, where l is the burst length.

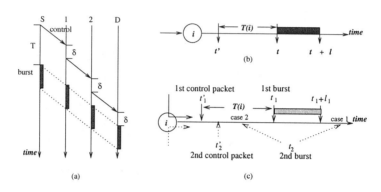

Figure 5.1   The use of an offset time and delayed reservation in Just-Enough-Time (JET).

As in any one-way reservation protocols such as TAG and RFD, if the requested bandwidth is not available, the burst is said to be blocked, and will be dropped if it cannot be buffered (a dropped burst may then be retransmitted later). The use of DR can reduce burst dropping (and increase bandwidth utilization) even without using any buffer, as illustrated in Fig. 5.1 (c). More specifically, when the 2nd control packet arrives, it knows that if either $t_2 > t_1 + l_1$ (case 1) or $t_2 + l_2 < t_1$ (case 2), bandwidth for the 2nd burst can be successfully reserved.

Note that, JET-based OBS can also take advantage of any FDLs available at an intermediate node by using the FDLs to delay a blocked burst until bandwidth becomes available (even though FDLs are not mandatory in JET). In fact, by taking advantage of the information on the duration of each reservation, DR can increase the effectiveness of the available FDLs, just as it can increase bandwidth utilization through scheduling[5] (for more detailed discussions, see [Yoo and Qiao, 1997; Yoo et al., 1997]). In addition, if the control delay is relatively large compared to the average burst length, then with the same FDLs, JET will achieve a better performance (e.g. a lower burst dropping probability) than optical packet/cell switching and other OBS protocols that do not use any offset time (i.e. $T = 0$). This is because JET can use 100% of the available FDLs for the purpose of resolving conflicts but these protocols cannot (due to the fact that some FDLs must be used to delay the burst while the header or control packet is being processed).

TAG-based OBS and Other Related Work

The TAG and terabit burst switching protocols studied in [Hudek and Muder, 1995; Turner, 1997] are example of TAG-based OBS protocols[6], in which the bandwidth is released using an explicit tear-down signal. Since loss of the tear-down signal during transmission will result in bandwidth waste, each source is required to periodically send out a refresh signal in order to keep a circuit alive, and if no refresh signals is received after a time-out period, each intermediate node will automatically release the circuit.

Although bandwidth can be released as soon as burst is transmitted in TAG, IBT and JET (or in general RFD), the lack of information on when existing reservations will expire (i.e. the duration of each reservation) in TAG and IBT prevents them from scheduling the bandwidth reservations (and allocating FDLs, if any) as intelligently and efficiently as JET. For instance, if TAG or IBT were used in Fig. 5.1 (c), there is no way for the 2nd control packet to know that the bandwidth will be released before the 2nd burst will arrive (case 1), or that the length of the 2nd burst is short enough (case 2). In fact, it is not beneficial in terms of performance (end-to-end latency) and bandwidth utilization in TAG- (or IBT-) based OBS protocols to use an offset time between a control packet and its corresponding data burst.

Note that, like JET, both the CC [Hudek and Muder, 1995] and ERVC [Varvarigos and Sharma, 1995] protocols reserve bandwidth (or capacity) on fiber-optic links only till the end of a burst. In addition, protocols such as RIT [Hudek and Muder, 1995] and ERVC reserve bandwidth starting at the time it is actually needed (such a technique is often called *just-in-time switching*). However, these protocols use two-way reservation as in circuit-switching.

5.   OBS AND OPTICAL INTERNET

In this section, we address issues related to building the next generation Optical Internet and in particular, show how OBS can be used to provide quality-of-service (QoS) at the WDM layer.

In most of the existing networks, IP routers are interconnected with ATM switches, which in turn are interconnected with SONET/SDH ADMs. IP over ATM (see e.g. [White, 1998]) has been considered mainly because ATM can support integrated services with QoS. However, using ATM to transport IP packets incurs a high ATM "cell tax" as well as high ATM signaling overheads. An alternative is IP directly over SONET/SDH (see e.g. [Manchester et al., 1998]), but the cost of SONET/SDH ADMs at a high data rate (e.g. 10 Gbps or above) is extremely high, and in addition, many sophisticated functionalities related to network operation, administration and management (OAM) of SONET/SDH, which were defined for voice-dominant telephony traffic, are not necessary or simply do not make sense for data-dominant Internet traffic. In addition, with the line speed of IP routers reaching 2.5 Gbps, and especially the emergence of terabit IP routers, there is no longer a compelling reason for having either ATM or SONET/SDH to multiplex lower rate data streams onto wavelengths (currently operating at 2.5 Gbps or higher). Accordingly, there have been several initiatives in building the so-called Optical Internet where IP routers are interconnected directly with WDM links (see e.g. [Arnaud, 1998]).

In the rest of the section, we consider the *next generation* Optical Internet where IP runs over a WDM layer consisting of WDM switches and WDM links. Having the WDM layer will enable a huge amount of "through" traffic to be switched in the optical domain, and as a result, can reduce the number of expensive terabit routers and high-speed transceivers required at the IP layer (in addition to creating high-speed communication pipes that are transparent to bit-rate and coding format, as mentioned earlier).

We will address the issue of how to provide QoS support at the WDM layer. This issue is important because it is well known that current IP provides only *best-effort* services and thus lacks QoS support (although this keeps IP simple, robust and scalable). Given that some applications such as Internet telephony and video conferencing require a higher QoS than electronic mail and general web browsing, it becomes apparent that for the envisioned infrastructure to be truly ubiquitous, one must address, among other important issues, how the WDM layer can provide basic QoS support (e.g. a few priority levels). Such a WDM layer will facilitate as well as complement a QoS-enhanced version of IP (such as *diffserv* being developed in the QBone project as a part of the Internet2 [Hyperlink at http://www.internet2.edu/]). Furthermore, supporting basic QoS at the WDM layer is necessary not only for carrying some WDM layer traffic such as those for signaling and protection/restoration purposes, which require a higher priority than other ordinary traffic, but also for supporting certain applications directly (i.e. bypassing IP) or indirectly through other legacy or new protocols incapable of QoS support.

Even though a considerable amount of effort has been and is still being devoted to developing QoS schemes for ATM and IP, none of them has taken into account the unique properties of the WDM layer. Specifically, most QoS schemes such as fair queueing (FQ) and its variations (see e.g. [Varma and Stiliadis, 1997; Briem et al., 1998]) are based on the use of buffer (queues) and scheduling algorithms. These schemes are not applicable to the all-optical WDM layer because on one hand, optical RAM is not yet available, and only limited delay may be provided to optical packets via the use of FDLs; and on the other hand, the use of electronic buffer would necessitate O/E and E/O conversions at intermediate nodes, thus sacrificing the data transparency, for instance. How to develop an efficient QoS scheme that can achieve effective service differentiation (or traffic class isolation) at the WDM layer without requiring any buffer at intermediate nodes becomes a challenging problem.

Before we describe how the principles of RFD-based OBS (in particular JET) can be applied as a solution to the above problem, we note that when OBS is used in the next generation Optical Internet, the control packet will be processed at each and every intermediate IP entities to establish an all-optical path, but the corresponding burst (e.g. several IP packets) will go through only the pre-configured WDM switches at the intermediate nodes along the established all-optical path [Qiao and Yoo, 1999] (to some extent, this is similar to MPLS).

5.1. PRIORITIZED OBS

Consider a *prioritized OBS* protocol called pJET, where bursts are classified into multiple (e.g. two) classes, and differentiated services are to be provided. For example, class 0 corresponds to best-effort services and can be used for non-real-time applications such as email and FTP, while class 1 corresponds to priority services and can be used for delay sensitive applications such as real-time audio and video communications. Since a dropped class 0 burst may be retransmitted but not a dropped class 1 burst (due to its stringent delay constraint), it is desirable to assign class 1 bursts a higher priority than class 0 bursts when reserving bandwidth to ensure that class 1 bursts incur a lower blocking (dropping) probability.

The main idea of pJET is to assign an *extra* offset time, denoted by t_{offset}, to each class 1 burst (but only a "base" offset time T is used when sending each class 0 burst), while all control requests are still treated equal, i.e. processed in the first-come-first-served (FCFS) order. Intuitively, this extra offset time allows a control packet corresponding to a class 1 burst to make bandwidth reservation in much more advance, thus giving it a greater chance of success than the control packet for a class 0 burst, which can only "buy tickets at door".

To illustrate the principle of pJET using terms common to queueing systems, let t_{ai} and t_{si} be the arrival time and the service-start time respectively, of a class i request, denoted by $req(i)$, where $i = 0, 1$. Also, let l_i be the service time (i.e. burst length) requested by $req(i)$. To simplify the following presentation, let us assume no FDLs at any intermediate node. In addition, we will ignore the effect of the base offset time assigned to both classes of bursts and concentrate on that of the extra offset time assigned only to class 1 bursts.

Since no extra offset time is given to class 0 bursts, a class 0 request, $req(0)$, will try to reserve bandwidth immediate upon its arrival, and will be serviced right away if bandwidth is available (and dropped otherwise). In other words, $t_{a0} = t_{s0}$ when reservation is successful (see Fig. 5.2(b)). However, for a class 1 request, $req(1)$, a delayed reservation is made with an extra offset time,

t_{offset}, and hence, it will be serviced at $t_{s1} = t_{a1} + t_{offset}$ when reservation is successful (see Fig. 5.2(a)).

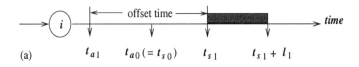

(a) t_{a1} $t_{a0}(=t_{s0})$ t_{s1} $t_{s1}+l_1$

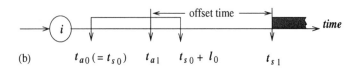

(b) $t_{a0}(=t_{s0})$ t_{a1} $t_{s0}+l_0$ t_{s1}

Figure 5.2   Priority scheme using the offset time combined with DR.

Fig. 5.2 illustrates why a class 1 request that is assigned t_{offset} can obtain a higher priority for reservation than a class 0 request that is not. Consider the following two cases where contention between two requests in different classes is possible. In the first case illustrated in Fig. 5.2(a), $req(1)$ arrives first and reserves the bandwidth (using delayed reservation), and $req(0)$ arrives afterwards. Clearly, $req(1)$ will succeed, but $req(0)$ will be blocked if $t_{a0} < t_{s1}$ but $t_{a0} + l_0 > t_{s1}$, or if $t_{s1} < t_{a0} < t_{s1} + l_1$. In the second case illustrated in Fig. 5.2(b), $req(0)$ arrives first, followed by $req(1)$. When $t_{a1} < t_{a0} + l_0$, $req(1)$ would be blocked *had* t_{offset} not been assigned to $req(1)$. However, such a blocking is avoided because of t_{offset} as long as $t_{s1} = t_{a1} + t_{offset} > t_{a0} + l_0$.

If $t_{a1} = t_{a0} + \sigma$, where $\sigma > 0$ is very small, t_{offset} needs to be longer than the maximum burst length over all class 0 bursts in order for $req(1)$ to completely avoid being blocked by any $req(0)$. With that much extra offset time, the blocking probability of class 1 bursts becomes independent of the offered load in class 0, that is, class 1 is completely (i.e. 100%) isolated from class 0.

Note that however, a reasonable t_{offset} can be used to achieve a sufficient degree of class isolation. For example, if the length of class 0 bursts is exponentially distributed with an average of L_0, then $t_{offset} = 3 \cdot L_0$ is sufficient to achieve at least 95% isolation [Yoo and Qiao, 1998]. In addition, with a reasonable degree of class isolation, class 1 bursts will have a blocking probability that is several orders of magnitude lower than class 0 bursts, although the overall blocking probability does not depend on the degree of isolation but rather other factors such as the overall traffic load and number of wavelengths [Yoo and Qiao, 1998]. The results in [Yoo and Qiao, 1998] have also shown that even with a link load of 0.8 and 40 wavelengths, the blocking probability of class 1 bursts can be as low as 10^{-7}.

5.2. PRE-TRANSMISSION DELAY

For real-time bursts, it is imperative to discuss their end-to-end delay. We first note that, the use of a (base) offset time as in JET does not increase the end-to-end delay of a burst as the offset merely substitutes for the total processing delay to be encountered by the corresponding control packet. Compared to circuit-switching, the end-to-end delay in OBS is about $2P$ shorter, where P is the total propagation delay between the burst's source and destination switches (typically around tens of milliseconds or ms coast-to-coast). In addition, compared to IP routing which uses store-and-forward, OBS can reduce the end-to-end delay because the data can cut-through the WDM switches as described earlier.

To assess the impact of using an extra offset time as in pJET, let the total processing delay be Δ (typically tens of microseconds or μs) and the length of a burst be l (typically a few μs or less). In addition, assume that in a real-time application, each byte (or bit) generated at the source has to reach its destination in D ms, which is often large compared to Δ or even P. For example, for today's voice and video communications, it may be acceptable for D to be as large as a few hundreds of ms. Nevertheless, if one takes into account the delays introduced by the higher-layer protocols at the source and destination, the available budget for the total delays in the WDM layer, denoted by B, could be significantly smaller.

Let $B = P + \Delta + b$, where b is the available budget for pre-transmission delay and could be as large as several ms[7]. Accordingly, if the extra offset time used in pJET is equal to a few times of L_0 (the average burst length), the increase in the end-to-end delay may not be significant (although the reduction in the blocking probability of higher priority bursts could be significant as discussed earlier).

Note that in RFD-based OBS (such as JET and pJET), it is required that a control packet specify the duration of the following burst, where a burst usually consists of all the IP packets belonging to the same data message. Such a requirement can be easily met when each burst can be assembled in less than a few ms (i.e. the burst assembly time is shorter than the tolerable pre-transmission delay b). Nevertheless, there may be cases where a message is long, the data belonging to the same message arrives (from the IP layer) at a slow rate, and/or each message is too short so that IP packets belonging to different messages (or flows) may need to be assembled into one burst in order to lower the percentage of the overhead introduced by the control packet. In such cases, the time to assemble a burst with all the desirable content, A, may be larger than b.

More specifically, let the time that the burst assembly starts be t_0. If $A > b^8$, an obvious solution is to stop accumulating data prior to (or at) time $(t_0 + b)$ (or in pJET where an extra offset time t_{offset} is used, before $(t_0 + b - t_{offset})$), and immediately send out a control packet specifying the length of the (partial) burst assembled so far, say l'.

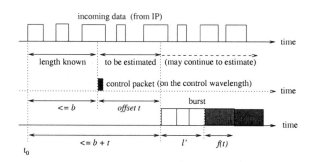

Figure 5.3   A control packet may be sent with an estimated burst length.

If the total offset time t (which may include both the base and extra offset time in pJET) to be used is relatively large (e.g. due to a large Δ), an alternative to the above solution is to continue to accumulate data for another t μs, but send a control packet prior to time $(t_0 + b)$ (or $t_0 + b - t_{offset}$ in pJET), which will specify the *expected* length of the (partial) burst to be transmitted. As illustrated in Fig. 5.3, the control packet will specify the length to be $l = l' + f(t)$, where $f(t)$ is the *estimated* average burst assembly rate during the next t μs (which may be calculated based on the actual rate observed so far). Note that if it is an over-estimation, some extra bandwidth reserved by the control packet will be wasted. On the other hand, if it is an under-estimation, additional data accumulated will have to be transmitted later as a separate (partial) burst. The potential advantage of this alternative is that more data can be sent, thus reducing the percentage of the overhead introduced by the control packets.

Note that we may extend the estimating period t to include the burst transmission time as well. Such a flexibility of OBS in sending out a control packet while a burst is still being assembled, combined with the fact that the line speed of IP routers is reaching OC-48, enables OBS to efficiently support real-time applications such as Voice-over-IP that generate data periodically. In addition, in some applications such as file transfers, WWW downloadings or video-on-demand (VOD), a server can determine the burst length as soon as a request for transferring/downloading a file from its client is processed. Accordingly, the server can send out a control packet specifying the exact burst duration even before the file is retrieved from a storage unit. This essentially overlaps the processing of the control packet at the intermediate nodes with the file retrieval operation, thus reducing the overall end-to-end delay.

Finally, we note that in order to reduce the overhead and delay due to the processing (and particularly, routing) of the control packets, one may use labels in OBS, resulting in what might be called labeled optical burst switching (or LOBS). More specifically, for a long flow (of bursts of IP packets), a label switching table can be set up at every node along the path either by the first few control packets or the network itself. Thereafter, every burst will be preceded by a control packet containing an appropriate label. Such an LOBS approach is perceived to be especially suitable for supporting WDM layer *multicast* of bursty traffic [Qiao et al., 1999], for which wavelength-routed WDM multicasting approaches will result in poor bandwidth utilization.

6. SUMMARY

In this paper, we have described various switching techniques used in today's voice and data networks as well as those showing promises for tomorrow's transparent optical networks. These switching techniques may be classified into three categories, with *circuit-* and *packet-switching* as the two extremes and *burst-switching* in the middle. Their correspondence in the optical (WDM) layer are wavelength-routing, optical packet-switching and optical burst switching (OBS), respectively, although the distinctions between the latter two are becoming more and more obscure as optical packet-switching starts to employ *switching cut-through* and *out-of-band* control to switch *variable-length* packets.

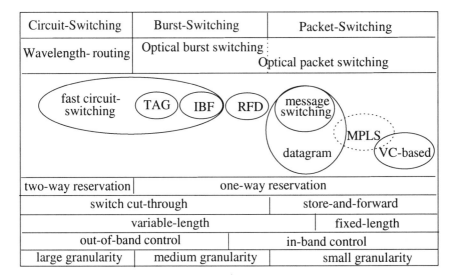

Figure 5.4 Major switching techniques.

Fig. 5.4 summarizes several major concepts and their relationship. We note that based on the current and near future technologies and applications, the question of whether there will be an single switching technique (and if so, which one) in use in the electronic layer and/or the optical (WDM) layer is wide open for debate. A way to enable a WDM network to utilize multiple switching techniques so that it can provide multiple services has been proposed in [Qiao et al., 1998]. However, *if* there will be a single switching technique in use in both layers, we expect that circuit-switching is not likely to be the one in the electronic layer, but consider optical burst switching (OBS) as the most promising one for the WDM layer.

Notes

1. In a backbone network, *source* and *destination* will refer to *ingress* and *egress* nodes, respectively.

2. This is why circuit-switching may also be referred to as *Synchronous Transfer Mode* or STM.

3. In IP routing, one needs to find the longest substring in the routing table that matches the packet's destination address.

4. Recently, the ATM Forum has been working on a specification called Fast (Framed ATM over SONET Transport), which will allow up to 64 Kbytes of data for every header of 4 bytes.

5. Scheduling is also useful for RFD-based OBS without an offset time, or optical packet switching with variable-length packets which, when combined with out-of-band control and switch cut-through, can be quite similar to OBS as mentioned earlier.

6. More recent publications by Turner [Turner, 1999] described a RFD-based OBS protocol similar to JET instead.

7. The assumption here is that $b < 2P$ so circuit-switching is out of the question. Note that even if b is as large as $2P$, OBS may still be preferred since retransmissions (of a set-up request and a burst) may be possible when using OBS, but not when using circuit-switching.

8. Since A is not the same as the burst length, l, this does not necessarily mean that $l > b$ as well.

References

Amstutz, S. (1983). Burst switching - an introduction. *IEEE Communications Magazine*, 21:36–42.

Arnaud, B. (1998). Architectural and engineering issues for building an optical internet. In *SPIE Proceedings, All optical Communication Systems: Architecture, Control and Network Issues*, volume 3531, pages 358–377.

Bannister, J., Touch, J., Willner, A., and Suryaputra, S. (1999). How many wavelengths do we really need in an optical backbone network. In *IEEE Gigabit Networking Workshop (GBN)*. (see related links at http://www.isi.edu/touch/pubs/gbn99/).

Bertsekas, D. and Gallager, R. (1992). *Data Networks*. Prentice Hall.

Blumenthal, D., Prucnal, P., and Sauer, J. (1994). Photonic packet switches - architectures and experimental implementations. *Proceedings of the IEEE*, 82:1650–1667.

Boyer, P. E. and Tranchier, D. P. (1992). A reservation principle with applications to the ATM traffic control. *Computer Networks and ISDN Systems*, 24:321–334.

Briem et al., U. (1998). Traffic management for an ATM switch with per-VC queuing: Concept and implementation. *IEEE Communications Magazine*, 36(1):88–93.

Callon et al., R. (1997). A framework for multiprotocol label switching. *IETF Draft*. draft-ietf-mpls-framework-02.txt.

Chang, G. (1998). Optical label switching. In *DARPA/ITO Next Generation Internet PI Meeting*.
(see www.dyncorp-is.com/darpa/meetings/ngi98oct/agenda.html).

Chlamtac, I., Ganz, A., and Karmi, G. (1992). Lightpath communications: an approach to high-bandwidth optical WANs. *IEEE Transactions on Communications*, 40:1171–1182.

Haselton, E. (1983). A PCM frame switching concept leading to burst switching network architecture. *IEEE Communications Magazine*, 21:13–19.

Hudek, G. and Muder, D. (1995). Signaling analysis for a multi-switch all-optical network. In *Proceedings of Int'l Conf. on Communication (ICC)*, pages 1206–1210.

Hui, J. (1990). *Switching and traffic theory for integrated broadband networks*. Kluwer Academic Publishers.

Hyperlink at http://www.internet2.edu/.

ITU-T Rec. I.371 (1995). Traffic control and congestion control in B-ISDN. Perth, U.K. Nov. 6-14.

Kermani, P. and Kleinrock, L. (1979). Virtual cut-through : A new computer communication switching technique. *Computer Networks*, 3:267–286.

Manchester et al., J. (1998). IP over SONET. *IEEE Communications Magazine*, 36(5):136–142.

Masetti, F., Gavignet-Morin, P., Chiaroni, D., and Loura, G. D. (1993). Fiber delay lines optical buffer for ATM photonic switching applications. In *Proceedings of IEEE Infocom*, volume 3, pages 935–942.

McDysan, D. and Spohn, D. (1994). *ATM: theory and application*. McGraw-Hill.

Mei, Y. and Qiao, C. (1997). Efficient distributed control protocols for WDM optical networks. In *Proc. Int'l Conference on Computer Communication and Networks (IC3N)*, pages 150–153.

Moy, J. (1998). *OSPF: Anatomy of an Internet Routing Protocol*. Addison-Wesley.

Mukherjee, B. (1997). *Optical Communication Networks*. McGraw-Hill.

Newman, P., Monshall, G., and Lyon, T. (1998). IP switching – ATM under IP. *IEEE/ACM Transactions on Networking*, 6:117–129.

Qiao, C. (1997). Optical burst switching - a new paradigm. In *Optical Internet Workshop*. (see related links at http://www.isi.edu/ workshop/oi97/).

Qiao, C. and Mei, Y. (1996). Wavelength reservation under distributed control. In *IEEE/LEOS Broadband Optical Networks*, pages 45–46.

Qiao, C. and Mei, Y. (1999). Off-line permutation embedding and scheduling in multiplexed optical networks with regular topologies. *IEEE/ACM Transactions on Networking*, pages 241–250.

Qiao, C., Mei, Y., Yoo, M., and Zhang, X. (1998). Polymorphic control for cost-effective design of optical networks. In *NSF DIMACS Workshop on Multichannel Optical Networks: Theory and Practice*, pages 157–179.

Qiao, C. and Yoo, M. (1999). Optical burst switching (OBS) - a new paradigm for an Optical Internet. *J. High Speed Networks (JHSN)*, 8(1):69–84.

Qiao et al., C. (1999). Multicasting in IP over WDM networks. Technical Report 99-05, CSE Dept, University at Buffalo (SUNY).

Ramaswami, R. and Segall, A. (1996). Distributed network control for wavelength routed optical networks. In *Proceedings of IEEE Infocom*, pages 138–147.

Ramaswami, R. and Sivarajan, K. (1998). *Optical Networks: A Practical Perspective*. Morgan Kaufmann.

Rekhter et al., Y. (1997). Tag switching architecture overview. *IEEE Proceedings*, 82:1973–1983.

Sengupta et al., A. (1997). On an adaptive algorithm for routing in all-optical networks. In *SPIE Proceedings, All Optical Communication Systems: Architecture, Control and Network Issues*, volume 3230, pages 288–297.

Shimonishi, H., Takine, T., Murata, M., and Miyahara, H. (1996). Performance analysis of fast reservation protocol with generalized bandwidth reservation method. In *Proceedings of IEEE Infocom*, volume 2, pages 758–767.

Spragins, J., Hammond, J., and Pawlikowski, K. (1994). *Telecommunications protocols and design*. Addison-Wesley.

Stern, T. and Bala, K. (1999). *Multiwavelength Optical Networks: A Layered Approach*. Addison-Wesley.

Turner, J. S. (1997). Terabit burst switching. Technical Report WUCS-97-49, Department of Computer Science, Washington University.

Turner, J. S. (1999). Terabit burst switching. *J. High Speed Networks (JHSN)*, 8(1):3–16.

Varma, A. and Stiliadis, D. (1997). Hardware implementation of fair queuing algorithms for asynchronous transfer mode networks. *IEEE Communications Magazine*, 35(12):74–80.

Varvarigos, E. and Lang, J. (1996). Performance analysis of deflection routing with virtual circuits in a Manhattan Street network. In *IEEE Globecom*, pages 1544–1548.

Varvarigos, E. and Sharma, V. (1995). The ERVC protocol for the Thunder and Lightning network: operation, formal description and proof of correctness. Technical Report CIPR 95-05, ECE Dept, UC Santa Barbara.

Varvarigos, E. and Sharma, V. (1997). The ready-to-go virtual circuit protocol : A loss-free protocol for multigigabit networks using FIFO buffers. *IEEE/ACM Transactions on Networking*, 5(5):705–718.

White, P. (1998). ATM switching and IP routing intergration: The next stage in Internet evolution. *IEEE Communications Magazine*, 36(4): 79–83.

Yoo, M., Jeong, M., and Qiao, C. (1997). A high-speed protocol for bursty traffic in optical networks. In *SPIE Proceedings, All Optical Communication Systems: Architecture, Control and Network Issues*, volume 3230, pages 79–90.

Yoo, M. and Qiao, C. (1997). Just-enough-time(JET): a high speed protocol for bursty traffic in optical networks. In *Digest of IEEE/LEOS Summer Topical Meetings on Technologies for a Global Information Infrastructure*, pages 26–27.

Yoo, M. and Qiao, C. (1998). A new optical burst switching protocol for supporting quality of service. In *SPIE Proceedings, All Optical Networking: Architecture, Control and Management Issues*, volume 3531, pages 396–405.

Yuan, X., Gupta, R., and Melhem, R. (1996). Distributed control in optical WDM networks. In *IEEE MILCOM*, pages 100–104.

Chapter 6

WDM NETWORKS WITH LIMITED OR NO WAVELENGTH CONVERSION FOR WORST CASE TRAFFIC[*]

Galen Sasaki
Department of Electrical Engineering
University of Hawaii
2540 Dole Street, Honolulu, HI 96822
sasaki@spectra.eng.hawaii.edu

Abstract   WDM networks are considered that have limited or no wavelength conversion, and they accommodate all traffic as long as the offered *load* on any link does not exceed some parameter L. Wavelength requirements W are given that are functions of L for *static* and *dynamic* traffic. It is shown that wavelength requirements can be decreased significantly by small amounts of wavelength conversion.

1.   INTRODUCTION

Wavelength division multiplexing (WDM) is an important and practical technology to exploit the wide communication bandwidths in single mode optical fibers. WDM is basically *frequency division multiplexing*. A fiber carries multiple optical communication channels, each at different *wavelengths*, which are essentially the inverse of the carrier frequencies. In this way, WDM transforms a fiber that carries a single signal into multiple *virtual fibers*, each carrying its own signal.

We consider *wavelength routed networks*. These networks support end-to-end optical communication connections referred to as *lightpaths*. Fig. 6.1 shows a network with two lightpaths. Lightpath $L1$ is made up of WDM channels along its path that are all at wavelength w_0. The intermediate nodes

---

[*]The work was sponsored in part by NSF grant 9612846.

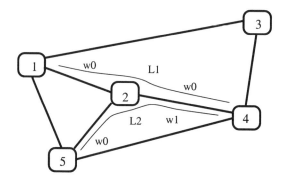

Figure 6.1   Two lightpaths $L1$ and $L2$.

have switching that forwards the optical signals between channels. A lightpath performs as a *virtual fiber* carrying its optical signal.

In the case of $L1$, the signals are forwarded while preserving their wavelengths. However, lightpath $L2$ is made up of channels of different wavelengths. At the intermediate node 2, a *wavelength conversion* device is required that will shift the lightpath's signal from w_0 to w_1. A network with wavelength conversion can support more lightpaths than one without. However, wavelength converter technology can be difficult or expensive.

Wavelength conversion can be done *all-optically* or by receiving the signal, switching it electronically and retransmitting it on another wavelength, i.e., *optical-electrical-optical* (O-E-O) conversion. The all-optical approach keeps the signal optical during conversion. Technologies, such as *four-wave mixing* [Zhou et al., 1994], has conversion efficiency that is a strong function of the input and output wavelengths, naturally leading to limited conversion capability. In particular, it may lead to a limit on the range of wavelength shifting.

We consider the relationship between the amount of wavelength conversion and bandwidth efficiency. In the next few subsections we present the network and traffic models, objectives, and the organization of the rest of the paper.

1.1.   NETWORK MODEL

The network is composed of a set of nodes and links, where there is at most one link between any pair of nodes. The number of wavelengths (and channels) per link is denoted by W, and the wavelengths are denoted by $\{\omega_0, \omega_1, ..., \omega_{W-1}\}$, where $\omega_0 < \omega_1 < ... < \omega_{W-1}$. Each link and channel are *bidirectional*, i.e., full duplex. Such links may be implemented with two unidirectional fibers, each carrying W wavelengths in opposite directions, and each bidirectional channel will be composed of two unidirectional channels. Our assumption that a bidirectional channel is at a particular wavelength implies

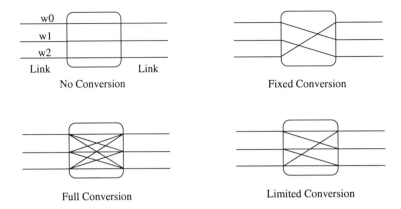

Figure 6.2   Types of wavelength conversion.

that both its unidirectional channels have the wavelength. In practical systems, this may not be a restriction. We will comment on this in Section 4.

Each node has switching and conversion capability to connect WDM channels in different links to form lightpaths. We will refer to two channels that may be connected through a node as being *compatible* at the node. A node with *no* conversion has compatible channels if the channels are at the same wavelength. A node with *full* conversion has all channels compatible. A node with *limited* conversion is somewhere in between. Fig. 6.2 shows the different types of conversion for nodes with two links and three wavelengths. The lines within the nodes show which channels are compatible. The figure also shows *fixed* conversion, which is where each channel is compatible to exactly one channel in each other link.

We characterize the amount of switching and conversion at a node by a parameter we call the *wavelength degree*. A node has wavelength degree d (for some integer d) if for each pair of incident links, each channel is compatible to at most d channels in all other links. Thus, if a node has k incident links then each channel is compatible with a total of at most $(k - 1) \cdot d$ channels. Notice that nodes with no and fixed conversion have wavelength degree of 1 and nodes with full conversion has wavelength degree W.

The network model is fairly general and can model networks considered in [Yates et al., 1996; Subramaniam et al., 1996]. In [Yates et al., 1996], four-wave mixing was considered, so wavelengths could only be shifted within some range. *Sparse* wavelength conversion was studied in [Subramaniam et al., 1996], where networks are comprised of a mix of nodes with no or full conversion. However, the wavelength degree parameter is not the only measure of wavelength conversion complexity. In [Subramaniam et al., 1996], the number of full conversion nodes is the appropriate measure. In [Lee and Li, 1993a; Lee and Li, 1993b], the number of converting devices is the appropriate measure, where a wavelength converter device can shift an optical signal between arbitrary wavelengths.

1.2.   TRAFFIC MODEL

In the traffic model, *requests* are made for lightpaths to be set up in the network. A lightpath request comes with a pre-computed route that the lightpath must follow. Thus, the network is only required to determine an assignment of channels for the lightpath along the route.

The traffic is *static* if all requests are made at one time. In this way the network may find the channel assignments together. The traffic is *dynamic* if requests may arrive and lightpaths may terminate at arbitrary times. A special case of dynamic traffic is *incremental* traffic, which is where lightpaths never terminate. This models the fact that lightpaths support wide-band end-to-end connections carrying large volumes of tributary traffic. It is likely that lightpaths will have very long life times.

The traffic is modeled by a parameter called its *load L*, which is a measure of the link bandwidth requirements. In the case of static traffic, it is the maximum number of lightpath requests that are offered over any link. In the case of dynamic traffic, an arriving lightpath request will find each link along its route having less than L existing lightpaths. Notice that $L \leq W$ is a necessary condition for no lightpath blocking. It is also a sufficient condition if each node has full conversion.

Another traffic parameter we consider is the *maximum length of a lightpath*, which we denote by H. The parameter can affect the bandwidth requirements of a network since typically shorter lightpaths are easier to set up than longer ones.

The traffic model will be used to define the wavelength requirements of networks that do not block lightpaths, as we shall see in the next subsection. It is a worst case traffic model with few constraints. This is in contrast to standard statistical traffic models (e.g., Poisson process of arrivals and random independent holding times of lightpaths) used in [Barry and Humblet, 1996; Ramaswami and Sivarajan, 1995; Yates et al., 1996; Birman, 1996; Kovačević and Acampora, 1996] for networks that allow blocking. The disadvantage of the worst case traffic model is that the resulting wavelength requirements are conservative. However, the same can be said for typical results of statistical traffic models because networks are evaluated assuming low blocking probabilities, which implies low link utilization. An advantage of the worst case traffic model is that it leads to formulas for wavelength requirements that shows the dependence on traffic and network parameters.

1.3. GOALS AND ORGANIZATION

We will determine networks that need minimal numbers of wavelengths and limited wavelength conversion to support the traffic. We consider two types of networks:

Static: These are non-blocking for all static traffic with load L. Note that they are also non-blocking for dynamic traffic with load L if existing lightpaths can rearrange their channels.

Wide-sense non-blocking: In this case, once a lightpath is given a channel assignment it cannot change. The networks are non-blocking for dynamic traffic given the restriction. The restriction is a practical one to insure high quality of service requirements.

We require the network to be non-blocking with very little assumptions on the traffic. As mentioned before, resulting wavelength requirements will be for the "worst case" traffic.

In Section 2, we survey our results for the ring topology. The ring is important because optical network architectures such as FDDI and SONET/SDH use it to support fault-recovery. The simplicity of the ring also leads to analytical results.

Fig. 6.3 summarizes the wavelength requirements of the ring for different traffic types and wavelength conversion. We can make two observations from the figure. First, a small amount of conversion can go a long way. For example, in the case of static traffic, no conversion leads to wavelength requirements of $2L - 1$, while fixed conversion can almost halve this to $L + 1$. Another example is the case of incremental traffic, where no conversion leads to a wavelength requirement of $3L - 2$, while wavelength degree $d = 2$ can lead to a requirement of $2L - 2$. Second, wavelength requirements depend on how much control a network will have in managing lightpaths. The network has the most control with static traffic since it can set up all lightpaths together. It has the least control with dynamic traffic, since it cannot reconfigure existing lightpaths and lightpaths can terminate arbitrarily, leaving free channels that are difficult to use.

In Section 2, we also present results for trees. In Section 3, we discuss results for general topologies. Finally, in Section 4, we have our final remarks.

2. RINGS AND TREES

We will present wavelength requirements for ring and tree networks for different types of traffic. To simplify the discussion for ring networks, we will assume that they have a clockwise direction, and lightpath routes follow the direction. The direction does not imply anything about the physical transmission on the ring. The lightpaths and channels are still full duplex.

Model	Conv:	No conversion	$d = 1$	$d > 1$	Full
Static	Lower	$2L - 1^*$	$L + 1^*$	L	L
	Upper	$2L - 1$	$L + 1$	L	L (trivial)
Incre-	Lower	$3L - 2$	L	L	↑ same
mental	Upper	$3L - 2$	L	$\max(L, 2L - d)$	↑ same
Dyna-	Lower	$0.5L \log_2 N + L$	← same	L	↑ same
mic	Upper	$L \log_2 N + L$	← same	$\min(L \log_2 L + 4L,$ $2L \log_2 \log_2 L + 4L)$ $(d = 2)$	↑ same
	(FF)	$1.63 L \log_2 N$	?	?	↑ same

Figure 6.3   Wavelength requirements for ring networks. Note that the lower bound L is a trivial lower bound. Also note that the starred lower bounds for static traffic apply in certain cases, but perhaps not all.

2.1.   STATIC TRAFFIC

In this subsection, we will discuss results for static traffic. For a ring with no conversion, the following results have been shown.

Theorem 1   *[Tucker, 1975] A ring network with no conversion at all nodes has wavelength requirement $W \leq 2L - 1$ for static traffic.*

There is a simple wavelength assignment algorithm to achieve the wavelength requirements. Choose a node x that has at most $L - 1$ lightpaths crossing through it, and assign distinct wavelengths to the lightpaths. For the remaining lightpaths, assign wavelengths using a greedy strategy and at most L wavelengths. The strategy is to scan nodes in the ring starting from x and going in the clockwise direction, and each new lightpath is assigned the lowest available wavelength among the L. In [Tucker, 1975] examples are given that may be extended to show that the upper bound in the theorem is tight. For instance, if $L \leq \frac{N+1}{2}$ and N is odd then there is a collection of $2L - 1$ lightpaths with load L, where each lightpath has length $\frac{N+1}{2}$. Then $W = 2L - 1$ because each lightpath requires its own wavelength.

For a tree with no conversion, there is the following.

Theorem 2   *[Raghavan and Upfal, 1994] A tree network with no conversion at all nodes has wavelength requirements of $W \leq 3L/2$ for static traffic.*

Fig. 6.4 shows an example when this wavelength requirement is also necessary.

The next set of results are for limited conversion networks, and are based upon the following observation about lightpath requests. Consider an arbitrary collection of lightpath requests where each link has exactly L requests. Note that having each link with L requests is not a real restriction since "dummy" lightpath requests of one-hop can be added.

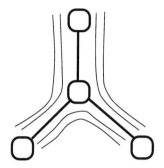

Figure 6.4 An example of lightpaths on a star network with wavelength requirements of $W \geq \frac{3}{2}L$ for $L = 4$.

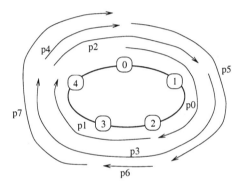

Figure 6.5 A spiral of lightpaths $p_0, p_1, ..., p_7$.

The collection can be partitioned into subsets, each forming a *spiral* of lightpaths. A spiral is a sequence of lightpaths where each lightpath ends where the next one begins, except for the last which ends at the beginning of the spiral. An example collection of lightpath requests that forms a spiral is shown in Fig. 6.5.

A spiral of lightpaths can be assigned to a subset of WDM channels that forms a *spiral* of channels. A spiral of channels is a sequence of channels where the consecutive channels are compatible, and the first channel is compatible to the last. Note that the spiral of channels must go around the ring the same number of times as the spiral of lightpaths. Fig. 6.6 shows the spiral of lightpaths in Fig. 6.5 assigned to a spiral of channels.

The next theorem uses the fact that any traffic with load L, with some additional dummy lightpaths, can form a lightpath spiral that goes around the ring $L + 1$ times. The ring network considered in the theorem has a spiral of channels that goes around the ring $L + 1$ times.

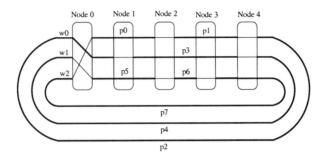

Figure 6.6   A spiral of channels.

Theorem 3 *[Ramaswami and Sasaki, 1998] There is a ring network with fixed conversion at one node and no conversion at the other nodes which has wavelength requirement $W = L + 1$ for static traffic.*

The above wavelength requirements are fairly tight.

Theorem 4 *[Ramaswami and Sasaki, 1998] Any ring network with fixed conversion at every node has wavelength requirement $W = L + 1$ for static traffic if it has a sufficiently large number of nodes.*

The theorem is true because a network with fixed conversion has a fixed collection of spirals of channels. For any fixed collection of spirals of channels, we can construct a spiral of lightpaths which will not fit. Having a large number of nodes helps the construction by having each lightpath be part of only one spiral of lightpaths.

With a bit more conversion, the wavelength requirements become L.

Theorem 5 *[Ramaswami and Sasaki, 1998] The following ring networks have the wavelength requirement $W = L$:*

- *One node has full conversion and the other nodes have no conversion.*

- *One node has wavelength degree 5 and the other nodes have no conversion. (See Fig. 6.7 for an example node.)*

- *Two nodes have wavelength degree 2 and the other nodes have no conversion. (See Fig. 6.8 for an example of two nodes.)*

These networks require exactly L wavelengths because they can form spirals of channels of any size.

2.2.   INCREMENTAL TRAFFIC

For incremental traffic, wavelength requirements for ring networks are higher than for static traffic because lightpaths are set up as they arrive and cannot

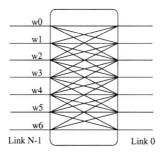

Figure 6.7 A node with wavelength degree 5.

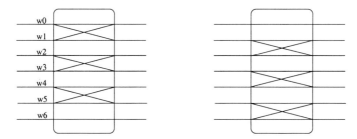

Figure 6.8 Two nodes with wavelength degree 2.

be rearranged thereafter. In the case of no conversion, we have the following result.

Theorem 6 *[Slusarek, 1995] For any N, there is a ring network with no conversion that requires $W = 3L - 2$ wavelengths for incremental traffic.*

In [Slusarek, 1995], the theorem was proven by using a wavelength assignment algorithm called $COLOR$. The algorithm assigns incoming lightpaths into L sets called *shelves*, which are denoted by SHELF(0), SHELF(1), ..., SHELF($L - 1$). Each SHELF(i) is assigned to its own pool of wavelengths, denoted by POOL(i). Each pool has three wavelengths, except POOL(0), which has one. This is sufficient for SHELF(0) since its lightpaths do not overlap. This is sufficient for other shelves because a lightpath will overlap with at most two other lightpaths in its shelf.

The next theorem considers a network with wavelength degree $d > 1$.

Theorem 7 *[Gerstel et al., 1996] For any N and $W \geq d > 1$, there is a ring network with wavelength degree d that requires $W = \max\{0, L - d\} + L$ wavelengths for incremental traffic.*

To prove the theorem, $COLOR$ is modified as shown in Fig. 6.9. There are now $L - d + 1$ shelves and pools. POOL(0) has d wavelengths, while all other

---

Input: A sequence of requests to add lightpaths $p_0, p_1, ...$, one at a time, where the load of the lightpaths is at most L.

Data Structure:

- A collection $\{SHELF(i)\}_{i=0}^{\max\{L-d,0\}}$, where for $i \geq 0$, $SHELF(i)$ is a set of lightpaths.

- A collection of pools $\{POOL(i)\}_{i=0}^{\max\{L-d,0\}}$ where $POOL(0)$ contains wavelengths $0, 1, ..., \min\{L, d\} - 1$ and each other pool contains a disjoint pair of consecutive wavelengths. (The lightpaths in $SHELF(i)$ will be accommodated by $POOL(i)$.)

Initialization: For each $i \geq 0$, set $SHELF(i) = \emptyset$

Processing a Request: Upon arrival of the next lightpath request whose route is p do:

   1 Set $i = 0$

   2 While $LOAD(p/SHELF(0) \cup ... \cup SHELF(i)) > i + d$, Set $i = i + 1$

   3 Set $SHELF(i) = SHELF(i) \cup \{p\}$

   4 Accommodate the request using wavelengths in $POOL(i)$.

Figure 6.9   Lightpath channel assignment for incremental traffic.

pools have 2 wavelengths. Channels are compatible if they have wavelengths in the same pool, so the network has wavelength degree d. These wavelengths are sufficient because the load of lightpaths in $SHELF(0)$ is at most d, and the load of lightpaths in all other shelves is at most 2.

To determine which pool to place a lightpath p, the algorithm in Fig. 6.9 uses the definition $LOAD(p/S)$, which is the maximum load on links along p of lightpath in some set S. In other words, it is the value $\max_{e \in p} LOAD(e/S)$, where $LOAD(e/S)$ denotes the number of lightpaths in S that traverse link e. Notice that the algorithm tries to place a lightpath into the shelf with the smallest index, but where the load in the shelf and lower shelfs is not too high.

2.3.   DYNAMIC TRAFFIC AND NO/FIXED CONVERSION

We will first present wavelength requirements for line networks with no conversion. The requirements lead to results for ring networks, because a ring network becomes a line network after removing a link. Then results for ring and tree networks with no and fixed conversion are presented.

Theorem 8   *[Gerstel et al., 1996] There is a line network with N nodes and no conversion with wavelength requirements of at most $W = L\lceil \log_2 N \rceil$ for dynamic traffic.*

This can be argued by induction. It is certainly true for $N = 2$. Suppose it is true for all $N \leq k$, where $k \geq 2$ is an arbitrary integer. Now suppose

$N = k+1$. We can pick a link e that divides the network in half, and dedicate L wavelengths for the lightpaths that use it. All other lightpaths are in one of the network's halves. By the theorem, each half requires at most $L(\lceil \log_2 N \rceil - 1)$ wavelengths. Since their lightpaths do not intersect, the halves can use the same wavelengths. Thus, the line network with $N = k + 1$ has wavelength requirements as stated in the theorem.

The wavelength requirement is fairly tight.

Theorem 9 *[Gerstel and Kutten, 1996] A line network with N nodes and no conversion has a wavelength requirement of at least $W > 0.5L\lfloor \log_2 N \rfloor + L$ for dynamic traffic.*

The theorem is proven by constructing a sequence of lightpath additions and terminations so that the lightpaths are forced to use more wavelengths while keeping the load within an upper bound.

Since a ring network becomes a line network after deleting a link, the next two theorems follow from Theorems 8 and 9.

Theorem 10 *[Gerstel et al., 1996] A ring network with N nodes and no conversion has a wavelength requirement of at most $W = L\lceil \log_2 N \rceil + L$ for dynamic traffic.*

Theorem 11 *[Gerstel and Kutten, 1996] A ring network with N nodes and fixed conversion has a wavelength requirement of at least $W > 0.5\lfloor \log_2 N \rfloor + L$ for dynamic traffic.*

Although the wavelength requirements of Theorem 10 are fairly tight, the channel assignment strategy used to prove the theorem may be inefficient for typical traffic. A channel assignment that does work well for randomly generated traffic is *first fit*, which is to assign to an arriving lightpath the lowest available wavelength. The strategy also works well for worst case traffic.

Theorem 12 *[Veeramachaneni and Berman, 1997] A ring network with no conversion and using first-fit has wavelength requirements of at most $1.63L \log_2 N$ and at least $0.74L \log_2 N - 0.56L$.*

Note that the Theorem is an improvement over the upper bound on wavelength requirements of $2.54L \log_2 N$ reported earlier in [Gerstel et al., 1996].

For trees with no conversion, we have the following result.

Theorem 13 *[Gerstel et al., 1996] A tree network with no conversion has wavelength requirements $W \leq (2L - 1)\lceil \log_2 N \rceil$ for dynamic traffic.*

This can be shown using similar inductive arguments to prove Theorem 8. Suppose the theorem is true for all $N \leq k$. Consider an arbitrary tree with

$N = k + 1$. One can find a central node x in the tree such that deleting the node leaves subtrees of size at most $N/2$ [Zelinka, 1969]. $2L - 1$ wavelengths are dedicated to the lightpaths that pass through x. From the theorem, each subtree requires at most $(2L - 1)(\lceil \log_2 N \rceil - 1)$ wavelengths. The subtrees can use the same wavelengths because their lightpaths do not intersect.

2.4.   DYNAMIC TRAFFIC AND LIMITED CONVERSION

Theorem 11 shows that if a ring network has wavelength degree of only one, then its wavelength requirements for dynamic traffic will grow with N. With a little more wavelength conversion the wavelength requirements can be independent of N.

Theorem 14 *[Gerstel et al., 1996] There is a ring network with wavelength degree two with wavelength requirement $W = L\lceil \log_2 L \rceil + 4L$ for dynamic traffic.*

We will now discuss the architecture of the network. The ring is divided into segments, having length of about L (between L and $2L$). For a segment, lightpaths can be classified as being *local* or *transit* depending on whether they are within a segment or cross segments, respectively. The wavelengths in the network are divided into two types: *local* and *transit*, where $L\lceil \log_2 L \rceil + L$ are local, while the other $3L$ are transit. Local and transit wavelengths are for local and transit lightpaths, respectively.

At the local wavelengths, channels with the same wavelengths are compatible. In this way, a segment with its local wavelengths can operate like a line network with no conversion. From Theorem 8, $L\lceil \log_2 L \rceil + L$ wavelengths are sufficient for local lightpaths.

At the transit wavelengths, channels are compatible so that a segment can operate as a crossbar interconnection network for transit traffic. In this way, transit traffic can be set up through a segment without disturbing existing lightpaths. Fig. 6.10 has a graph illustrating how channels are compatible within a segment. Edges represent channels, and edges are compatible if they are incident to a common vertex. Each stage of edges are channels in a link, and the clockwise direction is from left to right. Each stage has L *v-channels*, *u-channels*, and *shift-channels*. Notice that there are *J-vertices* which is where lightpaths must pass through to go from one segment to the next. (However, a J-vertex does not really exist in the network. Passing through a J-vertex means traversing channels that correspond to its incident edges.) A J-vertex is *busy* if a lightpath passes through it.

A transit lightpath is set up segment by segment following the clockwise direction of the ring. In the first segment, it is placed in v-channels that lead

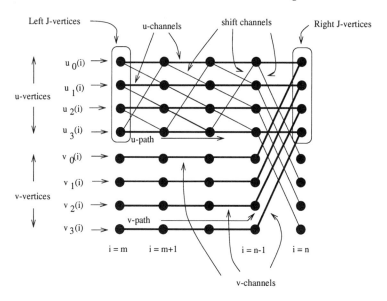

Figure 6.10 A graph illustrating how transit channels are compatible within a segment. Edges represent channels, and edges are compatible if they are incident to a common vertex.

to an idle J-vertex (this is possible since there are L J-vertices in a stage). In subsequent segments, the lightpath is set up as follows. If the lightpath crosses the segment to the next segment, then first an idle J-vertex on the right is found. Next, the lightpath is set up on shift-channels until it reaches the row of the J-vertex. Then it is set up on the u-channels to the J-vertex. On the other hand, if the lightpath does not completely cross the segment then the lightpath is set up on the shift channels.

Note that rather than using the u-channels and shift-channels to form a crossbar-like interconnection network, we could have them form another wide-sense nonblocking crossconnect. For example, the interconnection network in [Cantor, 1972] requires only $\lceil \log_2 L \rceil^2$ stages [1]. Then the segment sizes can be limited to at most $2\lceil \log_2 L \rceil^2$. Then the number of local wavelengths can be $L\lceil \log_2 (\lceil \log_2 L \rceil) \rceil + L \le 2L\lceil \log_2 \log_2 L \rceil + 4L$, while the number of transit wavelengths are still $3L$. The next result is implied.

Theorem 15 *[Gerstel et al., 1996] There is a ring network with wavelength degree two with wavelength requirement* $W = 2L\lceil \log_2 \log_2 L \rceil + 4L$ *for dynamic traffic.*

3. GENERAL TOPOLOGIES

In this section, we will consider networks with arbitrary topologies. We will first present wavelength requirements for dynamic traffic which also hold for static traffic. Then we will consider static traffic only, with the additional restriction that all lightpaths have at most two hops.

3.1.   DYNAMIC TRAFFIC

In this subsection, we consider dynamic traffic. The following is a simple result that comes from [Aggarwal et al., 1994] [2].

Theorem 16 *[Aggarwal et al., 1994] A network with no conversion has wavelength requirements of*

$$W \leq \min\{(L-1)H + 1, (2L-1)\sqrt{M} - L + 2\}$$

for dynamic traffic, where M is the number of links in the network and H is an upper bound on the length of any lightpath.

Proof.   Suppose $(L-1)H + 1 \leq (2L-1)\sqrt{M} - L + 2$. Then $W = (L-1)H + 1$ is sufficient because an arriving lightpath will share links with at most $(L-1)H$ other lightpaths. Therefore, we can find and use a wavelength unused by these other lightpaths.

Now suppose $(L-1)H + 1 > (2L-1)\sqrt{M} - L + 2$. Then $W = (2L-1)\sqrt{M} - L + 2$ is also sufficient. In this case, there are $L\sqrt{M}$ wavelengths dedicated to *long* lightpaths that have at least $\sqrt{M}$ hops. The rest of the $(L-1)(\sqrt{M} - 1) + 1$ wavelengths are dedicated to *short* lightpaths that have at most $\sqrt{M} - 1$ hops.

We can always find a wavelength for an arriving short lightpath because it has at most $\sqrt{M} - 1$ hops and so will intersect with at most $(L-1)(\sqrt{M} - 1)$ other lightpaths. We can always find a wavelength for an arriving long lightpath for the following reasons. Let K denote the number of long lightpaths including the arriving lightpath. Note that the average number of long lightpaths crossing any link is at least $\frac{K\sqrt{M}}{M}$ because each of the K lightpaths has at least $\sqrt{M}$ hops. Since one or more links have at least $\frac{K\sqrt{M}}{M}$ lightpaths, $\frac{K\sqrt{M}}{M} \leq L$. Thus, $K \leq L\sqrt{M}$, and we can give each long lightpath, and in particular the arriving one, its own wavelength.

The bound can be loose. For ring networks, $M = N$ and, assuming lightpaths take shortest-hop routes, $H \leq N/2$. The wavelength requirements then become $\min\{(L-1)N/2, (2L-1)\sqrt{N} - L + 2\}$ which is $O(L\sqrt{N})$. This is larger than the $O(L \log N)$ wavelengths requirements in Theorem 10.

The next result shows that with "limited wavelength conversion" the number of wavelengths required for dynamic traffic is only dependent on L and not on H, N, or number of links. What we mean by limited conversion is that the wavelength degree is a constant d.

Theorem 17 *[Gerstel et al., 1996] There is a fraction $\delta > 0$ and integer $d > 0$ such that for any network topology, there is a network with wavelength degree d with wavelength requirements $W \geq L/\delta$ for dynamic traffic with L sufficiently large.*

The result is impractical because the constants are too small (or too large). It relies upon *expander graphs*, which have been used to define wide-sense nonblocking interconnection topologies.

Definition 1 *Consider a bipartite graph (V_1, V_2, E) with each node having at most d incident edges. For each subset of nodes $S \subseteq V_1$, let $\Gamma(S)$ denote the subset of nodes in V_2 that are adjacent to a node in S (i.e., $\Gamma(S) = \{j \in V_2 : \exists i \in S, (i, j) \in E\}$). The graph is called an (α, β, d)-expander, for some $0 < \alpha < \frac{1}{2}$ and $\beta > 1$, if for each subset of nodes $S \subseteq V_1$ such that $|S| \leq \alpha|V_1|$, $|\Gamma(S)| \geq \beta|S|$.*

Lemma 1 *[Arora et al., 1990] There is a triple (α, β, d), where $0 < \alpha < \frac{1}{2}$ and $\beta > 1$, such that for each n that is sufficiently large, there is a symmetric (α, β, d)-expander with n nodes.*

The expander graph is used to define how WDM channels are attached at a node. $V_1 = \{u_0, u_1, ..., u_{W-1}\}$ represents WDM channels in one link, while $V_2 = \{v_0, v_1, ..., v_{W-1}\}$ represents WDM channels in another link. An edge in the expander graph indicates which channels are attached.

To see why Theorem 17 is true, consider a lightpath request with route corresponding to a sequence of links $e_1, e_2, ..., e_h$, where h denotes the route's length. For $i = 1, 2, ..., h$, label a channel in e_i *idle* if a lightpath can be set up from e_1 through e_i using it as the last channel. Notice that the idle channels in e_1 are the ones that are unused by lightpaths.

The theorem is true because the fraction of idle channels in a link never goes below α. This can be argued by induction. For e_1, the fraction of unused channels is at least $1 - \delta$ (here, $\delta = \frac{L}{W}$), and δ is chosen to be less than $1 - \alpha$. Now suppose that the fraction of idle channels in e_i ($i \geq 1$) is at least α. To show that the fraction of idle channels in e_{i+1} is at least α, first notice that these idle channels are unused and compatible with idle channels in e_i. The expander graph property implies that the fraction of channels in e_{i+1} that are compatible with idle channels in e_i is at least $\alpha\beta$. Since the fraction of used channels in e_i is at most δ, the fraction of idle channels in e_{i+1} is at least $\alpha\beta - \delta$. The fraction is at least α by the way the parameters are chosen.

3.2.   STATIC TRAFFIC

If we restrict the lightpaths to two hops, we can get tight wavelength requirements for static traffic.

Theorem 18 *[Ramaswami and Sasaki, 1998] For any network topology, there is a network with fixed conversion that has wavelength requirements*

$$W = \begin{cases} L & \text{if } L \text{ is even} \\ L+1 & \text{if } L \text{ is odd} \end{cases}$$

for static traffic with lightpaths restricted to at most two hops.

Note that without wavelength conversion, there are cases when the wavelength requirements are $W \geq \frac{3}{2}L$ even when the lightpaths are restricted to two hops (recall Fig. 6.4 for an example).

The theorem is a result of the following result for star topology networks.

Theorem 19 *[Ramaswami and Sasaki, 1998] For a star topology and L even, there is a network with fixed conversion that has wavelength requirements $W = L$ for static traffic with lightpaths restricted to at most two hops.*

The wavelength conversion for the star topology is described next. First, the wavelengths are divided in half into two subsets $A = \{a_0, a_1, ..., a_{W/2-1}\}$ and $B = \{b_0, b_1, ..., b_{W/2-1}\}$. The wavelengths are paired such that for all i, (a_i, b_i) are a pair. WDM channels are compatible if their wavelengths are distinct but belong to a common pair. We refer to this as *fixed-conversion wavelength-paired* (FCWP).

The procedure to find a channel assignment for a set of lightpath requests is as follows [Ramaswami and Sasaki, 1998]. First, assign logical *directions* to each lightpath route so that each link has at most $W/2$ lightpaths in either direction. Note that the lightpaths are still full duplex, and the *directions* are only used by the channel assignment algorithm. Assigning directions can be done using a greedy algorithm [Ramaswami and Sasaki, 1998] similar to the algorithm in Subsection 2.1 to find lightpath spirals. Second, define a bipartite graph with vertices $U = \{u_0, u_1, ..., u_{m-1}\}$ and $V = \{v_0, v_1, ..., v_{m-1}\}$, where m is the number of links in the star, and each vertex pair $\{u_i, v_i\}$ represents a link i. Each edge in the graph represents represents a two-hop lightpath, where an edge (u_i, v_j) represents the lightpath that first traverses link i and then link j. Now assign wavelength pairs to the edges of the bipartite graph such that at each vertex, the incident edges have distinct pairs. (This is possible since this is equivalent to *coloring* edges of a bipartite graph using a minimum number of colors [Berge, 1976].) Note that the assignment of wavelength pairs to edges is really an assignment to two-hop lightpaths. In particular, if a lightpath traversing links j then k is assigned the pair (a_i, b_i) then its channel assignment is to have wavelength a_i on link j and wavelength b_i on link k. Finally, to complete the procedure, the one-hop lightpaths are trivially assigned to the remaining channels.

Theorem 18 follows from Theorem 19. In particular, consider a general topology network as stated in Theorem 18 where the fixed conversion is FCWP. For simplicity, assume L is even. Suppose we have a collection of lightpath requests. Now let us rearrange the topology so it has the same links but is now a star, and where the lightpaths still use their same links. Since the topology is now a star, the procedure described above can find a channel assignment for the lightpaths. This channel assignment is valid for the lightpaths in the original general topology because all nodes have FCWP.

4.   SUMMARY

We have provided the numbers of wavelengths to insure no blocking of lightpaths for the ring, tree, and arbitrary topologies, and for static, incremental and dynamic traffic. Fig. 6.3 summarizes the wavelength requirements for ring networks. For tree networks, we have results for no conversion. In particular, the wavelength requirements for static traffic is $W = 3L/2$, and for dynamic traffic is $W \leq (2L - 1)\lceil \log_2 N \rceil$. For general topologies, we have the simple wavelength requirement bound given in Theorem 16 for dynamic traffic. This bound depends on the number of nodes and links there are in the network. If we allow limited conversion for some wavelength degree d then the wavelength requirement is $W \leq L/\delta$ for some fraction δ. If we consider static traffic and limits of two hops per lightpath then we have the tight wavelength requirements given in Theorem 18.

Notice that with very little conversion, wavelength requirements can drop substantially. We also found that wavelength requirements can increase substantially from static to dynamic traffic.

Note that the many of the results can be trivially extended to networks that have directed lightpaths and WDM channels. These include all the results on ring networks. For trees, the results may be different. For example, a tree network with directed lightpaths and no conversion requires $15L/8$ wavelengths [Mihail et al., 1995], where as the wavelength requirements for undirected lightpaths is $3L/2$, from Theorem 13.

Notes

1. The switching network in [Cantor, 1972] requires $8\lceil \log_2 L \rceil^2$ stages, but it is claimed there that the factor of 8 can be eliminated.

2. Actually, the result in [Aggarwal et al., 1994] is for static traffic but the proof works for dynamic traffic too.

References

Aggarwal, A., Bar-Noy, A., Coppersmith, D., Ramaswami, R., Scheiber, B., and Sudan, M. (1994). Efficient routing and scheduling algorithms for optical networks. In *Proc. Fifth Annual ACM-SIAM Symposium on Discrete Algorithms*, pages 412–423.

Arora, S., Leighton, T., and Maggs, B. (1990). On-line algorithms for path selection in a non-blocking network. In *Proc. ACM Symp. on Theory of Computering*, pages 149–158.

Barry, R. A. and Humblet, P. A. (1996). Models of blocking probability in all-optical networks with and without wavelength changers. *IEEE J. Sel. Areas Comm.*, 14(5):858–867.

Berge, C. (1976). *Graphs and Hypergraphs*. North Holland.

Birman, A. (1996). Computing approximate blocking probabilities for a class of all-optical networks. *IEEE J. Sel. Areas Comm.*, 14(5):852–857.

Cantor, D. (1972). On non-blocking switching networks. *Networks*, 1:367–377.

Gerstel, O. and Kutten, S. (1996). Dynamic wavelength allocation in WDM ring networks. Research Report RC 20462, IBM.

Gerstel, O., Sasaki, G., and Ramaswami, R. (1996). Dynamic channel assignment for WDM optical networks with little or no wavelength conversion. In *1996 Allerton Conference*, pages 32–43, Monticello, IL.

Kovačević, M. and Acampora, A. S. (1996). Benefits of wavelength translation in all-optical clear-channel networks. *IEEE J. Sel. Areas Comm.*, 14(5):868–880.

Lee, K. C. and Li, V. O. K. (1993a). Routing and switching in a wavelength convertible lightwave network. In *Proc. INFOCOM*, pages 578–585.

Lee, K. C. and Li, V. O. K. (1993b). A wavelength-convertible optical network. *IEEE/OSA J. Lightwave Tech.*, 11(5/6):962–970.

Mihail, M., Kaklamanis, C., and Rao, S. (1995). Efficient access in all-optical networks. In *Proc. IEEE Symp. Foundations of Com. Sci.*, pages 548–557.

Raghavan, P. and Upfal, E. (1994). Efficient routing in all-optical networks. In *Proc. ACM Symp. Theory of Computing*, pages 134–143.

Ramaswami, R. and Sasaki, G. H. (1998). Multiwavelength optical networks with limited wavelength conversion. *IEEE/ACM Transactions on Networking*, 6(6):744–754.

Ramaswami, R. and Sivarajan, K. N. (1995). Routing and wavelength assignment in all-optical networks. *IEEE/ACM Trans. Networking*, 3(5):489–500.

Slusarek, M. (1995). Optimal online coloring of circular arc graphs. *Informatique Theoretique et Applications*, 29(5):423–429.

Subramaniam, S., Azizoğlu, M., and Somani, A. K. (1996). Connectivity and sparse wavelength conversion in wavelength-routing networks. In *Proc. INFOCOM '96*, pages 148–155.

Tucker, A. (1975). Coloring a family of circular arcs. *SIAM J. Appl. Math.*, 29(3):493–502.

Veeramachaneni, V. and Berman, P. (1997). Lower and upper bounds on the performance of first-fit in dynamic WDM optical networks. Technical Report CSE-97-017, Pennsylvania State University, University Park, PA, USA.

Yates, J., Lacey, J., Everitt, D., and Summerfield, M. (1996). Limited-range wavelength translation in all-optical networks. In *Proc. INFOCOM*, pages 954–961.

Zelinka, B. (1969). Medians and peripherians on trees. *Arch. Math. (Brno)*, pages 87–95.

Zhou, J., Park, N., Vahala, K. J., Newkirk, M. A., and Miller, B. I. (1994). Broadband wavelength conversion with amplification by four-wave mixing in semiconductor travelling-wave amplifiers. *Electronics Letters*, 30(11):859–860.

IV

BROADCAST STAR NETWORKS

Chapter 7

MAC PROTOCOLS FOR WDM NETWORKS – SURVEY AND SUMMARY

Bo Li

Department of Computer Science
Hong Kong University of Science and Technology
Clear Water Bay, Hong Kong

Maode Ma

Department of Computer Science
Hong Kong University of Science and Technology
Clear Water Bay, Hong Kong

Mounir Hamdi

Department of Computer Science
Hong Kong University of Science and Technology
Clear Water Bay, Hong Kong

Abstract With the proliferation of the world-wide-web (WWW), current local and wide area networks can hardly cope with the huge demand for network bandwidth. As a result, there is a world-wide effort in upgrading current networks with high-capacity fiber-optic links that can potentially deliver Tera-bits/sec bandwidth. The *Wavelength-Division-Multiplexing* (WDM) is an effective technique for utilizing the large bandwidth of an optical fiber. By allowing multiple simultaneous transmission over a number of channels, WDM has the potential to significantly improve the performance of optical networks. The nodes in the network can transmit and receive messages on any of the available channels by employing one or more tunable transmitter(s) and/or tunable receiver(s). Several topologies have been proposed for WDM networks. Of particular interest to us in this chapter is the single-hop topology where a WDM optical network is configured as a broadcast-and-select network in which all the input nodes are combined using a passive star coupler, and the mixed optical information is broadcast to all destinations. To unleash the potential of single-hop WDM passive star networks, effective medium access control (MAC) protocols are needed to efficiently allocate and coordinate the system resources. In this chapter, we present a comprehensive survey of state-the-art MAC protocols for WDM networks so as to give the reader an overview of the research efforts conducted in this area for the past decade. In addition, it can serve as a starting point for further investigation into ways of coping with the current and the anticipated explosion of multimedia information transfer.

1.   INTRODUCTION

Wavelength Division Multiplexing (WDM) is the most promising multiplexing technology for optical networks. By using WDM, the optical transmission spectrum is divided into a number of non-overlapping wavelength bands. In particular, allowing multiple WDM channels to coexist on a single fiber, the task of balancing the opto-electronic bandwidth mismatch, can be implemented by designing and developing appropriate WDM optical network architectures and protocols.

WDM optical networks can be designed by using one of two types of architectures, Broadcast-and-Select networks or Wavelength-Routed networks. Typically, the former is used for local area networks, while the latter is used for wide-area networks. A local WDM optical network may be set up by connecting computing nodes via two-way fibers to a passive star coupler, as shown in Fig. 7.1. A node can send its information to the passive star coupler on one available wavelength by using a laser which produces an optical stream modulated with information. The modulated optical streams from multiple transmitting nodes are combined by the passive-star coupler and then the combined streams are separated and transmitted to all the nodes in the network. A destination's receiver is an optical filter and is tuned to one of the wavelengths to receive its designated information stream. Communication between the source and destination nodes is proceeded in one of the following two modes: single-hop, in which communication takes place directly between two nodes [Mukherjee, 1992a], or multihop, in which information from a source to a destination may be routed through the intermediate nodes of the network [Mukherjee, 1992b; Li and Ganz, 1992].

Based on the architectures of WDM optical networks, Medium Access Control (MAC) protocols are required to allocate and coordinate the system resources. The challenge of developing appropriate MAC protocols is to efficiently exploit the potentially vast bandwidth of an optical fiber to meet the

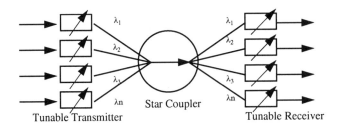

Figure 7.1 Structure of single-hop passive star coupled WDM optical network.

increasing information transmission demand under the constraints of the net-work resources and the constraints imposed on the transmitted information.

In this chapter, we review the state-of-art MAC protocols in passive star coupler-based WDM networks. Our focus is on the MAC protocols for the single-hop architecture. We will discuss several protocols in some detail to show their importance in the development of MAC protocols for the single-hop passive-star coupler based WDM networks. According to the network service provided to the transmitted information, we roughly divide the MAC protocols into three categories as follows: MAC protocols for packet transmission, MAC protocols for variable-length message transmission, and MAC protocols for real-time message transmission (i.e., MAC protocol with QoS concern).

The reminder of this chapter is organized as follows. Section 2 surveys the MAC protocols for packet transmission. Section 3 reviews the MAC protocols for variable-length message transmission. Section 4 discusses the MAC protocols for real-time service. Section 5 concludes the chapter with a summary.

2. MAC PROTOCOLS FOR PACKET TRANSMISSION

The MAC protocols for packet transmission in single-hop passive-star cou-pled WDM networks are so called "legacy" protocols, since they are dedicated for fixed length packet transmission, and are often adopted from legacy shared medium networks. In a single-hop network, significant amount of dynamic coordinations among nodes are required in order to access the network re-sources. According to the coordination schemes, the MAC protocols can be further classified into the following sub-categories.

2.1.   NON-PRETRANSMISSION COORDINATION PROTOCOLS

Protocols with non-pretransmission coordination do not need any channels for pretransmission coordination. All the transmission channels are either preassigned to transmitting nodes or accessed by transmitting nodes through contest. These protocols can be categorized accordingly in the following subgroups:

2.1.1   Fixed Assignment.   A simple approach, based on the fixed wavelength assignment technique, is a time-division-multiplexing (TDM) extended over a multichannel environment [Chlamtac and Ganz, 1988]. It is pre-determined that a pair of nodes is allowed to communicate with each other in the specified time slots within a cycle on the specified channel. Several extensions to the above protocol have been proposed to improve the performance. One approach, named weighted TDM, assigned different number of time slots to different transmitting nodes according to the traffic load [Rouskas and Ammar, 1995]. Another approach proposed a versatile time-wavelength assignment algorithm [Ganz and Gao, 1994]. Under the condition that a traffic demand matrix is given beforehand, the algorithm can minimize the tuning times and has the ability to reduce transmission delay. Some new algorithms based on [Ganz and Gao, 1994] investigated problems such as the performance of scheduling packet transmissions with an arbitrary traffic matrix; and the effect of the tuning time on the performance [Pieris and Sasaki, 1994; Borella and Mukherjee, 1996; Azizoglu et al., 1996].

2.1.2   Partial Fixed Assignment Protocols.   Three partial fixed assignment protocols have been proposed in [Chlamtac and Ganz, 1988]. The first one is the Destination Allocation (DA) protocol. In this protocol, the number of source and destination node pairs can be the same as the number of nodes. A Source Allocation (SA) protocol is also defined in which the control of access to transmission channels is further relaxed. Similar to the SA protocol, Allocation Free (AF) protocol has been proposed, in which all source-destination pairs of computing nodes have full rights to transmit packets on any channel over any time slot duration.

2.1.3   Random Access Protocols.   Two slotted-ALOHA based protocols been were proposed in [Dowd, 1991]. In the first protocol, transmissions on different channels are synchronized into slots; while in the second protocol, the transmission is synchronized into mini-slots, and each packet transmission can take multiple such mini-slots. Two similar protocols were presented in [Ganz and Koren, 1991].

2.2.   PRETRANSMISSION COORDINATION PROTOCOLS

For protocols that require pretransmission coordination, transmission channels are grouped into control channels and data channels. These protocols can be categorized according to the ways to access the control channels into the following subgroups:

2.2.1   Random Access Protocols.   The architecture of the network protocols in this subgroup is as follows. In a single-hop communication network, a control channel is employed. Each node is equipped with a single tunable transmitter and a single tunable receiver.

In [Kavehrad and Sundberg, 1987], three random access protocols including ALOHA, slotted-ALOHA, and CSMA are proposed to access the control channel. ALOHA, CSMA, and N-server switch scheme can be the sub-protocols for the data channels. Under a typical ALOHA protocol, a node transmits a control packet over the control channel at a randomly selected time, after which it immediately transmits a data packet on a data channel, which is specified by the control packet.

In [Mehravari, 1990], an improved protocol named slotted-ALOHA/delayed-ALOHA has been proposed. The characteristics of this protocol is that it requires that a transmitting node delays transmitting data on a data channel until it receives the acknowledgment that its control packet has been successfully received by the destination node. The probability of data channel collision can be decreased. And the performance in terms of throughput can be improved.

In [Sudhakar et al., 1991b], one set of slotted-ALOHA protocols and one set of Reservation-ALOHA protocols have been proposed. Further performance gains are observed when comparing to the the protocols in [Kavehrad and Sundberg, 1987].

In [Sudhakar et al., 1991a], a so-called Multi-Control-Channel protocol is proposed, which aims at improving Reservation-ALOHA-based protocols. All channels are used to transmit control information as well as data information. Control packet transmission is contention-based operation; while data transmission follows it.

A different protocol was proposed in [Li et al., 1995], in which it does require separate control channel for control packet dissemination. The reservation packet are transmitted along with data packet using the same channels in different time slot. This is referred as in-band signaling protocol in [Li et al., 1995].

One potential problem in the above protocols is that there can be conflict at the destination node if multiple transmissions from different source nodes and different wavelengths/channels reach the destination at the same time given that there is only one receiver at the destination node. This is referred as receiver collision, in which the issue was resolved in [Jia and Mukherjee, 1993].

2.2.2   Reservation Protocols.

In the Dynamic Time-Wavelength Division Multiple Access (DT-WDMA) protocol, a channel is reserved as control channel and it is accessed only in a pre-assigned TDM fashion. It requires that each node has two transmitters and two receivers [Chen et al., 1990]. One pair of the transceivers are fixed to the control channel, while another pair are tunable to all the data channels. If there are N nodes in the network, N data channels and one control channel are required. Although this protocol cannot avoid receiver collisions, it ensures that exactly one data packet can be successfully received when more than one data packet come to the same destination node simultaneously.

One proposal [Chlamtac and Fumagalli, 1991] to improve the DT-WDMA algorithm is to use an optical delay line to buffer the potentially collided packets, when more than one node transmit data packets to the same destination node at the same time. Its effectiveness depends on the relative capacity of the buffer. Another protocol [Chen and Yum, 1991] also tries to improve the DT-WDMA algorithm by making transmitting nodes remember the information from the previous transmission of control packet and combining this information into the scheduling of future packet transmission.

In [Chipalkatti et al., 1992] and [Chipalkatti et al., 1993], another two protocols aiming at improving the DT-WDMA algorithm are proposed. The first one is called Dynamic Allocation Scheme (DAS), where each node runs an identical algorithm based on a common random number generator with the same seed. The second protocol is named Hybrid TDM. Time on the data channels is divided into frames consisting of several slots. In a certain period of time, one slot will be opened for a transmitting node to transmit data packets to any destination receiver.

A reservation-based Multi-Control-Channel protocol can be found in [Humblet et al., 1993]. In this protocol, x channels ($1 < x < (N/2)$ can be reserved as control channels to transmit control information, where N is the number of the channels in the network. The value of x is a system design parameter, which depends on the ratio of the amount of control information and the amount of actual data information. The objective to reserve multiple control channels in the network is to decrease the overhead of control information processing time as much as possible.

The properties of the "legacy" MAC protocols are summarized as follows:

- Although the protocols using fixed-channel assignment approach can ensure the successful data transmission and reception, they are sensitive to the dynamic bandwidth requirements of the network and they are difficult to scale in terms of the number of nodes. The protocols using contention-based channel assignment approach introduce contention on data channels in order to adapt to the dynamic bandwidth requirements. As a result, either channel collision or receiver collision will occur.

■ The protocols with contention-based control channel assignment still have either data channel collision or receiver collision because contention is involved in the control channel. While some protocols proposed in [Jeon and Un, 1992; Lee and Un, 1996], have the capability to avoid both collisions by continuously monitoring the network states. The reservation-based protocols, which take fixed control channel assignment approach, can only ensure data transmission without collisions. However, by introducing some information to make the network nodes intelligent, it has potential to avoid receiver collisions as well. It also has potential to accommodate application traffic composed of variable-length messages.

3.   MAC PROTOCOLS FOR VARIABLE-LENGTH MESSAGE TRANSMISSION

The "legacy" MAC protocols are designed to handle and schedule fixed length packets. Using these MAC protocols, most of the application level data units (ADU) must be segmented into a sequence of fixed size packets in order to be transmitted over the networks. However, as traffic streams in the real world are often characterized as bursty, consecutive arriving packets in a burst are strongly correlated by having the same destination node. A new idea about this observation is that all the fixed size packets of a burst should be scheduled as a whole and transmitted continuously in a WDM network rather than schedule them on a packet-by-packet basis. Another way of looking at this is that the ADUs should not be segmented. Rather they should be simply scheduled as a whole without interleaving. The main advantages of using a burst-based or message transmission over WDM networks are: 1) To an application, the performance metrics of its data units are more relevant performance measures than ones specified by individual packets; 2) It perfectly fits the current trend of carrying IP traffic over WDM networks; and 3) Message fragmentation and reassembly are not needed.

The first two MAC protocols in [Sudhakar et al., 1991b] proposed for variable-length message transmission are protocols with contention-based control channel assignment. Another two Reservation-ALOHA-based protocols in [Sudhakar et al., 1991b] are presented in order to serve the long holding time traffic of variable-length messages. The first protocol aims to improve the basic slotted-ALOHA-based technique in [Sudhakar et al., 1991b]. The second protocol aims to improve the slotted-ALOHA-based protocol with asynchronous cycles on the different data channels. Data channel collisions can be avoided in the protocols presented in [Sudhakar et al., 1991b].

The protocol in [Bogineni and Dowd, 1991; Dowd and Bogineni, 1992; Bogineni and Dowd, 1992] tries to improve the reservation-based DT-WDMA protocol in [Chen et al., 1990]: The number of nodes is larger than the number of channels; the transmitted data is a variable-length message rather than a fixed length packet; data transmission can start without any delay. Both data collision and receiver collision can be avoided because any message transmission scheduling has to consider the status of the data channels as well as receivers.

There are two protocols called FatMAC in [Sivalingam and Dowd, 1995] and LiteMAC in [Sivalingam and Dowd, 1996] respectively. They try to combine reservation based and preallocation based techniques to schedule variable-length message transmission. FatMAC is a hybrid approach which reserves access preallocated channels through control packets. Transmission is organized into cycles where each of them consists of a reservation phase and a data phase. A reservation specifies the destination, the channel and the message length of next data transmission. LiteMAC protocol is an extension of FatMAC. By LiteMAC protocol, each node is equipped with a tunable transmitter and a tunable receiver rather than a fixed receiver in FatMAC. LiteMAC has more flexibility than FatMAC because of the usage of tunable receiver and its special scheduling mechanism. So that more complicated scheduling algorithms could be used to achieve better performance than FatMAC. Both FatMAC and LiteMAC have the ability to transmit variable-length messages by effect scheduling without collisions. Their performance have been proved to be better than the preallocation based protocols while less transmission channels are used than reservation based protocols. With these two protocols, low average message delay and high channel utilization can be expected.

3.1.   A RESERVATION-BASED MAC PROTOCOL FOR VARIABLE-LENGTH MESSAGES

In [Jia et al., 1995], based on the protocols in [Bogineni and Dowd, 1991; Dowd and Bogineni, 1992; Bogineni and Dowd, 1992], an intelligent reservation based protocol for scheduling variable-length message transmission has been proposed. The protocol employs some global information of the network to avoid both data channel collisions and receiver collisions while message transmission is scheduled. Its ability to avoid both collisions makes this protocol a milestone in the development of MAC protocols for WDM optical networks.

The network consists of M nodes and $W + 1$ WDM channels. W channels are used as data channels, the other channel is the control channel. Each node is equipped with a fixed transmitter and a fixed receiver for the control channel, and a tunable transmitter and a tunable receiver to access the data channels. The time on the data channels is divided into data slots. It is assumed that there is a network-wide synchronization of data slots over all data channels. The duration of a data slot is equal to the transmission time of a fixed-length data packet. A node generates variable-length messages, each of which contains one or more fixed-length data packets. On the control channel, time is divided into control frames. A control frame consists of M control slots. A control slot has several fields such as address of destination node, the length of the message, etc. A time division multiple access protocol is employed to access the control channel so that the collision of control packets can be avoided.

Before a node sends a message, it needs to transmit a control packet on the control channel in its control slot. After one round-trip propagation delay, all the nodes in the network will receive the control packet. Then a distributed scheduling algorithm is invoked at each node to determine the data channel and the time duration over which the message will be transmitted. Once a message is scheduled, the transmitter will tune to the selected data channel and transmit the scheduled message at the scheduled transmission time. When the message arrives at its destination node, the receiver should have been tuned to the same data channel to receive the message.

The data channel assignment algorithm determines the data channel and the time duration over which the message will be transmitted. The algorithm schedules message transmissions based on some global information in order to avoid the data channel collisions and the receiver collisions. The global information is expressed through two tables which reside on each node. One table is the Receiver Available Time table (RAT). RAT is an array of M elements, one for each node. $RAT[i] = n$, where $i = 1, 2, ..., M$, means that node i's receiver will become free after n data slots. If $n = 0$, then node i's receiver is currently idle, and no reception is scheduled for it as yet. RAT is needed for avoiding receiver collisions. Another table is named the Channel Available Time table (CAT). CAT is an array of W elements, one for each data channel. $CAT[k] = m$, where $k = 1, 2, ..., W$, means that data channel k will be available after m data slots. If $CAT[k] = 0$, data channel k is currently available. CAT is needed to avoid collisions on data channels. Local and identical copies of these two tables are on each node. They contain consistent information on the messages whose transmissions have been scheduled but not yet transmitted so far. The contents of the tables are relative to current time. Three data channel assignment algorithms have been proposed. The fundamental one is named Earliest Available Time Scheduling algorithm (EATS). This algorithm schedules the transmission of a message by selecting a data channel which is the earliest available.

This reservation-based protocol has been shown to have a quite good performance while it can avoid data channel collisions and receiver collisions.

3.2.   A RECEIVER-ORIENTED MAC PROTOCOL FOR VARIABLE-LENGTH MESSAGES

Some related protocols have been proposed to improve the performance of the network based on the same system architecture of [Jia et al., 1995]. In [Muir and Garcia-Luna-Aceves, 1996], the proposed protocol tries to avoid the head-of-queue blocking during the channel assignment procedure by introducing the concept of "destination queue " to make each node maintain M queues, where M is the number of nodes in the network. In [Hamidzadeh et al., 1999], the authors notice that the performance of the network could be further improved by the way of exploiting more existing global information of the network and the transmitted messages. From this point of view, a general scheduling scheme, which combines the message sequencing techniques with the channel assignment algorithms in [Jia et al., 1995], is proposed to schedule variable-length message transmission. In [Ma et al.,], as an example of the general scheduling scheme in [Hamidzadeh et al., 1999], a novel scheduling algorithm is proposed. The new algorithm, named Receiver-Oriented Earliest Available Time Scheduling (RO-EATS), decides the sequence of the message transmission by the information of the receiver's states to decrease message transmission blocking caused by avoiding receiver collisions.

The RO-EATS scheduling algorithm employs the same system structure and network service as those of the protocol in [Jia et al., 1995] to form a receiver-oriented MAC protocol, which is an extension to the protocol in [Jia et al., 1995]. The logic structure of the system model for the RO-EATS protocol can be expressed as in Fig. 7.2. The new protocol proceeds messages' transmission and reception is the same as that of the protocol in [Jia et al., 1995]. The difference between the two protocols is on the scheduling algorithm for message transmission.

The RO-EATS algorithm works as follows. It first considers the earliest available receiver among all the nodes in the network and then selects a message which is destined to this receiver from those which are ready and identified by the control frame. After that, a channel is selected and assigned to the selected message by the principle of EATS algorithm.

The scheme to choose a suitable message to transmit is based on the information of the states of the receivers presented in RAT. The objective of this scheme is to avoid lots of messages going to one or a few nodes at the same time and try to raise the channels utilization. The motivation of the algorithm comes from the observation that two consecutive messages with the same destination may not fully use the available channels when EATS algorithm is employed. The new algorithm enforces the idea of scheduling two consecutive messages away from going to the same destination node. The RO-EATS algorithm always checks the table of RAT to see which node is the least visited destination and

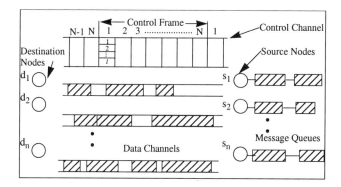

Figure 7.2   The system model of RO-EATS protocol.

to choose the message which is destined to this node to transmit. In this way, average message delay can be shown too be quite low and channel utilization can be shown to be high.

4.   MAC PROTOCOLS FOR REAL-TIME SERVICE

An important function of high-speed computer networks such as WDM optical networks is to provide real-time service to time-constrained application streams such as video or audio information. Most of the MAC protocols that provide real-time service on passive star-coupled WDM optical networks are protocols with reservation based pre-coordination. According to the type of the real-time service provided to the transmitted messages, the MAC protocols for real-time service can be classified into three types: protocols with best-effort service, protocols with deterministic guaranteed service, and protocols with statistical guaranteed service

4.1.   MAC PROTOCOLS FOR BEST-EFFORT REAL-TIME SERVICE

A protocol named Time-Deterministic Time and Wavelength Division Multiple Access (TD-TWDMA) presented in [Jonsson et al., 1997] provides services for both hard real-time messages and soft real-time messages for single destination, multicast, and broadcast transmissions. All channels can be accessed by a fixed-assignment method, which is a TDM approach. Using this approach, each channel is divided into time-slots. Each node has a number of slots for hard real-time message transmissions. Soft real-time messages can be transmitted if there is no hard real-time messages requiring service. Each node is equipped with one fixed-transmitter and tunable receivers. The transmitter is fixed to its assigned channel, while the receiver can be tuned over all channels in the network. Each node has a specified channel because the number of nodes, C, is equal to the number of channels, M in the network. At each node, there are $2 \times M$ queues, M queues for the hard real-time messages, another M queues for the soft real-time messages. For each type of queues, one queue is for broadcast and $M-1$ queues for the single destination. The messages in the broadcast queue can be either control information or data to be broadcast. The protocol works as follows: First it sends a broadcast slot containing the control information; then invokes the slot-allocation algorithm to determine the slots used to transmit the data information; at last, each node tunes to the specified channel to receive the data. The slot-allocation algorithm follows the static priority approach. The basic idea of the algorithm can be summarized as: 1) The M hard real-time message queues have higher priorities; while the M soft real-time message queues have lower priorities; 2) Each queue in each group has a fixed priority; while the queues for broadcast have the highest priority in each group; 3) Message transmission scheduling is based on the queue priority; and 4) For the hard real-time messages, if transmission delay is over their deadlines, these messages will be dropped; while for the soft real-time messages, they will be scheduled whether they are beyond their deadlines or not.

A reservation-based MAC protocol for best-effort real-time service can be found in [Ma et al., 1999]. This protocol is for the same network structure as that in [Jia et al., 1995]. Both hard real-time and soft real-time variable-length message transmissions have been considered. The scheduling algorithms for the protocol are based on the time related dynamic priority scheme, namely Minimum Laxity First (MLF) scheduling. This protocol employs global information of the network as well as the transmitted messages to ensure zero message loss rate caused by both data channel collisions and receiver collisions and decrease the message loss rate caused by network delay.

4.2.   MAC PROTOCOLS FOR DETERMINISTIC GUARANTEED SERVICE

In [Tyan et al., 1996], a pre-allocation based channel access protocol is proposed to provide deterministic timing guarantees to support time constrained communication in a single-hop passive star-coupled WDM optical network. This protocol takes a passive star-coupled broadcast-and-select network architecture in which N stations are connected to a passive star coupler with W different wavelength channels. Each of the W channels is slotted and shared by the N stations by means of a TDM approach. The slots on each channel are pre-assigned to the transmitters. A schedule specifies, for each channel, which slots are used for data transmission from node i to node j, where $1 \le i \le N$, $1 \le j \le N$, $i \le j$. Each node of the network can be equipped with a pair of tunable transmitters and tunable receivers which can be tuned over all the wavelengths. Each real time message stream with source and destination nodes specified is characterized with two parameters, relative message deadline D_i and maximum message size C_i that can arrive within any time interval of length D_i. A scheme called Binary Splitting Scheme (BSS) is proposed to assign each message stream sufficient and well-spaced slots to fulfill its timing requirement. Given a set of real-time message streams M specified by the maximum length of each stream C_i and the relative deadline of each stream D_i, this scheme can allocate time slots over as few channels as possible in such a way that at least C_i slots are assigned to M_i in any time window of size D_i slots. So that the real-time constraints of the message streams can be guaranteed.

In [Ma and Hamdi, 2000], a reservation-based MAC protocol for deterministic guaranteed real-time service can be found. This protocol is for the same network structure as that in [Jia et al., 1995]. In [Ma and Hamdi, 2000], a systematic scheme is proposed to provide deterministic guaranteed real-time service for application streams composed of variable-length messages. It includes admission control policy, traffic regularity, and message transmission scheduling algorithm. A traffic intensity oriented admission control policy is developed to manage flow level traffic. A g-regularity scheme based on the Max-plus algebra theory is employed to shape the traffic. An Adaptive Round-Robin and Earliest Available Time Scheduling (ARR-EATS) algorithm is proposed to schedule variable-length message transmission. All of the above are integrated to ensure that the deterministic guaranteed real-time service can be achieved.

4.3. MAC PROTOCOLS FOR STATISTICALLY GUARANTEED SERVICE

A reservation-based protocol to provide statistically guaranteed real-time services in WDM optical token LAN can be found in [Yan et al., 1996]. In the network, there are M nodes and $W + 1$ channels. One of the channels is the control channel, while the others are data channels. Unlike the structure in [Jia et al., 1995], the control channel is accessed by token passing. At each node, there is a fixed receiver and transmitter tuned to the control channel. There is also a tunable transmitter which can be tuned to any of the data channels. There are one or more receivers fixed to certain data channels.

The protocol provides transmission service to either real-time or non real-time messages. The packets in the traffic may have variable-length but bounded by a maximum value. At each node, there are W queues, each corresponds to one of the channels in the network. The messages come into one of the queues according to the information of their destination nodes and the information of the channels which connect to the corresponding destination nodes. The protocol works as follows: A token exists on the control channel to ensure collision free transmission on data channels. The token has a designated node K. Every node can read the contents of the token and updates its local status table by the information in the fields of the token. When node K observes the token on the network, it will check the available channels. If there are no channels available, node K gives up this opportunity to send its queued packets. Otherwise, the Priority Index Algorithm (PIA) is invoked to evaluate the priority of each message queue on node K and then use Transmitter Scheduling Algorithm (TSA) to determine the transmission channel. Also the Flying Target Algorithm (FTA) is used to decide the next destination of the control token. After all these have completed, node K's status and scheduling result will be written into the token. Then the token on the node K will be sent out. And the scheduled packets will be transmitted.

4.4.   A STATE-OF-ART MAC PROTOCOL FOR STATISTICALLY GUARANTEED SERVICE

A novel reservation-based MAC protocol is proposed in [Li and Qin, 1998] to support statistically guaranteed real-time service in WDM networks by using a hierarchical scheduling framework. It shows that such hierarchical scheduling is essential for achieving scalability such that larger input-output ports can be accommodated.

The major advantage of the protocol in [Li and Qin, 1998] over that in [Yan et al., 1996], is that it divides the scheduling issue into flow scheduling and transmission scheduling. The former is responsible for considering the order of traffic streams to be transmitted. The latter is to decide the order of the packets transmission. The packets involved in the transmission scheduling are those selected from the traffic streams by the flow scheduling scheme. Compared with the protocol in [Yan et al., 1996], this protocol is expected to diminish the ratio of the packets which are over their deadlines. Another advantage of the protocol is that it employs a multiple retransmission scheme to alleviate the result of unsuccessful packet transmissions. This work is developed for a similar network structure as that in [Jia et al., 1995]. It has further enriched the structure of the network transmitting node to support the proposed protocol. The detailed architecture of the network and transmitting node is presented in Fig. 7.3.

4.4.1   Network Configuration.   An $N \times N$ packet-switching system configuration with single hop connection is considered. The switching and

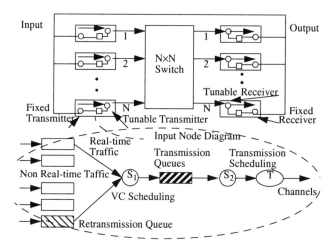

Figure 7.3 The structure of the networks and transmitting node.

routing can possibly be implemented by a recently designed switch, which can support link capacity between $2.4 - 10Gb/s$. In this architecture, a small delay line is employed to separate the data packet and header for allowing the header processing. The queues shown in Fig. 7.3 can be implemented by optical fiber loops. There are $W + 1$ wavelengths in the system, in which W channels are data channels, another channel is control channel. Each input node is equipped with one transmitter fixed to the control channel for control packet transmission, and one data packet transmitter for data packet transmission. Each output node has one control packet receiver fixed to the control channel for control packet reception, and one data packet receiver for data packet reception. Data transmitter and data receiver are tunable, and are assumed to be able to tune to all the channels in the system. Control channels are assumed to be synchronized with data channels. Data channels are divided into data slots and control channel is divided into control slots. The control slot is further divided into mini-slots with each holding a control packet from a node. The number of mini-slots is assumed to equal the number of input/output nodes in the switch, i.e., each input node has its dedicated mini-slot in each control slot then some kind of mini-slot assignment based on either a round robin or contention can be adopted. The hierarchical scheduling frame work is readily applicable to handle variable length packets.

Each input port is fed with multiple traffic streams, and it is noted as a Virtual Circuit. Each VC has its own separate queue. Two generic types of VCs are considered in the switch: one is real-time traffic, the other is non-real-time datagram type of traffic. Real-time traffic has QoS requirements such as delay bound, in-sequence delivery, and packet loss rate, and non-real-time traffic either has no QoS requirement such as the "best effort" UBR or certain minimum QoS such as minimum bandwidth guarantee like ABR.

4.4.2   Scheduling Mechanisms.   The are two stages in scheduling procedure: the VC scheduling and the transmission scheduling. Each input node has one transmission queue for scheduling the packet transmission. VC scheduling is responsible for scheduling packets from VC queues to the transmission queue, the transmission scheduling handles the coordination of packet transmissions among multiple input nodes. A simple round robin scheme is adopted in the VC scheduling and a random scheduling with age priority is used in the transmission scheduling.

One of the major complications in an on-demand scheduling algorithm is that the reservation might fail due to either output conflict or channel conflict, thus re-scheduling would be necessary. The outcome of the reservation will be known to all nodes at the end of one round-trip propagation delay. If a scheduling fails, there is a decision that has to be made whether re-scheduling the same packet or scheduling a new packet from another VC. If the failed reservation is from a real-time VC, it certainly makes sense to re-schedule the very same packet as soon as possible. The very same RT reservation packet will be re-transmitted immediately in the next control slot, thus no other new reservation either from RT VCs or non-RT VCs of the same source node can be initiated. Some kind of randomized algorithms, such as the exponential backoff algorithm might be needed in the scheme preferred in which an immediate re-transmission of the reservation packet will take place once a real-time reservation packet fails.

The more intriguing part of the scheduling algorithm is the re-scheduling of non-RT traffic. If real-time traffic has more stringent QoS requirements, in particular the delay, it is worth trading the complication of switch design and the non-real-time best effort traffic performance for the real-time traffic performance. Two scenarios, non-preemptive and preemptive, are considered. Under the non-preemptive case, a failed reservation of non-RT VCs will be re-transmitted at the next control slot, exactly like RT VCs. This is non-preemptive in nature in that once a non-RT VC reservation is made, it can not be interrupted by any other RT VCs. Under the preemptive case, however, at the time when the reservation of a non-RT VC fails (at the end of one propagation delay after the reservation is sent), since there might exist packet(s) from a real-time VC of the same input node ready for transmission, i.e., an RT VC with non empty queues and no HOL blocking, in order to avoid the non-RT traffic blocking the RT traffic, one additional queue called re-transmission queue is introduced for temporarily holding the non-RT packet for later re-transmission if the previous reservation of a non-RT packet is unsuccessful. A number of immediate observations can be drawn on the behavior of the re-transmission queue:

- No real-time reservation packet will be put on the re-transmission queue.

- The non-RT reservation packet on the re-transmission queue will be sent in the next available control slot when there is no RT VC reservation packet available. Notice, this includes a new reservation packet from an RT VC, or a previous failed RT reservation packet. Thus, the non-RT reservation packet on the re-transmission queue might have to wait for more than one reservation cycle.

- No new non-RT reservation can be initiated if the re-transmission queue is non empty.

- The queue length of the re-transmission queue is finite, which can be shown to be bounded by the pipelining depth, defined as ratio between the propagation delay and the the size of control packet.

VC Scheduling Algorithm.   The VC scheduling essentially selects a proper candidate if there is any from multiple application VC queues, and moves it into the transmission queue ready for transmission scheduling. Notice there are differences in handling real-time and non-real time traffic. For real-time traffic, the in-sequence delivery mandates that a new transmission cannot be initiated until the successful reservation of current packet transmission. On the other hand, the non-real-time VC does not have this constraint. Depending on the VC type, different scheduling at the VC scheduling level can be used. The non-real-time traffic are assumed as "best effort" UBR type. This results in the real-time traffic having strict priority over the non-real-time traffic. Within each type, a simple round robin policy is employed.

Transmission Scheduling Algorithm.   During the transmission scheduling stage, a control packet if any is sent in the corresponding control mini-slot. Depending on whether the transmission queue or/and the re-transmission queue is empty or not, and depending on the VC type, a number of scenarios are possible. The following summarizes the priority order of reservation packet transmission.

- A failed real-time reservation packet will be re-transmitted in the next control slot.

- A new non-blocked real-time VC's reservation packet will be transmitted.

- A non-real-time reservation packet from the re-transmission queue will be transmitted.

- A new non-real-time VC's reservation packet will be transmitted.

Within one control slot, each node has one control mini-slot for sending out its reservation. The following is the order for transmission scheduling to avoid both channel conflict and output conflict.

- Real-time packet has higher priority than non-real-time packet.

- Real-time (non-real-time) traffic with earlier arrival has higher priority than real-time (non-real-time) packet of late arrival (age priority).

- Random selection with equal probability further breaks the tie.

5.   SUMMARY

This chapter has summarized state-of-the art medium access control protocols for wavelength division multiplexing (WDM) networks. Depending on the characteristics, complexity, and capabilities of these MAC protocols, we have classified them under non-pretransmission coordination protocols, pretransmission coordination protocols, variable-length transmission protocols, and quality-of-service oriented protocols. Detailed architectural, qualitative and quantitative descriptions of various protocols within each category have been provided. Most of these protocols are targeted towards local and metropolitan area environments. This chapter should serve as a good starting point for researchers working on this area so as to give them an overview of the research efforts conducted for the past decade. In addition, it presents the fundamentals for further investigation into ways of coping with the current and the anticipated explosion of multimedia information transfer.

References

Azizoglu, M., Barry, R. A., and Mokhtar, A. (1996). Impact of tuning delay on the performance of bandwidth-limited optical broadcast networks with uniform traffic. *IEEE J. Sel. Areas Comm.*, 14(6):935–944.

Bogineni, K. and Dowd, P. W. (1991). A collisionless media access protocols for high speed communication in optically interconnected parallel computers. *SPIE*, 1577:276–287.

Bogineni, K. and Dowd, P. W. (1992). A collisionless multiple access protocol for a wavelength division multiplexed star-coupled configuration: Architecture and performance analysis. *IEEE/OSA J. Lightwave Tech.*, 10(11):1688–1699.

Borella, M. S. and Mukherjee, B. (1996). Efficient scheduling of nonuniform packet traffic in a WDM/TDM local lightwave network with arbitrary transceiver tuning latencies. *IEEE J. Sel. Areas Comm.*, 14(6):923–934.

Chen, M. and Yum, T.-S. (1991). A conflict-free protocol for optical WDM networks. In *IEEE GLOBECOM '91*, pages 1276–1291, Phoenix, AZ.

Chen, M.-S., Dono, N. R., and Ramaswami, R. (1990). A media access protocol for packet-switched wavelength division multiaccess metropolitan area networks. *IEEE J. Sel. Areas Comm.*, 8(8):1048–1057.

Chipalkatti, R., Zhang, Z., and Acampora, A. S. (1992). High-speed communication protocols for optical star networks using WDM. In *Proc. INFOCOM '92*, pages 2124–2133, Florence, Italy.

Chipalkatti, R., Zhang, Z., and Acampora, A. S. (1993). Protocols for optical star-coupler network using WDM: Performance and complexity study. *IEEE J. Sel. Areas Comm.*, 11(4):579–589.

Chlamtac, I. and Fumagalli, A. (1991). Quadro-stars: High performance optical WDM star networks. In *Proc. IEEE GLOBECOM '91*, pages 1224–1229, Phoenix, AZ.

Chlamtac, I. and Ganz, A. (1988). Channel allocation protocols in frequency – time controlled high speed networks. *IEEE Trans. Comm.*, 36(4):430–440.

Dowd, P. W. (1991). Random access protocols for high speed interprocessor communication based on an optical passive star topology. *IEEE/OSA J. Lightwave Tech.*, 9(6):799–808.

Dowd, P. W. and Bogineni, K. (1992). Simulation analysis of a collisionless multiple access protocol for a wavelength division multiplexed star-coupled configuration. In *Proceedings of the 25th Annual Simulation Symposium*, Orlando, FL.

Ganz, A. and Gao, Y. (1994). Time-wavelength assignment algorithms for high performance WDM star based systems. *IEEE Trans. Comm.*, 42(2-3-4):1827–1836.

Ganz, A. and Koren, Z. (1991). WDM passive star protocols and performance analysis. In *Proc. INFOCOM '91*, pages 991–1000, Bal Harbour, FL.

Hamidzadeh, B., Ma, M., and Hamdi, M. (1999). Message sequencing techniques for on-line scheduling in WDM networks. *IEEE/OSA J. Lightwave Tech.*, 17(8):1309–1319.

Humblet, P. A., Ramaswami, R., and Sivarajan, K. N. (1993). An efficient communication protocols for high- speed packet-switched multichannel networks. *IEEE J. Sel. Areas Comm.*, 11(4):568–578.

Jeon, H. and Un, C. (1992). Contention-based reservation protocols in multiwavelength optical networks with a passive star topology. In *Proc. ICC*, pages 1473–1477.

Jia, F. and Mukherjee, B. (1993). The receiver collision avoidance (rca) protocol for a single-hop lightwave network. *IEEE/OSA J. Lightwave Tech.*, 11(5/6):1052–1065.

Jia, F., Mukherjee, B., and Iness, J. (1995). Scheduling variable-length messages in a single-hop multichannel local lightwave network. *IEEE/ACM Trans. Networking*, 3(4):477–487.

Jonsson, M., Borjesson, K., and Legardt, M. (1997). Dynamic timedeterministic traffic in a fiber optic WDM star network. In *Proceedings. Ninth Euromicro Workshop on Real Time Systems*, pages 25–33, Toledo, Spain.

Kavehrad, I. M. I. H. M. and Sundberg, C.-E. W. (1987). Protocols for very high speed optical fiber local area networks using a passive star topology. *IEEE/OSA J. Lightwave Tech.*, 5(12):1782–1794.

Lee, J. H. and Un, C. K. (1996). Dynamic scheduling protocol for variable-sized messages in a WDM-based local network. *IEEE/OSA J. Lightwave Tech.*, 14(7):1595–1600.

Li, B. and Ganz, A. (1992). Virtual topologies for WDM star LANs - the regular structure approach. In *Proc. INFOCOM*, Florence, Italy.

Li, B., Ganz, A., and Krishna, M. (1995). A novel transmission coordination scheme for single hop lightwave networks. In *Proc. GLOBECOM*, Singapore.

Li, B. and Qin, Y. (1998). Traffic scheduling in a photonic packet switching system with qos guarantee. *IEEE / OSA J. Lightwave Tech.*, 16(12):2281–2295.

Ma, M. and Hamdi, M. (2000). Providing deterministic quality-of-service guarantees on WDM optical networks. Submitted to ICC.

Ma, M., Hamidzadeh, B., and Hamdi, M. A receiver-oriented message scheduling algorithm for WDM lightwave networks. Accepted by Computer Networks and ISDN Systems.

Ma, M., Hamidzadeh, B., and Hamdi, M. (1999). Efficient scheduling algorithms for real-time service on WDM optical networks. *Photonic Network Communications*, 1(2).

Mehravari, N. (1990). Performance and protocol improvements for very highspeed optical fiber local area networks using a passive star topology. *IEEE / OSA J. Lightwave Tech.*, 8(4):520–530.

Muir, A. and Garcia-Luna-Aceves, J. J. (1996). Distributed queue packet scheduling algorithms for WDM-based networks. In *Proc. INFOCOM '96*.

Mukherjee, B. (1992a). WDM-based local lightwave networks – Part I: Single-hop systems. *IEEE Network*, 6(3):12–27.

Mukherjee, B. (1992b). WDM-based local lightwave networks – Part II: Multihop systems. *IEEE Network*, 6(4):20–32.

Pieris, G. R. and Sasaki, G. H. (1994). Scheduling transmissions in WDM broadcast-and-select networks. *IEEE/ACM Trans. Networking*, 2(2):105–110.

Rouskas, G. N. and Ammar, M. H. (1995). Analysis and optimization of transmission schedules for single-hop WDM networks. *IEEE/ACM Trans. Networking*, 3(2):211–221.

Sivalingam, K. M. and Dowd, P. W. (1995). A multilevel WDM access protocol for an optically interconnected multiprocessor system. *IEEE/OSA J. Lightwave Tech.*, 13(11):2152–2167.

Sivalingam, K. M. and Dowd, P. W. (1996). A lightweight media access protocol for a -based distributed shared memory system. In *Proc. INFOCOM '96*, pages 946–953.

Sudhakar, G. N. M., Georganas, N., and Kavehrad, M. (1991a). Multi-control channel for very high-speed optical fiber local area networks and their interconnections using passive star topology. In *Proc. IEEE GlOBECOM '91*, pages 624–628, Phoenix, AZ.

Sudhakar, G. N. M., Kavehrad, M., and Georganas, N. (1991b). lotted aloha and reservation aloha protocols for very high-speed optical fiber local area networks using passive star topology. *IEEE/OSA J. Lightwave Tech.*, 9(10):1411 – 1422.

Tyan, H.-Y., Hou, C.-J., Wang, B., and Han, C.-C. (1996). On supporting time-constrained communications in WDMA-based star-coupled optical networks. In *Proceedings. 17th IEEE Real-Time Systems Symposium*, pages 175–184, Los Alamitos, CA, USA.

Yan, A., Ganz, A., and Krishna, C. M. (1996). A distributed adaptive protocol providing real-time services on WDM-based LANs. *IEEE/OSA J. Lightwave Tech.*, 14(6):1245–1254.

Chapter 8

SCHEDULING ALGORITHMS FOR UNICAST, MULTICAST, AND BROADCAST

George N. Rouskas
Department of Computer Science
North Carolina State University
Raleigh, NC 27695-7534
rouskas@csc.ncsu.edu

Abstract   In this chapter we present a survey of algorithms for scheduling packet traffic in broadcast optical WDM networks. We first describe the context and motivations of the scheduling problem. We then review the current literature in the field with an emphasis on scheduling techniques for providing best-effort service as well as guaranteed service for both unicast and multi-destination traffic. We provide alternative formulations of the problem, and we compare the formulations and theoretical results, as well as algorithms and heuristics.

1.   INTRODUCTION

The broadcast WDM network architecture has been widely studied as an approach to building optically interconnected networks. Under one widely adopted scenario for the evolution of the optical network infrastructure [Kam et al., 1998], broadcast WDM subnetworks will be used to provide local access, and the subnetworks will be interconnected through wavelength routed MANs and WANs. One of the potentially difficult issues that arise in a broadcast WDM environment, is that of coordinating the various transmitters/receivers. Some form of coordination is necessary because (a) a transmitter and a receiver must both be tuned to the same channel for the duration of a packet's transmission, and (b) a simultaneous transmission by one or more nodes on the same channel will result in a collision. The issue of coordination is further complicated by the fact that tunable transceivers may need a non-negligible amount of time to switch between wavelengths. Thus, at the heart of media access control

Transmitting side   Receiving side

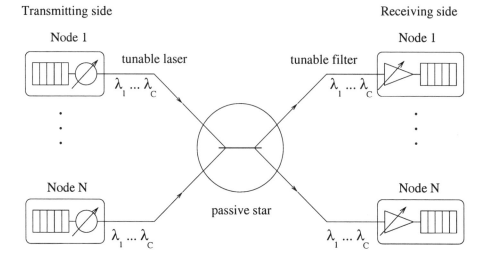

Figure 8.1   A broadcast WDM network with N nodes and C channels.

(MAC) protocols for broadcast WDM subnetworks is a scheduling algorithm responsible for coordinating access to the available channels (wavelengths).

In this chapter we survey a number of algorithms for traffic scheduling in a packet-switched broadcast WDM with N nodes and C channels, $N \geq C$, as shown in Fig. 8.1. Unless otherwise specified, it is assumed that each node has exactly one tunable transmitter and one tunable receiver. We let Δ denote the transceiver tuning latency, i.e., the time it takes a transmitter or receiver to tune from one wavelength to another. Packets in the network are of fixed size. Time is slotted, with a slot time equal to the packet time plus some guard band, which depends on the MAC protocol and the corresponding scheduling algorithm.

The chapter is organized as follows. In Section 2 we discuss the role of scheduling within the context of broadcast WDM networks, and we examine its relationship to reservation protocols and load balancing. In Section 3 we review scheduling algorithms for both best-effort and guaranteed-service unicast traffic. In Section 4 we discuss approaches to scheduling multi-destination traffic. We conclude the chapter in Section 5.

2.   LOAD BALANCING, RESERVATIONS, AND SCHEDULING

While in this chapter we are mainly concerned with scheduling algorithms, one should keep in mind that, in ensuring an acceptable level of network performance, scheduling is only one piece of the puzzle. In this section we briefly review two other components critical to the operation of broadcast

WDM networks, namely, load balancing and reservation protocols, and we discuss their relationship to scheduling.

We distinguish two levels of network operation, differing mainly in the time scales at which they take place. At the *media access control* level, connectivity among the network nodes is provided by a reservation protocol, whose main function is to collect information regarding traffic demands, and a scheduling algorithm whose objective is to provide collision-free communication among the nodes while optimizing some performance measure of interest (e.g., schedule length). At the *network dimensioning* level, which takes place at significantly longer time scales, the objective is to allocate resources in a way that optimizes network performance. In this context, the shared resource of interest is bandwidth, and load balancing algorithms are needed to ensure good performance and fairness at this level of network operation.

2.1. LOAD BALANCING AND RECONFIGURATION

In optical WDM networks, each channel will have to be shared by multiple receivers, and the problem of assigning receive wavelengths arises. A wavelength assignment(hereafter referred to as WLA) implies an allocation of the bandwidth to the various network nodes. Intuition suggests that if the traffic load is not well balanced across the available channels, the result will be poor network performance. A recent study on the performance of the HiPeR-ℓ reservation protocol [Sivaraman and Rouskas, 1997] has confirmed this intuition. Let us define parameter ϵ_b such that no channel carries more than $\frac{(1+\epsilon_b)}{C}$ times the total traffic offered to the network. In other words, ϵ_b is a measure of the *degree of load balancing* of the network; under perfect load balancing, $\epsilon_b = 0$. It was shown in [Sivaraman and Rouskas, 1997] that the maximum sustained throughput γ (i.e., the number of packets successfully transmitted per packet time) is directly affected by ϵ_b through the following stability condition:

$$\gamma < \frac{C}{(1 + \epsilon_b)(1 + \epsilon_s)}. \tag{8.1}$$

It can be seen from (8.1) that the higher the degree of load balancing (i.e., the lower the value of ϵ_b is), the higher the overall arrival rate γ that the network can accommodate, and vice versa. Parameter ϵ_s is the guarantee on the schedule length and depends on the scheduling algorithm used. In other words, even an optimal algorithm (for which $\epsilon_s = 0$ in (8.1)) will achieve a very low throughput if the load is not well balanced. Although the stability condition (8.1) was derived specifically for HiPeR-ℓ, we believe that load balancing has a similar effect on the performance of any protocol for WDM broadcast networks.

Hence, the time-varying conditions expected in this type of environment call for mechanisms that periodically adjust the bandwidth allocation to ensure that each channel carries an almost equal share of the corresponding offered load.

The problem of dynamic load balancing by re-tuning a subset of receivers in response to changes in the overall traffic pattern was studied in [Baldine and Rouskas, 1998; Baldine and Rouskas, 1999b]. Assuming an existing WLA and some information regarding the new traffic demands, this work studied two approaches to obtaining a new WLA such that (a) the new traffic load is balanced across the channels, and (b) the number of receivers that need to be re-tuned is minimized. The latter objective is motivated by the fact that tunable receivers take a non-negligible amount of time to switch between wavelengths during which parts of the network are unavailable for normal operation. Since this variation in traffic is expected to take place over larger time scales (i.e., re-tuning will be a relatively infrequent event), employing slowly tunable devices can be a cost effective solution. An approximation algorithm for the load balancing problem was presented that provides for tradeoff selection, using a single parameter, between two conflicting goals, namely, the degree of load balancing and the number of receivers that need to be re-tuned.

The issues arising in the reconfiguration phase of broadcast networks were also studied in [Baldine and Rouskas, 1999a], where reconfiguration policies to determine *when* to reconfigure the network were developed, and an approach was presented to carry out the network transition by describing a class of strategies that determine *how* to re-tune the optical receivers. The problem was formulated as a Markovian Decision Process, yielding a systematic and flexible framework in which to view and contrast reconfiguration policies, and it was shown how an appropriate selection of reward and cost functions can be used to achieve the desired balance among various performance criteria of interest.

2.2.   RESERVATION PROTOCOLS

Reservation protocols collect information about the short-term traffic demands between source-destination pairs, by having nodes periodically transmit reservation packets. The reservation packets are usually sent on a separate control channel dedicated to carrying signaling information. To prevent the control channel from becoming a bottleneck, it is also possible to use multiple control channels [Humblet et al., 1993] or employ in-band reservations, as in HiPeR-ℓ [Sivaraman and Rouskas, 1997], in which case all channels in the network may be used to carry both data and control packets. Reservation packets can contain information about the head-of-line packet only, as in PROTON [Levine and Akyildiz, 1995], or about the depth of all queues in a node, as in HiPeR-ℓ. The reservation information is used to build an identical (delayed) snapshot of the state of the queues in the network at each node [Foo and Robertazzi,

1995; Muir and Garcia-Luna-Aceves, 1996]. This snapshot is the main input to the scheduling algorithms discussed in the remainder of this chapter.

Reservation protocols may employ pipelining techniques to mask the effects of the tuning latency [Tridandapani et al., 1994] or the propagation delay [Sivaraman and Rouskas, 1997], both of which may otherwise have severe impact on overall performance in ultra high-speed WDM environments. Some reservation protocols organize time in distinct reservation and data phases [Sivalingam and Dowd, 1995], while in others [Sivaraman and Rouskas, 1997] there is no separate reservation phase and reservation information is multiplexed with data packets. The latter approach has the advantage that data transmission is not interrupted for large periods of time (i.e., during a separate reservation phase). It also makes it possible to transmit data while at the same time computing the schedule for the next transmission phase, effectively masking the computation time of the schedule.

3.   SCHEDULING OF UNICAST TRAFFIC

3.1.   BEST-EFFORT TRAFFIC

The scheduling problem in a broadcast optical network with N nodes and C, $N \geq C$, data channels is typically formulated as a matrix clearing problem. Specifically, it is assumed that there exists a $N \times C$ *traffic demand matrix* $\mathbf{M} = [m_{ic}]$, where integer m_{ic} represents the number of packets to be transmitted from node i, $i = 1, \cdots, N$, on channel λ_c, $c = 1, \cdots, C$. The traffic demands m_{ic} may be derived as: $m_{ic} = \sum_{j \in R_c} a_{ij}$, where the number a_{ij} of packets to be transmitted from node i to node j can be obtained using a reservation protocol such as the ones discussed in the previous section, and the sets of receivers R_c listening on a certain wavelength λ_c, $c = 1, \cdots, C$, (i.e., the WLA) are determined by the load balancing algorithm.

Given a traffic demand matrix $\mathbf{M}$, the objective most commonly considered in the literature is to construct an *optimal finish time* (OFT) schedule, i.e., one which has the least finish time among all schedules for matrix $\mathbf{M}$. An OFT schedule is highly desirable since it both minimizes average packet delay and maximizes the aggregate network throughput (recall the effect of parameter ϵ_s in the stability condition (8.1)). However, if no restriction is imposed on the number of reservations submitted by the nodes (that is, the values that quantities m_{ic} may take), the length of even the OFT schedule can become very large under high loads. In this case, while the average packet delay and aggregate network throughput will be optimal, the length and variability of the schedules make it impossible to provide guarantees (e.g., on delay and/or delay jitter) to individual packet flows. Thus, this approach is appropriate for best-effort traffic, but not well-suited to support real-time services.

When the transceiver tuning latency Δ is assumed to be small relative to the packet transmission time, a padding equal to Δ time units can be included within each slot to allow the transceivers sufficient time to switch between wavelengths, with minimal effects on overall performance [Bogineni et al., 1993; Humblet et al., 1993]. In this case, the matrix clearing problem is equivalent to the *open shop scheduling* problem where *preemption* is allowed [Gonzalez and Sahni, 1976]. The open shop scheduling problem formulation prevents channel, transmitter, and receiver collisions by including constraints which guarantee that, within each time slot (a) two or more sources do not transmit on the same channel, (b) a given source transmits on at most one channel, and (c) a given receiver listens on at most one channel. In the context of broadcast WDM networks, a preemptive schedule is such that there may exist a node-channel pair (i, λ_c) for which the m_{ic} packets are not transmitted in contiguous slots. In other words, the node i of any such pair (i, λ_c) will have to tune to channel λ_c multiple times in order to transmit all its traffic demands m_{ic}. Preemptive OFT schedules for the open-shop scheduling problem can be constructed in polynomial time using techniques for maximum matching on bipartite graphs [Rouskas and Ammar, 1995; Gonzalez and Sahni, 1976].

For networks where the value of the tuning latency is comparable to, or greater than, the packet transmission time, including a padding of Δ time units within each slot would be highly inefficient in terms of both throughput and delay. A better approach is to keep the slot time equal to the packet time, and introduce a new set of constraints to account for the time it takes a transceiver to tune from one wavelength to another, during which it is taken off-line and is not available for transmitting or receiving packets. The objective, then, becomes that of minimizing the impact of the tuning requirements on the length of the schedule. However, adding the new constraints introduces significant difficulty to the scheduling problem, and it makes it impossible to obtain an OFT schedule in polynomial time.

One approach to alleviating the effects of the tuning latency is to insist on *non-preemptive* schedules, whereby each node i tunes its transceiver to each channel λ_c exactly once during the schedule, and remains at that channel until all m_{ic} packets have been transmitted. The non-preemptive open shop scheduling problem with $\Delta = 0$ admits a polynomial-time solution when the number of channels $C = 2$, but it is NP-complete for $C \geq 3$ [Gonzalez and Sahni, 1976]. When the tuning latency Δ is non-zero, however, the problem becomes NP-complete even when $C = 2$ [Rouskas and Sivaraman, 1997]. We now discuss several heuristics and approximation algorithms for constructing near-optimal schedules.

In [Rouskas and Sivaraman, 1997], the design of non-preemptive open-shop schedules for broadcast WDM networks with non-uniform traffic demands and arbitrary tuning latencies was undertaken. Two distinct regions of network operation were identified. The *tuning-limited* region is such that the schedule length is determined by the transceiver tuning requirements. When the network operates in the *bandwidth-limited* region, the length of the schedule is determined by the traffic demands. The point at which the network switches between the two regions was also identified in terms of system parameters such as the number of nodes and channels and the tuning latency. A special class of schedules was then introduced such that the order in which the various transceivers tune to each channel is the same for all channels. This class of schedules permits an intuitive formulation of the scheduling problem, and, under uniform traffic (i.e., when $m_{ic} = m \; \forall \; i, c$), an OFT schedule within this class can be readily constructed. Based on the new formulation, polynomial-time algorithms were developed to construct OFT schedules when the elements of the traffic demand matrix $\mathbf{M}$ satisfy certain optimality conditions. In essence, the optimality conditions impose an upper bound on the "degree of non-uniformity" of matrix $\mathbf{M}$ for the algorithm to construct an OFT schedule within this class. A set of heuristics was also developed which, in the general case (that is, when matrix $\mathbf{M}$ does not satisfy the optimality conditions), where shown to construct schedules of length very close to the lower bound. An important outcome of this work was the realization that algorithms which work well within the bandwidth-limited region may not work well within the tuning-limited region, and vice versa. Consequently, optimality conditions, optimal algorithms, and heuristics were developed for both bandwidth-limited and tuning-limited networks.

An important feature of the algorithms and heuristics in [Rouskas and Sivaraman, 1997] is that their running-time complexity is a function only of system parameters, namely, the number N of nodes and the number C of wavelengths, and is independent of the actual length of the schedule. This property makes it possible to allow the transmission of variable-length packets over the broadcast WDM network without any extra control overhead. This can be accomplished by letting the slot time be equal to one byte, and having the nodes send reservation requests for the number of *bytes* they wish to transmit, rather than the number of *fixed-size packets*. The algorithms will then schedule each node's transmission in a number of contiguous bytes (slots). Having the nodes make reservations in terms of number of bytes eliminates the problem of selecting the length for the fixed-size packets, and it also eliminates the overhead for segmenting and then reassembling the upper layer variable-length packets (e.g., IP datagrams). The problem of determining the "best" fixed length for packets is a difficult one since it strongly depends on the (mostly unknown) mix of applications that will be carried over the network, and may lead to non-optimal compromises (e.g., as in the size of ATMcells). On the other hand, it would be inefficient to use algorithms whose running time is a function of the schedule length (in slots) in a network where the slot size is equal to one byte.

An approximation algorithm for constructing non-preemptive open shop schedules was developed in [Choi et al., 1996]. The algorithm is based on the well-known concept of list scheduling . Specifically, as soon as a transmitter completes its transmissions on a given wavelength, it tunes to the wavelength in which it can start transmitting at the *earliest time*. It was shown that the length of a schedule constructed by this algorithm is at most twice the length of the OFT schedule for a given matrix **M**, for any value of the tuning latency Δ.

A different two-phase heuristic for the same problem was derived in [Borella and Mukherjee, 1996]. In the first phase, nodes are assigned to transmit in contiguous slots on a given channel, in decreasing order of their demands for that channel. The assignment ensures that all channel, transmitter, and tuning constraints are satisfied. Since this approach may result in unused slots in which no node has been assigned to transmit, the second phase of the algorithm attempts to fill these slots. Specifically, for each unused slot, the nodes that are assigned to transmit in slots immediately before or after the unused slot are examined. If the tuning constraints allow, one of these nodes is assigned to transmit in the previously unused slot. The transmissions allocated during the second pass are in addition to the demands specified by matrix **M**, and can greatly increase the efficiency of the schedule by decreasing the number of unused slots.

Special cases of the general open shop scheduling problem have also been addressed in [Pieris and Sasaki, 1994; Azizoglu et al., 1996; Choi et al., 1996]. The *all-to-all* scheduling problem is a special case such that each transmitter has exactly one packet to send to each receiver. Under such uniform traffic, and assuming that the number N of nodes is a multiple of the number C of wavelengths, the optimal WLA is one in which exactly N/C receivers are tuned to each channel. Consequently, the traffic matrix **M** is such that $m_{ic} = N/C \ \forall \ i, c$. For this traffic matrix, lower and upper bounds on the schedule length were derived in [Pieris and Sasaki, 1994], and a scheduling algorithm was presented. This algorithm was shown in [Choi et al., 1996] to be optimal. It is interesting to note that the schedules constructed by this algorithm fall within the class of schedules considered in [Rouskas and Sivaraman, 1997]. A different special case was studied in [Azizoglu et al., 1996]. Specifically, the traffic demands were such that each transmitter has either one packet or no packet to send to each receiver (representing the existence or not, respectively, of a head-of-line packet at the various queues), and the value of the tuning latency Δ was restricted to be at most equal to the packet transmission time. A heuristic based on a variation of the list scheduling algorithm in [Choi et al., 1996] was analyzed through simulations and was shown to exhibit good average case behavior for this problem.

All the algorithms discussed so far have the same objective, namely, they attempt to construct OFT non-preemptive open shop schedules. In such schedules, the tuning and transmission periods are interleaved so as to minimize the overall finish time for a given traffic matrix. (A performance analysis of non-preemptive open-shop schedules under packet traffic has been carried out in [McKinnon et al., 1998b; McKinnon et al., 1999; McKinnon et al., 1998a].) Another approach to scheduling packet transmissions in WDM networks is to construct schedules satisfying the *tune-transmit separability constraint* [Pieris and Sasaki, 1994]. Specifically, time is divided into alternating periods of transmission and tuning. Each transceiver operates on a fixed channel during a transmission period; no packets are transmitted during the tuning periods, which are reserved to re-tune transceivers to be ready for the next transmission period.

This version of the problem is closely related to the well-known scheduling problem in satellite-switched time division multiple access (SS / TDMA) [Gopal and Wong, 1985; Inukai, 1979]. More formally, the problem can be defined as follows. Given a traffic demand matrix $\mathbf{M}$, the objective is to decompose it into sub-matrices $\mathbf{M}_k$, $k = 1, \cdots, K$, such that (a) each row and each column of each sub-matrix $\mathbf{M}_k$ has at most one non-zero element, (b) $\sum_{k=1}^{K} \mathbf{M}_k = \mathbf{M}$, (c) the total time to sequentially transmit the individual sub-matrices is minimized, and (d) the number K of sub-matrices is minimized. In this formulation, each sub-matrix $\mathbf{M}_k$ corresponds to a transmission period within the schedule. Requirement (a) ensures that there are no receiver or transmitter collisions within each transmission period, while requirement (b) ensures that all the traffic demands of matrix $\mathbf{M}$ will be met by following the transmissions indicated by the sub-matrices. Objective (c) reflects the desire to minimize the time it takes to clear matrix $\mathbf{M}$, while objective (d) is necessary to keep the time spent tuning the transceivers between transmission as short as possible; together, the two objectives ensure that the total time spent transmitting and tuning is minimized. As defined, this problem is NP-hard [Gopal and Wong, 1985]. Next, we discuss various heuristics in the context of broadcast WDM networks.

In [Ganz and Gao, 1994], the network was viewed as a bipartite graph of N sources and N destinations (in other words, the starting point for the decomposition is not the matrix $\mathbf{M}$ we have considered so far, but rather the $N \times N$ matrix of traffic demands between each source-destination node). A bipartite matching algorithm was used to decompose this graph into a number K of bipartite matchings, where each matching is constrained to have at most C arcs. No bounds on the performance of the algorithm were derived. In [Choi et al., 1996], a network with demand matrix $\mathbf{M}$ was modeled as a bipartite multi-graph with N sources and C destinations. The bipartite multi-graph is first edge-colored, and then decomposed into subgraphs consisting of edges of the same color. The transmissions corresponding to the edges of a subgraph can all take place simultaneously, similar to the transmissions in a sub-matrix $\mathbf{M}_k$ in the above formulation. A bound on the length of the schedule constructed by the algorithm was derived, and the average case behavior of this approach was studied through simulations. Also, a lower bound on the length of all-to-all schedules satisfying the tune-transmit separability constraints was derived in [Pieris and Sasaki, 1994].

A different approach to matrix decomposition was taken in [Sivalingam and Wang, 1996], where the focus was on the running-time efficiency and ease-of-implementation of the scheduling algorithm. Another important feature of the techniques developed in [Sivalingam and Wang, 1996] is that the schedule is built using partial information as it becomes available to the reservation protocol, without the need to wait until the entire traffic demand matrix $\mathbf{M}$ is complete. Specifically, as soon as a transmitter's reservation requests are received, all nodes in the network use the same greedy strategy to schedule the requests within an existing sub-matrix, if one that can accommodate the requests is found. Otherwise, a new sub-matrix is created for the transmitter's requests. A bound on the number K of sub-matrices generated by the algorithm was derived, and its performance in terms of average packet delay and network throughput was studied for both uniform and client-server traffic.

While most algorithms that have appeared in the literature attempt to minimize the finish time of a schedule, a different objective in scheduling packet transmissions was considered in [Kam et al., 1998]. Achieving *max-min fairness* was the primary concern of this work, taking precedence over maximizing throughput or minimizing delay. The algorithm developed uses information on whether a source is back-logged or idle, and builds the schedule slot by slot using the following greedy approach. To determine the transmissions in a given slot, each source is considered in increasing order of the number of packets sent by the source so far. If the source is back-logged, it is assigned to transmit in the slot if no transmitter, receiver, or channel constraints are violated by doing so. Otherwise, the next source is considered until either all channels have been assigned transmissions or no back-logged sources remain. Simulation studies presented in [Kam et al., 1998] demonstrate that the algorithm has good fairness properties while also achieving high throughput.

3.2.   GUARANTEED-SERVICE TRAFFIC

Packet-switched WDM networks will need to support a range of applications with varying quality of service (QoS) requirements, such as bandwidth and delay guarantees. The algorithms discussed in the previous section focus on minimizing the finish time of the schedule, and do not address the issue of supporting time-constrained communication. From the point of view of scheduling , the requirement to provide QoS guarantees necessitates algorithms which can transmit packets in some priority order, e.g., according to deadlines, virtual finish times, eligibility times, or other time-stamps associated with a packet [Liu and Layland, 1973].

Under the assumption of negligible tuning latency, the problem of scheduling real-time packet flows in WDM networks is related to the problem of scheduling periodic tasks in a real-time multiprocessor system [Dertouzos and Mok, 1989]. In [Wang et al., 1997], the problem of scheduling *isochronous* message streams was considered, where each stream l is characterized by its deadline D_l, and the maximum number C_l of packets that can arrive in any time interval of length D_l (the "computation time" in multiprocessor scheduling terminology). A *feasible* schedule for a set of message streams is such that exactly C_l slots are allocated to stream l in any time window of size D_l. An algorithm based on the *rate-monotonic* principle [Liu and Layland, 1973] was applied to schedule a static set of isochronous message streams. The algorithm may not be successful in constructing feasible schedules when the deadlines D_l are not multiples of a basic value $D \geq 2$, however, no sufficient condition for schedulability was derived. The dynamic problem was also considered, and a set of algorithms was presented to schedule transmissions of new message streams, as well as to deallocate slots assigned to terminating streams.

The problem of optimally scheduling periodic tasks on multiprocessors was studied in [Jackson and Rouskas, 1998]. The existence of a feasible schedule for this problem when the total task density $\rho = \sum_l (C_l/D_l) = C$, where C is the number of processors (channels in the corresponding WDM problem), has been an open problem since the work in [Dertouzos and Mok, 1989]. It was shown in [Jackson and Rouskas, 1998] that the condition $\rho \leq C$ is both necessary and sufficient for the existence of a feasible schedule. A network flow formulation was also presented, based on which an algorithm to construct a feasible schedule was developed. In addition to broadcast networks, this algorithm can have applications to scheduling packet traffic on WDM point-to-pointlinks between routers.

An algorithm which provides a minimum bandwidth guarantee to packet flows was presented in [Kam et al., 1998; Kam and Siu, 1998]. This algorithm is in fact an extension of the max-min fair algorithm discussed in the previous section. The main difference is that traffic flows are considered for slot allocation in increasing order of the *excess* bandwidth they have used beyond their guaranteed bandwidth. By letting the guaranteed bandwidth of best-effort traffic be zero, this algorithm can be used to provide both bandwidth guarantees and max-min fairness. Thus, this approach represents a first step towards supporting integrated services in a broadcast WDM environment.

4.   SCHEDULING OF MULTI-DESTINATION TRAFFIC

Many applications and telecommunication services, including teleconferencing, distributed data processing, and video distribution, require some form of multipoint communication. Traditionally, without network support for multicasting, a multi-destination message is replicated and transmitted individually to all its recipients. This method, however, consumes more bandwidth than necessary. Bandwidth consumption constitutes a problem since most multipoint applications require a large amount of bandwidth. An alternative solution is to broadcast a multi-destination message to all nodes in the network. The problem in this case is that nodes not addressed in the message will have to dedicate resources to receive and process the message. Thus, the ability to efficiently transmit messages addressed to multiple destinations has become increasingly important, and the issues associated with providing network support for multipoint communication have been widely studied within a number of different networking contexts [Ammar et al., 1997].

In WDM broadcast networks, information transmitted on any channel is broadcast to the entire set of nodes, but it is only received by those with a receiver listening on that channel. The broadcast feature, coupled with tunability at the receiving end, makes it possible to design scheduling algorithms [Rouskas and Ammar, 1997; Borella and Mukherjee, 1995] such that a *single* transmission of a multicast packet can reach all receivers in the packet's destination set simultaneously. The high degree of efficiency in using the network resources makes this approach especially appealing for transmitting multicast traffic. However, the design of appropriate receiver tuning algorithms is complicated by the fact that (a) tunable receivers take a non-negligible amount of time to switch between channels, and (b) different multicast groups may have several receivers in common. On the other hand, waiting until all receivers become available before scheduling a multicast packet may result in low wavelength throughput (i.e., low average number of packets transmitted per unit time), especially for medium to large size multicastgroups. To improve the situation, it was proposed in [Jue and Mukherjee, 1997] to partition a multicastgroup into several sub-groups, and to transmit a packet once to each sub-group. This approach leads to higher wavelength throughput despite the fact that each packet is transmitted multiple times, indicating the existence of a tradeoff between wavelength throughput and the degree of efficiency in using the bandwidth.

An approach similar to the one in [Jue and Mukherjee, 1997] was presented in [Modiano, 1998]. Specifically, a packet is also transmitted multiple times, until it is received by all members of its multicast group. Instead of partitioning the multicast group in advance, however, each receiver follows a set of rules to listen to a packet transmission in each slot. An analytical model for obtaining the average packet delay was developed for two schemes, one employing persistent transmissions and one that introduces a random back-off delay. Under the first scheme, a packet is continuously transmitted until it is received by all members of its multicast group. The random back-off scheme eliminates the head-of-line problem of persistent transmissions by retransmitting packets not received by all intended receivers after a random delay, and results in better performance. Also, it was shown that the algorithm used by the receiver to select one among multiple packets addressed to it can have a significant impact on performance. Specifically, an algorithm where the receiver selects the packet with the smallest number of intended receivers remaining outperforms one in which packets are selected based on the time of their initial transmission (i.e., a first-come first-served discipline).

The problem of scheduling multicast traffic was also considered in [Ortiz et al., 1999; Ortiz et al., 1997]. Let a *multicast completion* denote the completion of the transmission of a multicast packet to all receivers in its multicast group. The *multicast throughput*, defined as the average number of multicast transmissions per slot, was introduced as the performance measure of interest, and it was shown that it depends on two measures that have previously been considered in isolation, namely, the degree of efficiency in using the channel bandwidth and wavelength throughput. Then, a new technique was presented for the transmission of multicast packets based on the concept of a *virtual* receiver, a set of physical receivers which behave identically in terms of tuning. It was demonstrated that the number of virtual receivers naturally captures the performance of the system in terms of multicast throughput. By partitioning the set of all physical receivers into virtual receivers, a multicast packet must be transmitted to each virtual receiver containing a physical receiver in the packet's multicast group, and the original network with multicast traffic is transformed into a new network with unicast traffic. This approach decouples the problem of determining how many times each multicast packet should be transmitted, from the problem of scheduling the actual packet transmissions. Thus, rather than developing new scheduling algorithms for multicast traffic, one may take advantage of the algorithms discussed in the previous section. Consequently, the focus of the work in [Ortiz et al., 1997] was on the problem of optimally selecting the virtual receivers to maximize multicast throughput, and it was proven that it is NP-complete. Finally, four heuristics of varying degree of complexity were presented for obtaining a set of virtual receivers that provide near-optimal performance in terms of multicast throughput.

In [Ortiz et al., 1998] the performance of various strategies for scheduling a combined load of unicast and multi-destination traffic was studied. The performance measure of interest was schedule length. Three different scheduling strategies were presented, namely: separate scheduling of unicast and multicast traffic, treating multicast traffic as a number of unicast messages, and treating unicast traffic as multicasts of size one. A lower bound on the schedule obtained by each strategy was first obtained. Subsequently, the strategies were compared against each other using extensive simulation experiments in order to establish the regions of operation, in terms of a number of relevant system parameters, for which each strategy performs best. The main conclusions were as follows. Multicast traffic can be treated as unicast traffic under very limited circumstances. On the other hand, treating unicast traffic as multicast traffic produces short schedules in most cases. Alternatively, scheduling and transmitting each traffic separately is also a good choice.

5.   CONCLUDING REMARKS

We have reviewed algorithms for scheduling unicast and multi-destination traffic in broadcast WDM networks. A classification of the scheduling algorithms is presented in Tables 8.1 and 8.2.

Scheduling of best-effort traffic is a well-researched problem, and many efficient algorithms have been developed that give optimal or near-optimal results. More work is needed in the area of scheduling algorithms for providing QoS guarantees to real-time traffic, especially when the tuning latency must be taken into account. With the current interest on packet (especially IP) over WDM architectures, it would also be important to develop integrated approaches that fit within the Internet's differentiated services framework. The main challenge, however, is in the deployment of optical WDM packet network testbeds, such as the WDM LAN extension to the wideband all-optical network [Kaminow et al., 1996] at MIT and the Helios IP over WDM joint testbed between MCNC and North Carolina State University, which will provide opportunities for extensive experimentation with, and validation and extension of the proposed scheduling algorithms and heuristics.

References

Ammar, M., Polyzos, G., and Tripathi, S. (Eds.) (1997). Special issue on network support for multipoint communication. *IEEE Journal Selected Areas in Communications*, 15(3).

Azizoglu, M., Barry, R. A., and Mokhtar, A. (1996). Impact of tuning delay on the performance of bandwidth-limited optical broadcast networks with uniform traffic. *IEEE Journal on Selected Areas in Communications*, 14(5):935–944.

Baldine, I. and Rouskas, G. N. (1998). Dynamic load balancing in broadcast WDM networks with tuning latencies. In *Proceedings of INFOCOM '98*, pages 78–85. IEEE.

Table 8.1   Classification of scheduling algorithms for unicast traffic.

Objective	Δ	Schedule Class	Algorithm
QoS Guarantees		Periodic Task	[Jackson and Rouskas, 1998] [Wang et al., 1997]
Max-Min Fairness	≈ 0	Open Shop with preemption	[Kam and Siu, 1998] [Kam et al., 1998] [Rouskas and Ammar, 1995] [Choi et al., 1996, Sec. III.A]
Minimize Length	small	Open Shop without preemption	[Azizoglu et al., 1996] [Rouskas and Sivaraman, 1997] [Borella and Mukherjee, 1996] [Pieris and Sasaki, 1994]
	arbit-rary	Tune Transmit Separability	[Choi et al., 1996, Sec. III.B] [Sivalingam and Wang, 1996] [Ganz and Gao, 1994]

Table 8.2   Classification of scheduling algorithms for multi-destination traffic.

Δ	Approach	Algorithm
≈ 0	Repeated transmissions	[Modiano, 1998]
	Single transmission of multicast packet	[Rouskas and Ammar, 1997] [Borella and Mukherjee, 1995]
arbit-rary	Partition groups	[Jue and Mukherjee, 1997]
	Virtual Receivers (VR)	[Ortiz et al., 1997]
	VR for unicast & multicast	[Ortiz et al., 1998]

Baldine, I. and Rouskas, G. N. (1999a). Dynamic reconfiguration policies for WDM networks. In *Proceedings of INFOCOM '99*, pages 313–320. IEEE.

Baldine, I. and Rouskas, G. N. (1999b). Reconfiguration and dynamic load balancing in broadcast WDM networks. *Photonic Network Communications*, 1(1):49–64.

Bogineni, K., Sivalingam, K. M., and Dowd, P. W. (1993). Low-complexity multiple access protocols for wavelength-division multiplexed photonic networks. *IEEE Journal on Selected Areas in Communications*, 11(4):590–604.

Borella, M. and Mukherjee, B. (1995). A reservation-based multicasting protocol for WDM local lightwave networks. In *Proceedings of ICC '95*, pages 1277–1281. IEEE.

Borella, M. S. and Mukherjee, B. (1996). Efficient scheduling of nonuniform packet traffic in a WDM/TDM local lightwave network with arbitrary transceiver tuning latencies. *IEEE Journal on Selected Areas in Communications*, 14(5):923–934.

Choi, H., Choi, H.-A., and Azizoglu, M. (1996). Efficient scheduling of transmissions in optical broadcast networks. *IEEE/ACM Transactions on Networking*, 4(6):913–920.

Dertouzos, M. L. and Mok, A. K.-L. (1989). Multiprocessor on-line scheduling of hard-real-time tasks. *IEEE Transactions on Software Engineering*, 15(12):1497–1506.

Foo, E. M. and Robertazzi, T. G. (1995). A distributed global queue transmission strategy for a WDM optical fiber network. In *Proceedings of INFOCOM '95*, pages 154–161. IEEE.

Ganz, A. and Gao, Y. (1994). Time-wavelength assignment algorithms for high performance WDM star based networks. *IEEE Transactions on Communications*, 42(4):1827–1836.

Gonzalez, T. and Sahni, S. (1976). Open shop scheduling to minimize finish time. *Journal of the Association for Computing Machinery*, 23(4):665–679.

Gopal, I. and Wong, C. (1985). Minimizing the number of switchings in an SS/TDMA system. *IEEE Transactions on Communications*, 33(6):497–501.

Humblet, P. A., Ramaswami, R., and Sivarajan, K. N. (1993). An efficient communication protocol for high-speed packet-switched multichannel networks. *IEEE Journal on Selected Areas in Communications*, 11(4):568–578.

Inukai, T. (1979). An efficient SS/TDMA time slot assignment algorithm. *IEEE Transactions on Communications*, 27(10):1449–1455.

Jackson, L. E. and Rouskas, G. N. (1998). Optimal scheduling of periodic tasks on multiple identical processors. Technical Report TR-98-14, North Carolina State University, Raleigh, NC.

Jue, J. and Mukherjee, B. (1997). The advantages of partitioning multicast transmissions in a single-hop optical WDM network. In *Proceedings of ICC '97*. IEEE.

Kam, A. C. and Siu, K.-Y. (1998). A real-time distributed scheduling algorithm for supporting QoS over WDM networks. In *Proceedings of SPIE*, volume 3531, pages 181–193.

Kam, A. C., Siu, K.-Y., Barry, R. A., and Swanson, E. A. (1998). Toward best effort services over WDM networks with fair access and minimum bandwidth guarantee. *IEEE Journal Selected Areas in Communications*, 16(7):1024–1039.

Kaminow, I. P., Doerr, C. R., Dragone, C., Koch, T., Koren, U., Saleh, A. A. M., Kirby, A. J., Ozveren, C. M., Schoffield, B., Thomas, R. E., BArry, R. A., Castagnozzi, D. M., Chan, V. W. S., Hemenway, B. R., MArquis, D., Parikh, S. A., Stevens, M. L., Swanson, E. A., Finn, S. G., and Gallager, R. G. (1996). A wideband all-optical WDM network. *IEEE Journal Selected Areas in Communications*, 14(5):780–799.

Levine, D. A. and Akyildiz, I. F. (1995). PROTON: A media access control protocol for optical networks with star topology. *IEEE/ACM Transactions on Networking*, 3(2):158–168.

Liu, C. L. and Layland, J. W. (1973). Scheduling algorithms for multiprogramming in a hard-real-time environment. *Journal of the ACM*, 20(1):46–61.

McKinnon, M. W., Perros, H. G., and Rouskas, G. N. (1999). Performance analysis of broadcast WDM networks under IP traffic. *Performance Evaluation*, 36-37:333–358.

McKinnon, M. W., Rouskas, G. N., and Perros, H. G. (1998a). Performance analysis of a photonic single-hop ATM switch architecture with tunable transmitters and fixed frequency receivers. *Performance Evaluation*, 33(2):113–136.

McKinnon, M. W., Rouskas, G. N., and Perros, H. G. (1998b). Queueing-based analysis of broadcast optical networks. In *Proceedings of ACM SIGMETRICS/PERFORMANCE '98*, pages 121–130. ACM.

Modiano, E. (1998). Unscheduled multicasts in WDM broadcast-and-select networks. In *Proceedings of INFOCOM '98*. IEEE.

Muir, A. and Garcia-Luna-Aceves, J. J. (1996). Distributed queue packet scheduling algorithms for WDM-based networks. In *Proceedings of INFOCOM '96*, pages 938–945. IEEE.

Ortiz, Z., Rouskas, G. N., and Perros, H. G. (1997). Scheduling of multicast traffic in tunable-receiver WDM networks with non-negligible tuning latencies. In *Proceedings of SIGCOMM '97*, pages 301–310. ACM.

Ortiz, Z., Rouskas, G. N., and Perros, H. G. (1998). Scheduling of combined unicast and multicast traffic in broadcast WDM networks. In *Proceedings of PICS '98*, pages 137–150. Chapman & Hall.

Ortiz, Z., Rouskas, G. N., and Perros, H. G. (1999). Maximizing multicast throughput in WDM networks with tuning latencies using the virtual receiver concept. *European Transactions on Telecommunications.* (To appear).

Pieris, G. R. and Sasaki, G. H. (1994). Scheduling transmissions in WDM broadcast-and-select networks. *IEEE/ACM Transactions on Networking*, 2(2):105–110.

Rouskas, G. N. and Ammar, M. H. (1995). Analysis and optimization of transmission schedules for single-hop WDM networks. *IEEE/ACM Transactions on Networking*, 3(2):211–221.

Rouskas, G. N. and Ammar, M. H. (1997). Multi-destination communication over tunable-receiver single-hop WDM networks. *IEEE Journal on Selected Areas in Communications*, 15(3):501–511.

Rouskas, G. N. and Sivaraman, V. (1997). Packet scheduling in broadcast WDM networks with arbitrary transceiver tuning latencies.
IEEE/ACM Transactions on Networking, 5(3):359–370.

Sivalingam, K. and Dowd, P. (1995). A multi-level WDM access protocol for an optical interconnected multi-processor system. *IEEE/OSA Journal of Lightwave Technology*, 13(11):2152–2167.

Sivalingam, K. and Wang, J. (1996). Media access protocols for WDM networks with on-line scheduling. *IEEE/OSA Journal of Lightwave Technology*, 14(6):1278–1286.

Sivaraman, V. and Rouskas, G. N. (1997). HiPeR-ℓ: A High Performance Reservation protocol with ℓook-ahead for broadcast WDM networks. In *Proceedings of INFOCOM '97*, pages 1272–1279. IEEE.

Tridandapani, S., Meditch, J. S., and Somani, A. K. (1994). The MaTPi protocol: Masking Tuning times through Pipelining in WDM optical networks. In *Proceedings of INFOCOM '94*, pages 1528–1535. IEEE.

Wang, B., Hou, C.-J., and Han, C.-C. (1997). On dynamically establishing and terminating isochronous message streams in WDMA-based local area lightwave networks. In *Proceedings of INFOCOM '97*, pages 1263–1271. IEEE.

Chapter 9

DESIGN AND ANALYSIS OF A MEDIA ACCESS PROTOCOL FOR STAR COUPLED WDM NETWORKS WITH TT-TR ARCHITECTURE

Krishna M. Sivalingam
School of EECS
Washington State University
Pullman, WA 99164-2752
krishna@eecs.wsu.edu

Abstract   This paper presents the design and analysis of a media access protocol implemented in LIGHTNING, an optical WDM network testbed that has been designed for high-performance supercomputer interconnection. The architecture is based on a dynamically reconfigurable hierarchical WDM network to interconnect a large number of supercomputers and create a distributed shared memory (DSM) environment. This paper describes the network media access protocol, called LiteMAC, based on a single tunable transmitter and single tunable receiver (TT-TR) per node which exploits the bimodal traffic characteristics of a DSM system. The primary objectives of the protocol design are reduced average latency per packet, support of broadcast/multicast, and support of collision-less communication. The proposed approach is compared to an earlier protocol based on one tunable transmitter and one fixed receiver (TT-FR) per node. The performance of the protocol in terms of average latency and channel utilization is analyzed for varying system characteristics such as number of nodes, channels, and levels.

1. INTRODUCTION

This paper describes the media access protocol that has been implemented in an experimental optical WDM testbed known as "LIGHTNING" [Dowd et al., 1996]. The testbed is a joint project involving the State University of New York at Buffalo, University of Maryland (College Park and Baltimore County, MD), Laboratory for Physical Sciences (College Park, MD), Supercomputing Research Center (Bowie, MD), and David Sarnoff Research Center (Princeton,

NJ). The motivation behind the testbed is to interconnect a large number of supercomputer class machines to create a distributed shared memory (DSM) system. The primary objective is low-latency low-cost communication that is sufficiently scalable to support thousands of processors.

This paper describes an approach that combines the support of the cache coherence protocol, required for DSM, and the access protocol to achieve the low-latency requirement which exploits the natural traffic characteristics of this environment. The performance of the DSM organization and the cache coherence protocol is dependent on the media access protocol. A primary design constraint, needed to achieve the low-cost and scalability objectives, is the assumption that there will be only a *few* WDM channels (many less than the number of processors). Another objective is to develop an access protocol that is amenable to hardware implementation with reduced cost per node.

The DSM traffic, generated by the cache coherence protocol and the operating system, has two major forms: control (such as memory block requests, invalidations, cache-level acknowledgments) and data or memory blocks. Control packets are less than 64 bytes long while the memory blocks could be up to 8 Kbytes. Furthermore, there are multiple control packets for every memory block that needs to be transferred. The protocol described in this paper exploits this bimodal characteristic. The rest of the paper will assume fixed data packet lengths though the protocol described in the paper incorporates support of variable data packet lengths.

A protocol (referred to as FatMAC) that combined the advantages of reservation and receiver pre-allocation has been studied in [Dowd and Sivalingam, 1994; Sivalingam and Dowd, 1995]. Each node had a tunable transmitter and a fixed receiver (TT-FR) tuned to its *home channel*. Packet transmission was divided into two phases: reservation phase during which control packets and reservations for data packets were transmitted; and the subsequent data packet transmission phase. The principal drawbacks of this protocol are bandwidth loss due to unused slots in the data phase, lack of flexibility of data channel usage, and head-of-line effects with non-uniform traffic. On-line scheduling algorithms that allow multiple reservations per node per data phase have been studied as an alternative in [Sivalingam and Wang, 1996].

This paper proposes a protocol, referred to as LiteMAC, which attempts to alleviate the drawbacks of FatMAC. The protocol is based on a tunable transmitter and a tunable receiver (TT-TR) per node. There is no concept of a home channel and a node can receive on any free channel. The protocol design goals are collision-less communication, broadcast/multicast, variable packet lengths, and low latency communication. The protocol retains the reservation strategy of FatMAC and is composed of a reservation/control phase followed by the data phase. The data phase is divided into "slots" where the length of a slot is the data packet. Let M denote the number of nodes and C the number of

WDM channels. During each slot, up to C source-destination pairs may exist. The flexibility in the availability of data channels results in reduced data cycle lengths and reduced unused slots as will be seen in Section 4. This results in reduced average packet latency and improved channel utilization.

Unlike other reservation protocols, LiteMAC does not require either a dedicated control channel or a control transmitter/receiver. Protocols based on scheduled-TDM approaches that allow multiple reservations may provide better performance but are prohibitively expensive for real-time hardware implementation. The proposed protocol is considered lightweight since its goal is to obtain the performance advantages and flexibility of a scheduled-TDM type of protocol but (with the proposed protocol extensions) provide a limited schedule to reduce the complexity and preserve implementational feasibility.

The performance metrics of interest are average packet latency and overall channel utilization. The performance analysis using discrete-event simulation is conducted for varying number of nodes, channels, DSM traffic patterns, and multiple hierarchic levels.

The rest of the paper is organized as follows. Section 2 describes the network architecture. Section 3 describes the protocols. Section 4 considers the performance of LiteMAC and the comparison to FatMAC [Sivalingam and Dowd, 1995] and to TDMA-C [Bogineni and Dowd, 1992]. Section 5 summarizes the paper.

2.   NETWORK ARCHITECTURE

This section very briefly describes the network of the target architecture. Refer to [Dowd et al., 1996] for a detailed description of the network architecture.

A variety of optical channel topologies can be used to achieve a multiple access environment. Single-level interconnection can be accomplished using an optical passive star topology. Each node has a tunable transmitter and a tunable receiver. The star configuration has been chosen for its passive nature, high fault tolerance, high node fanout, and complete unity distance connectivity. This approach is able to support on the order of hundreds of nodes with the rapid advancement in optical amplifiers. The goal is to support thousands of processors with few multiple access channels. The system size limitation imposed by the star power budget constraints leads to the generalized space wavelength hierarchical architecture that can support thousands of processors. The following describes the generalized hierarchical architecture of Project LIGHTNING.

The LIGHTNING architecture is hierarchical as in Fat-Tree [Leiserson, 1985; Kannan et al., 1994], and is all-optical with spatial wavelength reuse at each hierarchic level [Dowd et al., 1996]. An example of a three level network is

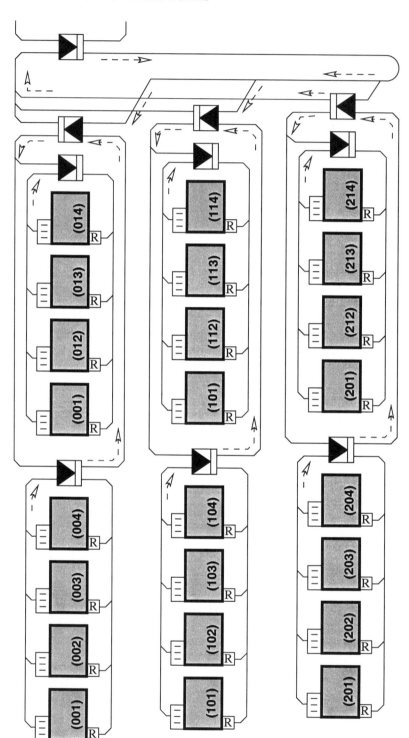

Figure 9.1 Example of the first three levels of the hierarchical architecture.

shown in Fig. 9.1. This system is based on a component called Λ-partitioner which is a space/wavelength switch. It is used as a wavelength-selective space-division switch to select a subset of wavelengths to switch.

An i-level Λ-partitioner couples a total of $m_i + 1$ input fibers and $m_i + 1$ output fibers: one fiber to/from its parent and m_i fibers to/from its children. The functionality of a Λ-partitioner can be described as a 2×2 switching element (the lower m_i links are passively coupled to a single fiber), as found in multistage interconnection networks, where the set of wavelengths below a partition point are bar-connected and the wavelengths above the partition point are cross connected. Functionally, this achieves wavelength re-use by retaining the wavelengths below the partition point to the local cluster thereby allowing all peer-level clusters to independently use the same set of wavelengths [Dowd et al., 1996]. It employs a wavelength partitioner (denoted as Λ-x) to partition the traffic between different levels of the hierarchy without electronic intervention.

The *processor* nodes are located at the leaves of the tree and each may actually be multiple processors. The organization of a single processor includes the Local Processor Cache (usually located on-chip), the Extended Cache (EC) and the Local Global Memory (LGM). The EC and the LGM are actually constructed from the same physical memory space.

Only the processors are equipped with transmitters and receivers (a tunable receiver per level and a laser array transmitter). The nodes at all higher levels are Λ-partitioners that provide wavelength/spatial switching. An r-level system partitions the wavelength channels into r non-overlapping subsets. The partition points are defined as $\mathbf{X} = \{X_0, X_1, X_2, \ldots, X_r\}$ where $X_r = C$, $X_0 = 0$ and $0 < X_i < X_j < C$ if $i < j$ for all $0 < i < r - 1$ and $1 < j < r$. Let $C_i = X_i - X_{i-1}$ denote the number of channels allocated for i-level communication, $1 \leq i \leq r$, such that $C = \sum_{i=1}^{r} C_i$.

The mixed radix system for node numbering is used [Bhuyan and Agrawal, 1984]. Let M_j denote the number of j-level nodes in this hierarchy, given by the equation $M_j = \prod_{i=j+1}^{r} m_i$ for all $0 \leq j \leq r - 1$ and $M = M_0$. Since there are a total of M_j j-level nodes, a total of $C_j M_j$ channels are provided to each j-level cluster. Let N_i denote the number of processor nodes under an i-level node, for $1 \leq i \leq r$, where $N_i = \prod_{j=1}^{i} m_i$, or $N_i = M/M_i$. An r-level architecture provides a total of $C = \sum_{j=1}^{r} C_j \prod_{k=j+1}^{r} m_k$ separate channels that may be concurrently accessed due to the combination of spatial and wavelength multiplexing.

Consider a small system of $M = 4 \times 4$ with $C = 4$. If the partition is set so that 3 channels are allocated for level one and one channel is allocated to level 2, the total number of effective channels increases to a maximum of $C = 13$ through wavelength re-use.

3.   PROTOCOL DEFINITION

This section describes the proposed media access protocol. The protocol is first defined for the single level system followed by the generalization to multiple level systems.

3.1.   DSM TRAFFIC

The traffic characteristics of the system play a crucial role in the access protocol design. Define two classes of traffic:

Class A: small amounts of data, such as control information generated by the cache control mechanism and the operating system.

Examples are memory block requests, invalidations, cache-level acknowledgments, application-level low-latency messages, operating system control information, bandwidth reconfiguration (for the multi-level network) and other network management packets.

Class B: large amounts of data, such as a memory block. *Typically, a memory block is two orders of magnitude larger than a control packet.*

This class of traffic generates a reservation control packet for media access if a reservation based protocol is used. In this case, the reservation for a Class B transmission is scheduled through a reservation piggybacked on a Class A packet.

The packet format is illustrated in Fig. 9.2. The payload in the Class A packet is used for transmission of the low-latency control information and application *hot-packets* such as semaphores and other application-level data structures that require extremely low latency for overall effective performance.

Memory block length is expressed in terms of control packet length: let L denote the ratio of memory block packet length to control packet length.

Often only a control packet needs to be sent to transmit control signals. Block transfers occur only a fraction of the time. This provides the motivation to develop a protocol that can reduce communication latency and transmit unfragmented memory block packets. Define γ as the fraction of Class B packets generated in the system, where $0 \leq \gamma \leq 1$. The case $\gamma = 0$ corresponds to no memory block packets and $\gamma = 1$ to all packets being memory block packets. The distribution of packet types is determined by *both* the media access protocol and the cache coherence protocol.

The ratio γ has a significant impact on system performance and in identifying the performance advantages of either protocol. A simulation study of snooping-based and directory-based cache coherence schemes was used to estimate γ.

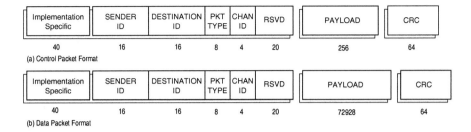

Figure 9.2 Format of control and data packets.

The *FFT* and *SPEECH* multiprocessor traces [Chaiken et al., 1990] were used with traces from the SPLASH benchmark suite [Singh et al., 1991] to estimate γ.

Snooping-based coherence schemes, which generally require a broadcast facility for requests and invalidations, used with the *mp3d* and *water* SPLASH benchmarks had a value of $\gamma \approx 0.3$. Directory-based coherence schemes, studied with the FFT and SPEECH traces, usually use explicit invalidations and the measured value of γ ranged between 0.06 and 0.03, depending on the amount of sharing. This shows that traffic is mostly composed of control packets in the studied applications even with a snooping-based coherence strategy. This led to the primary motivation for the development of LiteMAC to exploit the bi-modal DSM traffic characteristic.

3.2. WDMA PROTOCOLS

Reservation type protocols have been studied earlier for the satellite multiple access environment. The satellite access protocol studied in [Jacobs et al., 1978] used a time multiplexed reservation cycle followed by a data cycle, overlapping the long satellite propagation delays through interleaving reservation and data cycles. FatMAC and LiteMAC generalize this two-phase approach to a multi-channel environment. Typical WDM based reservation access protocols allocate one of the channels as a dedicated control channel [Mukherjee, 1992]. This is not attractive in the target environment since the number of channels available is usually small.

3.2.1 FatMAC . Transmission in FatMAC is organized into cycles where each cycle consists of a control phase and a data phase. Each node consists of a tunable transmitter and a fixed receiver always tuned to the node's *home channel*. The control phase operates in a broadcast environment.

A control packet sent during the Reservation/Control Block (RCB) phase of a cycle may perform up to two functions:

1 Transmit a memory block reservation, and/or

2 Include a Class A packet waiting to be transmitted through a small multi-purpose payload (see Fig. 9.2).

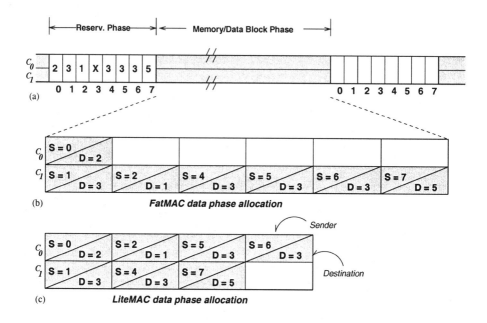

Figure 9.3 Assignment map for single level FatMAC and LiteMAC for $M = 8$ and $C = 2$: (a) The cycle consisting of reservation/control phase followed by data phase. The numbers in the boxes indicate the destination node for a sample cycle. X denotes that the node has no reservation; (b) FatMAC allocation map; and (c) LiteMAC allocation map.

A reservation specifies the destination, the data channel and the packet size (if variable sized packets are supported). Access during the control phase is TDM-based mainly to preserve a collision-less environment (and implementational simplicity) and other access strategies are possible. FatMAC exploits the orientation of the transmitter. Since it is constructed as a laser array, all channels can be simultaneously driven to achieve broadcast.

A control packet may contain only cache control information such as an invalidation and not need to reserve a data slot. This allows the average cycle length to be significantly reduced. A fast broadcast facility is achieved through this multi-purpose control packet.

3.2.2   LiteMAC .   Each node in LiteMAC has a tunable transmitter and a tunable receiver. Data transmission is organized in cycles as in FatMAC. Each cycle consists of a reservation/control phase followed by the data phase. The data phase is divided into *slots* where up to C simultaneous source-destination pairs may exist in one slot. The length of each data phase slot equals memory block packet transmission time.

The control phase may be designed to operate in a broadcast manner as in FatMAC. This is not strictly required since the nodes possess a tunable receiver. An alternative strategy would be to tune all the transmitters and receivers to one channel during the control phase. This broadcast approach is attractive in the situation where tuning latency is significant so that re-tuning of all the transceivers at the beginning of the control cycle may be avoided. The downside of broadcast when tunable receivers are available is the fact that effectively only one out of C channels is being utilized during this period. However, this period is typically more than two orders of magnitude shorter in duration than the data block phase. This restriction also arises primarily because of the fact that each node has only one receiver per level. Other approaches such as PROTON [Levine and Akyildiz, 1995] may achieve better utilization during the control phase since they possess a dedicated transmitter and receiver for control. However, that approach was not followed with LIGHTNING due to cost constraints.

Each unit of time in FatMAC and LiteMAC is equal to control packet transmission time. Each node has two queues, one for control packets (Class A) and the other for data packets (Class B).

This definition of FatMAC varies from the original definition in [Sivalingam and Dowd, 1995] which used a single queue for both control and data packets. Fig. 9.3 shows the allocation maps for single level FatMAC and LiteMAC for $M = 8$ nodes and $C = 2$ channels. The transmitter behavior can be described as follows:

1 When a packet is ready for transmission, the node waits for its turn on the reservation phase (cycle synchronization delay).

- Access during the reservation phase is based on TDM to support collision-less transmission.
- The service mechanism in this case is first-come first-serve though a general scheduling algorithm may be used.
  A reversing service during the control phase could improve fairness.
- Access to the reservation phase could be based on other strategies, such as Slotted Aloha, to further reduce packet latency. With SA, acknowledgments are not needed for the reservation phases but the sensitivity of the system to propagation delay is increased [Sivalingam, 1995].

2 The control packet is broadcast to all nodes during the reservation phase.

The control packet contains a Class A packet, and/or a block reservation if a block transfer is scheduled to follow the control packet. The reservation packet includes source and destination node information used to compute the data cycle offsets. Details of the control cell layout is shown in Fig. 9.2, where the payload and control information fields can be observed.

3 The node waits until the end of the current control phase before it attempts to transmit the corresponding memory block if any.

4 If the node has a data block to transmit, it determines the channel and its offset within the data phase as determined by the algorithm given below. Variable size blocks could be supported, but it is assumed in the analysis to follow that all nodes in a system use the same block size. Furthermore, it simplifies the network interface design if the block size remains constant.

Both protocols are collision free and do not require acknowledgment support under normal circumstances. However, there is always the possibility of corrupted or dropped packets, or loss of slot synchronization. So, although explicit MAC-level acknowledgments are not needed, LIGHTNING uses transaction-level acknowledgments provided by the cache coherence protocol. Furthermore, network synchronization and clock distribution are supported through a suite of synchronization protocols not described in this paper.

Offset Computation Algorithms.   A major difference between LiteMAC and FatMAC lies in the determination of the channel on which a packet is transmitted. FatMAC transmits a packet only on the home channel of the destination. LiteMAC eliminates this fixed allocation and provides flexibility in allocation of data channels. The following paragraphs describe how the source nodes compute the transmission channel and offset within the data phase. The data phase is composed of data slots where up to C simultaneous transmissions may take place – one per channel.

Each node in FatMAC maintains a set of channel counters $C_j, 0 \leq j \leq C-1$ initialized to zero at the beginning of each control phase. Consider a reservation from source M_s for destination M_d with home channel C_d. Node M_s will transmit on channel C_d in slot C_d. The counter C_d is incremented by one after the allocation.

LiteMAC uses the following counters per node to accomplish offset calculation:

1 a set of counters denoted by $\mathcal{N}_i, 0 \leq i \leq M-1$.

$\mathcal{N}_d$ keeps track of the number of nodes that have requested node M_d prior to the reservation under consideration. The counters are initialized to zero at the start of the control phase.

2 A counter S to keep track of the number of data slots created based on the reservations received so far.

S is initialized to zero at the start of the control phase.

3 For each slot t, F_t denotes the identifier of the first free channel in slot t where $0 \leq F_t \leq C - 1$ and $1 \leq t \leq S$.

Given a reservation by source M_s for destination M_d:

1 Find the first slot t such that $(t > \mathcal{N}_d)$ AND $(t \leq S)$ AND $(F_t < C - 1)$.

This ensures that destination conflicts to node M_d are avoided.

2 IF no such slot t exists THEN $S = S + 1; t = S; F_t = 0$

IF slot t exists THEN $F_t = F_t + 1$

A new data slot is created if the given request cannot be satisfied with the existing set of slots.

3 Increment $\mathcal{N}_d$.

4 Reserve channel F_t in slot t to source node M_s for destination M_d.

The data phase length (S) is the maximum number of slots reserved on the individual channels within the channel set (the channels allocated to that level). In general for $M > C$, the data cycle length of LiteMAC is less than that of FatMAC. Fig. 9.3 shows an example reservation set and the allocation maps for FatMAC and LiteMAC. For this case, the cycle length for FatMAC is 6 slots while the cycle length for LiteMAC is 4 slots. There are five unused slots in this example with FatMAC for a utilization of 0.58. With LiteMAC, there is one unused slot and utilization is 0.875. For $M = C$, both protocols result in the same data cycle length and hence identical system performance.

LiteMAC also provides better performance over FatMAC when the traffic is non-uniform such as client-server traffic. Consider a client-server system where a significant fraction of the traffic is targeted to the server node. With FatMAC, the server might share the home channel with other client nodes. As a result, traffic destined to these client nodes suffer from longer delays due to serialization of traffic on server's channel. One possible solution is dedicate a home channel for server traffic. This results in valuable loss of bandwidth especially when the number of channels is small or when server-directed traffic tends to occur in bursts. LiteMAC eliminates this head-of-line effect for the client-destined traffic since there are no fixed reception channels for the nodes.

With both protocols, slots may be unused in the data cycle as shown in Fig. 9.3. In general, LiteMAC reduces the number of wasted slots due to shorter cycle lengths. However, it does not completely eliminate unused slots. The impact of unused slots is higher for non-uniform client-server traffic.

3.2.3   Protocol Extensions.   This section considers two extensions to minimize the number of unused slots incurred in each data cycle. Both allow multiple reservations per node per data cycle. They differ in the way the requests are satisfied. The extensions are denoted as MRX (Multiple Reservations eXtended Cycle) and MRR (Multiple Reservations Reclaimed Slots).

MRX: Each node may request slots for more than one destination in one reservation packet. This allows the nodes to schedule all the packets using a scheduling algorithm such as in [Ganz and Gao, 1992; Bannister et al., 1990]. However, this might increase the data cycle length which results in increased packet delay for control packets. Since more data packets are transmitted per cycle, the latency of data packets may be reduced resulting in increased utilization. The performance analysis section studies this extension for the case where each node is allowed to make at most two reservations per cycle. This scheme is suitable for traffic which generates many more data packets than control packets.

is equivalent to the

MRR: Each node may place multiple requests in a single reservation packet as above. Initial slot allocation is done to satisfy all the first reservation requests as with basic LiteMAC. The second reservation requests are satisfied only if the data cycle length is not extended. This attempts to fill in unused slots with the second requests. Not all the requests may be satisfied with this approach in which case the node is required to re-submit the reservation. The advantage of this scheme is no increase in the data cycle length while attempting to improve utilization and delay. This scheme is suitable for traffic which generates many more control packets than data packets.

Fig. 9.4 depicts the extensions for a system with $M = 4$ and $C = 2$ for a sample set of reservation requests. Fig. 9.4(a) shows the status of the transmitter data queues at the nodes. The numbers in the boxes indicate the destination of the data packet. From Fig. 9.4(b), the cycle length of basic LiteMAC is seen to three slots per channel; there are two unused slots; and the data cycle utilization is 0.67. Fig. 9.4(c) shows that for MRR, the cycle length is retained at three slots per channel; there are no unused slots; and the data cycle utilization is 1.0. Six of the seven reservations were satisfied. The reservation by node M_2 for destination M_3 was not satisfied. Fig. 9.4(d) shows that for MRX, the cycle length is four slots per channel; there is one unused slot; and the data cycle utilization is 0.875.

Note that with both extensions, the second request may be transmitted in a slot ahead of the first request's allocated slot. To avoid this, an one-packet

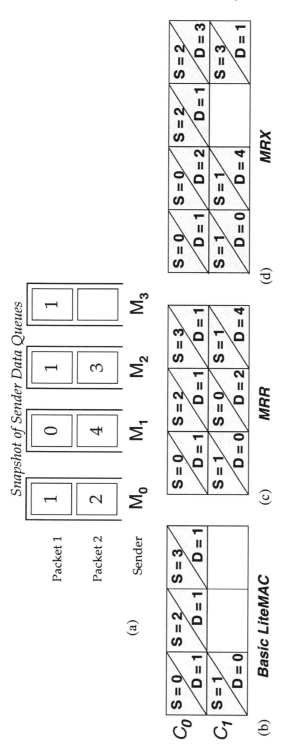

Figure 9.4   An example allocation for $M = 4$ and $C = 2$ for basic LiteMAC protocol and the two extensions MRX and MRR: (a) A snapshot of the data queues. The numbers in the boxes indicate the destination node; (b) Allocation with basic LiteMAC; (c) Allocation with MRR; and (d) Allocation with MRX. In the boxes, S denotes the source and D denotes the destination.

capacity additional queue may be used. To extend the schemes to support up to $M - 1$ reservations per node per cycle, $M - 1$ separate virtual queues may need to be used as in [Weller and Hajek, 1994].

The next section studies the performance of the two protocols.

4.   PERFORMANCE ANALYSIS

This section presents a comparison of the performance of LiteMAC and FatMAC for single-level systems and an analysis of LiteMAC for multi-level systems.

4.1.   PERFORMANCE COMPARISON OF LiteMAC AND FatMAC

This and the subsequent sections analyze the performance of LiteMAC and FatMAC using discrete-event simulation models. The system parameters of interest are C, M, and the traffic generation rate. Other parameters include the number of levels (r), the fraction of class B packets (γ) and traffic fraction requiring level-i communication (p_i), where $1 \leq i \leq r$. For the rest of this analysis, we will assume that each transmitter queue has the capacity to hold up to 8 packets. This value has been empirically chosen to ensure that there is virtually no loss of throughput due to blocking.

The following assumptions are made for the simulation model:

$\mathcal{A}_1$: *i.i.d.* behavior of the nodes and channels.

$\mathcal{A}_2$: Time is slotted on control packet boundaries and packet generation per slot follows a Poisson process.

$\mathcal{A}_3$: The traffic distribution is largely determined by reference of locality. Typically, reference locality would be exhibited within a cluster (level).

Under the uniform traffic model, a packet generated at node m_i is targeted to node m_j with probability $\mathcal{P}_{ij}$ where $\mathcal{P}_{ij}$ is defined as:

$$\mathcal{P}_{ij} = \begin{cases} \frac{1}{M-1} & \text{if } i \neq j \\ 0 & \text{if } i = j \end{cases}$$

$\mathcal{A}_4$: All memory block packets are of fixed length – L times the length of a control packet.

The performance metrics of interest are average packet latency and utilization. Packet latency is defined as the time between packet generation at source and reception at destination. Utilization is defined as the fraction of the time spent in transmitting useful information per time slot where time is slotted on

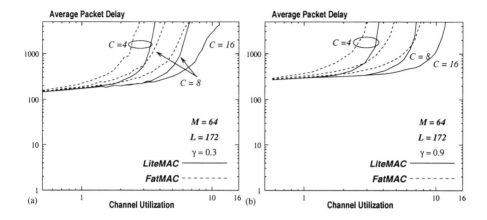

Figure 9.5 Comparison of LiteMAC and FatMAC under uniform traffic model for $M = 64$, $L = 172$, $C \in \{4, 8, 16\}$, and $\gamma \in \{0.3, 0.9\}$.

control packet boundaries. The theoretical maximum utilization for a single level system with C channels is C. In the following discussion, Γ indicates channel utilization and D indicates average packet delay.

The two metrics are dependent on the average data cycle length, which has been mathematically derived in [Sivalingam and Dowd, 1996] for both protocols. These equations have not been included here due to space constraints.

The performance of the protocols has been studied through discrete-event simulation. Simulation results have been obtained using the stochastic self-driven discrete-event models, written in C with YACSIM [Jump, 1992]. YACSIM is a C based library of routines that provides discrete-event and random variate facilities. Steady state transaction times and utilization were measured. Simulation convergence was obtained through the replication/deletion method [Law and Kelton, 1991], with a 95% confidence in a less than 5% variation from the mean.

The following paragraphs compare the performance of LiteMAC and Fat-MAC for a single level system under the uniform traffic model.

Fig. 9.5 compares the performance of LiteMAC and FatMAC with varying C for $M = 64$, $L = 172$, $C \in \{4, 8, 16\}$, and $\gamma \in \{0.3, 0.9\}$. At light loads, there is little improvement in latency with LiteMAC due to the small number of reservations in each cycle. The improvements are observed from intermediate to heavy loads as seen from the graph. The reduction in latency is due to the flexibility in the choice of data channels. With FatMAC, requests to two nodes sharing the home channel cannot be serviced in parallel. This is eliminated in LiteMAC which does not allocate home channels to receivers. For $M = 64$, $C = 4$, and $\gamma = 0.3$, the reduction in delay with LiteMAC is 67% at $\Gamma = 2$.

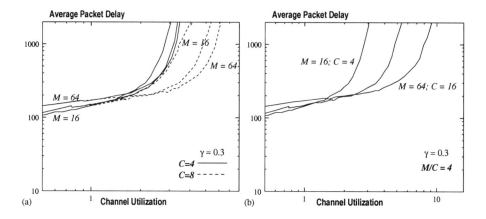

Figure 9.6 Performance of LiteMAC for varying M under uniform model for $L = 172$, and $\gamma \in \{0.3, 0.9\}$: (a) $C \in \{4, 8\}$ and $M \in \{16, 32, 64\}$; (b) $M/C = 4$ and $C \in \{4, 8, 16\}$.

However, the reduction in delay with LiteMAC for $\Gamma = 8$ is seen to be 85%. In conjunction with latency reduction, channel utilization improvement with LiteMAC is observed for all cases. The increase in maximum utilization for $M = 64$ with LiteMAC over FatMAC is 47% for $C = 16$, 43% for $C = 8$, and 25% for $C = 4$. This is primarily due to the reduced datacycle lengths.

The graphs indicate that LiteMAC takes advantage of an increase in the number of channels. Considering the delay aspects of LiteMAC alone, the reduction in latency for $M = 64$ and $\gamma = 0.3$ with respect to latency for $C = 4$ at $\lambda = 0.5$ is 51% for $C = 8$ and 73% for $C = 16$. Likewise, the maximum utilization for $M = 64$ with increasing C is: $\Gamma_{max}(4) = 3.9$, $\Gamma_{max}(8) = 7.3$, $\Gamma_{max}(16) = 11.9$. As the ratio M/C increases, note that the maximum utilization approaches C.

A similar comparison of the performance of LiteMAC and FatMAC was done for $M = 16$, $L = 172$, $C \in \{4, 8, 16\}$, and $\gamma \in \{0.3, 0.9\}$. When the number of channels is increased from $C = 4$ to $C = 8$ for $M = 16$, LiteMAC offers reduced latencies. However, when the the number of channels is increased to $C = 16$, the performance improvement observed with LiteMAC is small. This phenomenon has been observed with the TDMA-C protocol studied in [Bogineni and Dowd, 1992]. There is little performance improvement when the M/C ratio is reduced from 2 to 1 by increasing the number of channels.

To summarize the comparison of LiteMAC and FatMAC, the reduced data cycle lengths result in improved utilization and reduced latency. This has been shown using both simulation and equations derived in the previous section. The rest of the paper will consider only the performance of LiteMAC.

4.2.   PERFORMANCE OF LiteMAC

This section further evaluates the performance of LiteMAC with respect to scalability and multiple-level systems.

4.2.1   Varying M.   Fig. 9.6 studies the performance of LiteMAC with varying M for $L = 172$ and $\gamma \in \{0.3, 0.9\}$. Fig. 9.6(a) shows performance with varying M for $C \in \{4, 8\}$ and $M \in \{16, 32, 64\}$. Increasing M keeping C constant results in increase in latency due to longer control and data cycle lengths. Under light loads, the increase in latency is proportional to control slot length since not many data packets are transmitted. Under heavy loads, data cycle length increase is proportional to M and data cycle length resulting in increased latency. As the ratio M/C increases, data cycle lengths is approximately $\dfrac{ML}{C}$ and utilization approaches C as seen in the previous section.

Fig. 9.6(b) studies the performance of LiteMAC with varying M for $M/C = 4$. Regarding the scalability of the system, the system is not scalable with respect to delay for small values of M/C: the delay characteristics are not preserved by increasing C in proportion to increase in M. For larger values of M/C such as $M/C \geq 16$, system scalability shows improvement.

This section presented the key performance analysis for the access protocol. A more detailed analysis may be found in [Sivalingam and Dowd, 1996].

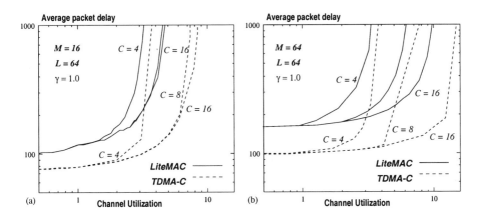

Figure 9.7   Comparison of TDMA-C [Bogineni and Dowd, 1992] with LiteMAC for uniform traffic for $M \in \{16, 64\}$, $C \in \{4, 8, 16\}$, $L = 64$, and $\gamma = 0.1$: (a) $M = 16$ and (b) $M = 64$.

4.3.   COMPARISON WITH A CONTROL CHANNEL PROTOCOL

In this section, the performance of LiteMAC is compared to a protocol based on a dedicated control channel (referred to as TDMA-C) studied in [Bogineni and Dowd, 1992].

TDMA-C is based on one control channel and $C - 1$ data channels. Each node is equipped with a tunable transmitter used for both data and control packets, a fixed receiver for control packets, and a tunable receiver for data packets. Packet collision due to destination node or data channel contention is eliminated through status tables at each node. Time multiplexing on the control channel provides each node a chance to transmit per control cycle. The basic protocol operation is as follows: The source node transmits a reservation packet containing the destination node address on the control channel in its allocated slot. All nodes receive and process all control packets transmitted on the control channel. The nodes maintain two status tables: one to track data channel availability, and one to track destination node availability. Each node updates the two tables in every control slot.

The main differences between LiteMAC and TDMA-C are as follows:

- Reserving one channel for control purposes as in TDMA-C reduces utilization, especially when number of available channels is small.

- Updating status tables in TDMA-C at the end of every control slot places a significant processing burden on the receiver subsystem.

- Broadcasting data packets requires more processing and synchronization and is not easily achievable with TDMA-C.

- Implementing TDMA-C tends to be more complex in comparison to LiteMAC.

Fig. 9.7 quantifies the performance comparison of LiteMAC and TDMA-C for $M \in \{16, 64\}$, $C \in \{4, 8, 16\}$, $L = 64$, and $\gamma = 1.0$. In general, the latency offered by TDMA-C is less than that offered by LiteMAC. For $M = 16$, $C = 4$ and $L = 64$, the reduction in latency with TDMA-C under light loads is approximately 20%. This is mainly due to the following reason: If a data packet is generated during LiteMAC's data phase, the node has to wait until the end of the data phase to transmit the reservation packet. In TDMA-C, reservation/control frames are consecutive on the control channel. Therefore, the node has to wait only for the destination node and a data channel to be available. Control packet starvation is aggravated in LiteMAC for systems with smaller γ (more control than data packets). This can be reduced to a certain extent by extending the length of each data slot to include a Class A packet.

TDMA-C offers better latency since the control packets can be transmitted in the next control frame.

For small values of C, maximum channel utilization achieved by TDMA-C is lower than LiteMAC. This is because for $\gamma = 1$, the control channel is used only for reservation overhead. For $M = 64$, $C = 4$, and $L = 64$, the maximum utilization achieved by LiteMAC is higher than that of TDMA-C by 25%. For higher values of C, the maximum utilization achieved by TDMA-C is higher (by 30% for $M = 64$, $C = 16$ and $L = 64$). For smaller γ, TDMA-C control channel will also be used for Class A packets thus increasing control channel utilization. An interesting feature common to both protocols is increasing C beyond $C = M/2$ does not significantly improve performance. For LiteMAC, maximum utilization closer to C is achieved for ratios $M/C \geq 4$.

Despite the performance disadvantages, the primary advantage of LiteMAC is simplicity and reasonable performance. This is the main reason for selecting LiteMAC as the prototype access protocol for Project LIGHTNING.

5.   CONCLUSIONS

This paper studied the design and analysis of a media access protocol for a WDM based hierarchical network architecture. The goal of the protocol is to provide low-cost, low-latency communication for a distributed shared memory multiprocessor system. The design objectives include support for broadcast/multicast and collision-less communication. The protocol is based on a system with one tunable transmitter and tunable receiver per node. Reservations are used to provide access to the data channels. The protocol does not use a separate transmitter/receiver pair for control processing. The performance of the protocol has been analyzed in detail through simulation. The performance of the protocol has been shown to be better than that of a similar protocol with tunable transmitter and fixed receiver per node.

Acknowledgments

The author is grateful to Dr. Patrick Dowd of the University of Maryland and all members of the LIGHTNING group.

References

Bannister, J. A., Fratta, L., and Gerla, M. (1990). Topological design of the wavelength-division optical network. In *Proc. IEEE INFOCOM'90*, pages 1005–1013.

Bhuyan, L. and Agrawal, D. P. (1984). Generalized hypercube and hyperbus structures for a computer network. *IEEE Transactions on Computers*, c-33:323–333.

Bogineni, K. and Dowd, P. W. (1992). A collisionless multiple access protocol for a wavelength division multiplexed star-coupled configuration: Architecture and performance analysis. *IEEE/OSA Journal of Lightwave Technology*, 10(11):1688–1699.

Chaiken, D., Fields, C., Kurihara, K., and Agarwal, A. (1990). Directory-based cache coherence in large-scale multiprocessors. *IEEE Computer*, 23(6):49–58.

Dowd, P., Perreault, J., Chu, J., Hoffmeister, D., Minnich, R., Hady, F., Chen, Y.-J., Dagenais, M., and Stone, D. (1996). Lightning Network and Systems Architecture. *IEEE/OSA Journal of Lightwave Technology*, 14(6):1371–1387.

Dowd, P. and Sivalingam, K. M. (1994). A multi-level WDM access protocol for an optically interconnected parallel computer. In *Proc. IEEE INFOCOM'94*, pages 400–408, Toronto, Ontario Canada.

Ganz, A. and Gao, Y. (1992). Efficient algorithms for SS/TDMA scheduling. *IEEE Transactions on Communications*, 40:1367–1374.

Jacobs, I., Binder, R., and Hoverstein, E. (1978). General purpose packet satellite networks. *Proc. IEEE*, 66:1448–1468.

Jump, J. R. (1992). *YACSIM Reference Manual*. Rice University, Department of Electrical and Computer Engineering, 1.2 edition.

Kannan, B., Fotedar, S., and Gerla, M. (1994). A two level optical star WDM metropolitan area network. In *Proceedings of GLOBECOM'94*, pages 563–566, San Francisco, CA.

Law, A. M. and Kelton, W. D. (1991). *Simulation Modeling and Analysis*. McGraw Hill.

Leiserson, C. E. (1985). Fat-trees: Universal networks for hardware-efficient supercomputing. *IEEE Transactions on Computers*, c-34:892–901.

Levine, D. and Akyildiz, I. (1995). PROTON: A Media Access Control protocol for optical networks with star topology. *IEEE/ACM Transactions on Networking*, 3(2):158–168.

Mukherjee, B. (1992). Architectures and protocols for WDM-based local lightwave networks Part I: Single-hop systems. *IEEE Network*, 6(3):12–27.

Singh, J. P., Weber, W. D., and Gupta, A. (1991). Splash: Stanford parallel applications for shared memory. Technical Report CSL-TR-91-469, Stanford University.

Sivalingam, K. M. (1995). Hybrid media access protocols for a DSM system based on optical WDM networks. In *Proc. Fourth IEEE International Symposium on High-Performance Distributed Computing (HPDC-4)*, pages 40–47.

Sivalingam, K. M. and Dowd, P. W. (1995). A Multi-Level WDM Access Protocol for an Optically Interconnected Multiprocessor System. *IEEE/OSA Journal of Lightwave Technology*, 13(11):2152–2167.

Sivalingam, K. M. and Dowd, P. W. (1996). A Lightweight Media Access Protocol for WDM-Based Distributed Shared Memory System. In *Proc. IEEE INFOCOM*, volume 2, pages 946–953, San Francisco, CA.

Sivalingam, K. M. and Wang, J. (1996). Media access protocols for WDM networks with on-line scheduling. *IEEE/OSA Journal of Lightwave Technology*, 14(6):1278–1286.

Weller, T. and Hajek, B. (1994). Scheduling NonUniform Traffic in a Packet Switching System with Small Propagation Delay. In *Proc. IEEE INFOCOM*, pages 1344–1351, Toronto, Canada.

Chapter 10

A MULTICASTING PROTOCOL FOR MULTICHANNEL LOCAL OPTICAL NETWORKS

Michael S. Borella

3Com Corp.
1800 W. Central Rd.
Mount Prospect, IL 60056
mike_borella@3com.com

Biswanath Mukherjee

Department of Computer Science
University of California
Davis, CA 95616
mukherje@cs.ucdavis.edu

Abstract   A wavelength-division multiplexed (WDM) based multicasting protocol for a single-hop broadcast-and-select local optical network is proposed. In particular, our approach employs a control-channel-based media-access protocol that schedules multicast packets, while incorporating arbitrary transceiver tuning times and propagation delays. A number of data channels (W) supply communication bandwidth to N nodes, where $N \geq W$, while the control channel is used for synchronization and scheduling . Each node is equipped with one fixed transmitter and one fixed receiver on the control channel , as well as one tunable transmitter and one or more tunable receivers for data channel access. The protocol takes advantage of the broadcast nature of a control channel by storing information of the system state of the network at each node. This allows efficient distributed scheduling of multicast packets. The metric of receiver throughput is defined to measure the expected number of busy receivers at steady state. Simulation results suggest that WDM single-hop multicasting experiences highest receiver throughput performance when multicast size is either small or very large, and when nodes are equipped with multiple receivers.

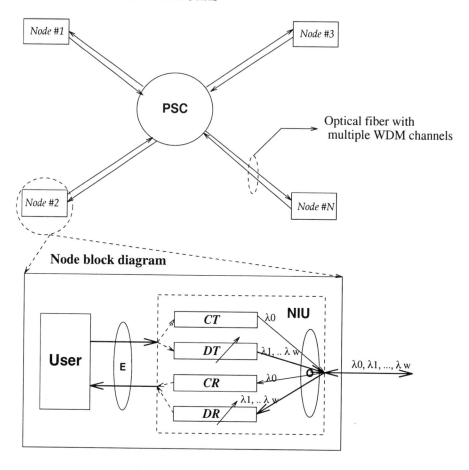

PSC -- Passive Star Coupler
NIU -- Network Interface Unit
CT -- Fixed Control Transmitter
DT -- Tunable Data Transmitter

O -- Optical inferface
E -- Electronic interface
CR -- Fixed Control Receiver
DR -- Tunable Data Receiver

Figure 10.1   A passive-star based local optical network.

1.   INTRODUCTION

The enormous bandwidth of optical fiber, approx. 25 THz in each of its low loss regions [Mukherjee, 1997], when compared to that of electronic communication media has led to the study of wavelength-division multiplexed (WDM) lightwave networks. While the aggregate information-carrying capacity of current high-speed networks, such as DQDB and FDDI is limited by the electronics at each network node, WDM carves the optical bandwidth into many channels that can each be operated at peak electronic speed. Thus, network throughput as a whole is no longer limited by the speed of electronics.

A popular method of building a local lightwave network is based on the passive star coupler(PSC). Fig. 10.1 shows the architecture of such a system. Each node is connected to the PSC by a transmitting and a receiving fiber. Nodes are equipped with some number of transmitters (lasers) and receivers (optical filters), some of which may be fixed to a particular wavelength (channel), while others may be tunable across a spectrum of wavelengths. The PSC is an $N \times N$ broadcast device. It has N inputs and N outputs, and the signal from each input is sent to each output. If one or more input signals using the same channel reach the PSC at the same time, a collision will occur, and none of those transmissions will be successful.

A connection is established between two nodes when the source node has a transmitter tuned to or fixed on a channel that the destination node has a receiver tuned to or fixed on. In a single-hop local lightwave network, all transmissions reach their destination in one pass through the PSC. This requires that the system be configured such that either: (1) each node has at least one tunable transmitter and one fixed receiver (TT-FR), (2) at least one tunable receiver and one fixed transmitter (FT-TR), or (3) one tunable transmitter *and* one tunable receiver (TT-TR)[1].

Once the hardware is chosen, a media-access protocol must be developed so that nodes can co-ordinate their transmissions. Random access protocols such as CSMA-CD tend to perform poorly due to the large ratio of propagation delay to packet size. For example, on a 1 Gbps channel, a 500 bit packet takes 500 ns to transmit. If the distance between two nodes is one kilometer, the propagation delay (given that the speed of light in a guided medium is approximately 2×10^8 m/s) will be 5 μs. It will take at least one round-trip propagation delay for a node to determine whether or not a collision has occurred.

WDM protocols based on synchronous TDM access to each channel have been proposed to eliminate collisions [Bogineni et al., 1993; Borella and Mukherjee, 1996; Rouskas and Sivaraman, 1997]. The bandwidth of each channel is divided into slots, which are globally synchronized over all channels. Node pairs are allocated slots over which to communicate. This allocation of slots forms a *transmission schedule.*

An alternative to random access and fixed-schedule protocols is deterministic control channel based arbitration for multichannel networks (also called reservation protocols) [Dono et al., 1990; Lu and Kleinrock, 1992; Jia and Mukherjee, 1993; Jia et al., 1995]. Nodes are allocated transmission bandwidth on demand (thus allowing nonuniform as well as time-varying traffic patterns). Some of these protocols (see [Lu and Kleinrock, 1992] and [Jia et al., 1995] for example) require extra transceivers per node for control channel access.

2.   PREVIOUS STUDIES ON MULTICASTING AND SCHEDULING

Multicast protocols for a single communication channel have been studied for fixed destination groups [Mase et al., 1983; Gopal and Jaffe, 1984; Wang and Silvester, 1988] (i.e., multicast transmission from a particular source has a fixed group of destinations) and multiple destination groups [Gopal and Rom, 1994] (i.e., multicast transmission from a particular source may have more than one destination group).

A multichannel, WDM single-hop protocol that incorporates multicasting was examined in [Rouskas and Ammar, 1994]. This study assumed a FT-TR architecture with one transmitter and one receiver per node. Channels are accessed via TDM on slot boundaries, with the length of a slot equal to the length of a data packet plus the receiver tuning latency. Each node's transmitter is fixed on a particular channel (which it may have to share with other nodes if $W < N$) and is allocated a number of multicast slots per frame. For one slot every F frames, all receivers tune to node i's transmitting wavelength. At this point, node i can start a new bulk-multicast[2] transmission. If a node $j \neq i$ is a member of the destination group for this transmission, then node j must tune to node i's channel during each of node i's multicast slots, until the transmission is completed.

The upper bound performance of multicasting on a WDM local network was studied in [Borella and Mukherjee, 1995]. Packet lengths were assumed to be exponentially distributed and the upper-bound throughput was examined by modeling the system as a W-server loss queue. It was found that, while multicasting decreases the transmitter (channel) throughput of the system, it drastically increases the receiver throughput of the system. In a unicasting system with W channels, the throughput of the system is asymptotically bounded by W. When multicasting is implemented on such a system, the amount of data received is greater than the amount of data transmitted, thus W is no longer a bound on receiver throughput. We found that, in the ideal case of a system with negligible tuning time, negligible propagation delay and a "perfect" media-access protocol, receiver-throughputs greater than W could be achieved. When nodes are equipped with multiple receivers, receiver-throughputs close to an order of magnitude greater than W could be achieved.

Unicast packet scheduling was examined in [Jia et al., 1995]. This single-hop WDM media access protocol assumes that each node is equipped with a tunable transmitter and tunable receiver, as well as a transmitter/receiver pair fixed to the control channel. Since every node receives all control transmissions, it is possible for each node to store the system state locally. Each node keeps track of the time at which each transmitter, receiver, and channel will be free, and the channel that each transmitter and receiver will be tuned to at the time they become free. When a control packet is received (uniform propagation delays are assumed so that control packets are received simultaneously at all nodes), all nodes execute a distributed, deterministic scheduling algorithm that returns the time at which the associated data packet will be transmitted, and the channel on which the transmission will occur. Then, each node updates its record of the system state. The system was studied under the assumption of bulk arrivals. The advantages of this protocol include the pipelining of propagation delay, high throughput, and collision-less transmission. The disadvantages include the large amount of information needed to be stored at each node and the processing overhead required with each control packet reception.

This article extends these studies by proposing a control channel based media access protocol that schedules multicast packets, while incorporating arbitrary transceiver tuning times and propagation delays.

3.   THE MULTICAST SCHEDULING ALGORITHM

We assume that each node is equipped with t tunable transmitters and r tunable receivers, each of which are wavelength-agile across all data channels. Tuning time is T_t for all transmitters and tuning time for all receivers is T_r. Each node is also equipped with one fixed transmitter and one fixed receiver that remain parked on the control channel. There are W data channels and one control channel to connect the N nodes. Data packet length is fixed at P bits, while control packets are P_c bits. The destination of a packet is a *multicast group*, which can be a set of 1 to $N - 1$ nodes. We assume that each node is equally distant from the passive star coupler, which makes the propagation delay (R) between all node pairs identical.

3.1.   CONTROL CHANNEL ACCESS

A control channel is used to share information between nodes. When a data packet arrives at a node, it is placed in that node's *arrival queue*. Once per frame, the node transmits a control packet (on the control channel) to schedule the transmission of the data packet at the head of its arrival queue. Access to the control channel is via round-robin TDM (see Fig. 10.2). The control frame (CF) is broken into N control slots ($CS\,0\,..\,CS$N-1), each of which consists of a source address field and a destination address field. The source address field encodes the source node's identity in $\log_2 N$ bits, and the destination address field is of N bits, such that the jth bit of this field is set if node j must receive the packet. Thus, the control packet length, P_c, equals $N + \log_2 N$ bits (which is the length of the control slot). Upon transmission of a control packet, the associated data packet is moved to the *waiting queue* until it is transmitted.

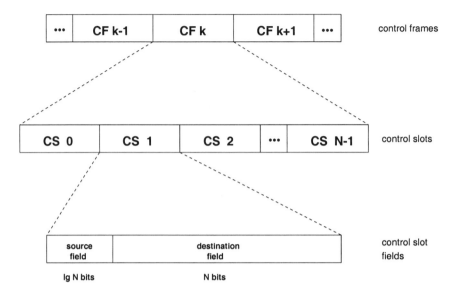

Figure 10.2   Control channel access via round-robin TDM.

3.2.   THE SCHEDULING PROTOCOL

After one propagation delay, the control packet is received by all nodes. Since all nodes are equipped with a fixed receiver on the control channel, all control information is received simultaneously by each node. Thus, if all nodes execute an identical media-access protocol upon a control packet arrival, data transmissions can be deterministically scheduled in a distributed fashion. In order to schedule data transmissions so that collisions will not occur, each node retains the following data described in Table 10.1. The transmitters and receivers are indexed from 0 to $Nt - 1$ and 0 to $Nr - 1$ respectively, where transmitter i belongs to node $\lfloor i/r \rfloor$ and receiver j belongs to node $\lfloor j/r \rfloor$.

Pseudo code of the Multicast Scheduling Algorithm (MSA) is presented in Fig. 10.3. Upon reception of a control packet, the MSA is run by all nodes in parallel. A description of the algorithm follows. Lines 3-13 examine each node in the multicast group to determine the earliest time at which the multicast packet can be received simultaneously by all nodes in the multicast group. Lines 7-12 find the receiver at each node that is free at the earliest time, and they store this information in *first_rec[i]* for each node i, $0 \le i \le N - 1$. Given this information, the earliest time that the packet can be received is *recfree[max_i(first_rec[i])]*, the value of which is stored in *last_node_free*. The current system time is stored in *CURRENT_TIME*. If all receivers are free before *CURRENT_TIME*, then *last_node_free* is set to *CURRENT_TIME*. The earliest system time at which the multicast packet can be received by all nodes in the multicast group is *last_node_free* $+T_r{}^3$.

```
 1 procedure MSA (source, dest[0..N-1],channel[0..Nt-1],chanfree[0..W-1],
 2                 transfree[0..Nt-1],recfree[0..Nr-1])
 3 last_node_free = -1
 4 for i=0 to N-1
 5     begin
 6     if dest[i] == 1 then
 7         begin
 8         first_rec[i] = the one of i's receivers that is free earliest
 9         first_rec_free = recfree[first_rec[i]]
10         if first_rec_free > last_node_free then
11             last_node_free = first_rec_free
12         end
13     end
14 if last_node_free < CURRENT_TIME then
15     last_node_free = CURRENT_TIME
16 earliest_rec_time = last_node_free + RecTuneTime
17 source_trans = the source node transmitter that is free the earliest
18 source_free = transfree[source_trans]
19 if source_free < CURRENT_TIME then
20     source_free = CURRENT_TIME
21     /* Case 1:  Try to schedule the packet on the channel that the
22     /* transmitter is currently on */
23 trans_chan_free = chanfree[channel[source_trans]]
24 if trans_chan_free <= source_free + TransTuneTime then
25     begin
26     earliest_tran_time = max(max(trans_chan_free,source_free),
       earliest_rec_time-R)
27     chan = channel[source_trans]
28     end
29     else begin
30     /* Case 2:  Find the earliest available channel to transmit on */
31     earliest_tran_time = max(source_free+TranTuneTime,earliest_rec_time-R)
32     chan = getfreechannel(earliest_tran_time)
33     end
34 tran_time = max(earliest_tran_time,chanfree[chan])
35 chanfree[chan] = tran_time+P
36 transfree[source_trans] = tran_time+P
37 for i=0 to N-1
38     recfree[first_rec[i]] = tran_time+P+R
39 end
```

Figure 10.3   The multicast scheduling algorithm (MSA).

A similar technique is used to find the earliest time at which the packet can be transmitted, viz. the time at which the transmitter of the source node that is free the earliest (lines 17-20). This transmitter's index is stored in *source_trans* and the system time at which it becomes free is stored in *source_free*. On line 23, *trans_chan_free* is assigned the time at which the channel that the transmitter is on becomes free (this information is stored in *chanfree[channel[source_trans]]*).

The MSA first attempts to schedule the packet to be transmitted on the channel that the transmitter is currently on, in order to eliminate having to tune the source transmitter. If *trans_chan_free* is less than or equal to *source_free* + T_t, the MSA chooses to transmit on *channel[source_trans]* (lines 25-28). The earliest time that both the channel and the transmitter is ready is given by max (*trans_chan_free, source_free*). The packet cannot be received before *earli-*

est_rec_time, which means that it cannot be transmitted before *earliest_rec_time* - *R*. Thus, the packet is scheduled to be transmitted at max(*trans_chan_free*, *source_free*, *earliest_rec_time* - *R*), which is stored in *earliest_tran_time*.

If the case that the channel that the transmitter is currently tuned to will not become free before *source_free* +T_t (lines 29-33), MSA calls the function *getfreechannel* with the parameter *earliest_tran_time* = *max(source_free* + *TranTuneTime, earliest_rec_time* - *R)*.

Function *getfreechannel* takes a time x as an argument, and returns the index of a channel free before x. If no channels will be free before than time, it returns the number of the channel that will be free soonest after x. Then, *earliest_tran_time* is redefined to be max(*earliest_tran_time, chanfree[y]*), where y = *getfreechannel (earliest_tran_time)*. Lines 34-39 update the data structures to reflect the addition to the schedule.

3.3.   RUNNING TIME

The running time of the MSA is derived as follows. The loop in lines 3-13 requires N iterations, each of which require $O(r)$ time to find the receiver that is free earliest. Lines 14-24 require $O(1)$ time. The *if* branch of line 24 requires $O(1)$ time, but the *else* branch of that lines requires $O(W)$ time. Lines 34-36 requires $O(1)$ time, and lines 37-38 require $O(N)$ time. This sums to a total worst case running time[4] of $O(Nr + t + W)$. The total storage requirements per node is based on the size of the arrays stored at each node. The sum of the sizes of these arrays (as shown by Table 10.1) is $N(2t + r + 1) + W$.

4.   ANALYSIS OF BOUNDS

This section derives the performance bounds of a WDM single-hop network using the protocol described in the previous section. A complete analysis of such a system seems to be intractable and such a study is still an open problem. Existing analyses assume either single packet buffers at all nodes [Lu and Kleinrock, 1992; Jia and Mukherjee, 1993] [Borella and Mukherjee, 1995] or an infinite number of channels [Jia et al., 1995]. An analysis of a finite-channel system which includes queueing was presented in [Bogineni et al., 1993], but the transceiver tuning time was assumed to be negligible, and the protocol analyzed was a deterministic round-robin scheduling scheme. Given that the upper bounds on throughput are provided in the analysis of [Borella and Mukherjee, 1995], we focus on providing a simple analysis of the lower bounds on mean packet delay.

The lower bound on the total packet delay is derived as follows. We assume light system load (queueing occurs for access to the control channel, but access to data channels is immediate. When a packet arrives at a node, it is put into that node's arrival queue. Access to the control channelis round-robin TDM, and can be modeled as an M/D/1 queue [Hammond and O'Reilly, 1986]

$$D_c = P_c + \frac{NP_c}{2} + \frac{N\rho_c P_c}{2(1 - \rho_c)}$$

where P_c is the length (in bits) of the control packet and D_c is the mean delay for access to and transmission on the control channel. The load offered to the control channelis ρ_c, which is defined to be $\frac{\rho W P_c}{P}$ where ρ is the load offered to the data channels and P is the length of a data packet in bits. After one propagation delay, R, all nodes have received the control packet. Assuming negligible processing delay and low system load, the earliest time that the MSA can transmit the associated data packet is $\min(0, T_r - R)$, and that packet will take P time units to complete transmission. Thus, a lower bound on the mean total queueing and transmission delay, D, is

$$D = D_c + 2R + \min(0, T_r - R) + P$$

5.   NUMERICAL RESULTS

Simulations of the MSA were conducted with $N = 50$, $W = 10$, channel bitrate = 1 Gb/s, round-trip propagation delay (R) of 5 μs, uniform traffic patterns and one tunable transmitter per node. The parameters of packet size (P), number of tunable receivers per node (r), transmitter and receiver tuning time (T_t and T_r, respectively), and mean multicast size were varied. In all configurations, it is assumed that an infinite number of messages may be queued at each node.

To measure the performance of a system in which the number of nodes receiving messages can be larger than the number of nodes transmitting, we define *transmitter throughput* (S_T) to be the mean number of transmitters in use at steady state (alternatively, transmitter throughput can be called *channel throughput*). *Receiver throughput* (S_R) is defined to be the mean number of receivers in use at steady state.

5.1.   SYSTEM PERFORMANCE WITHOUT MULTICASTING

Before we examine the numerical results of networks implementing multicasting, it is helpful to first take into account the performance of the MSA without multicasting[5]. Fig. 10.4 shows that, for a system with no multicasting, negligible tuning times and one receiver per node ($r = 1$), throughput approaches 10 as packet size increases to about 1000 bits, then slowly decreases as packet size gets larger. This can be explained by noting that a larger packet size will allow more information to be transmitted with the same scheduling

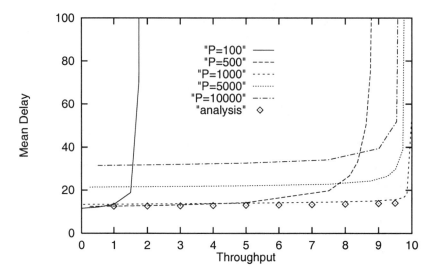

Figure 10.4   Mean delay versus transmitter throughput (S_T), no multicasting, different data packet sizes.

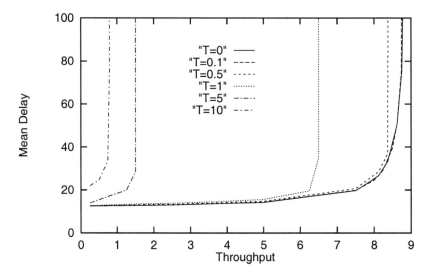

Figure 10.5   Mean delay versus transmitter throughput (S_T), varied transceiver tuning times.

overhead as a smaller packet. However, very large packets will require the transmitters and receivers servicing them to be busy for longer periods of time. Since the MSA reserves a transmitter, receiver(s), and a channel for the period of time up to and including the packet transmission, if one or more of these three resources are busy, the others are marked "reserved" even though they may not be in use. Thus, large packets may cause a schedule to not be filled up as efficiently as possible. However, as shown in Fig. 10.4, the reduction in throughput due to large packets is not great. The analysis points are for the system with $P = 1000$, using the lower bound formula given in the previous section. Another factor to consider is the bandwidth bottleneck that occurs when the ratio $\frac{P}{P_c}$ is small. The control channel is the main bottleneck of the system when P is small, which limits data channel throughput. This situation can be remedied by using a control channel with a higher bitrate – see [Mukherjee and Jia, 1992] for a discussion of this phenomenon.

Fig. 10.5 displays the mean delay versus transmitter throughput for the same system, except that the packet size is fixed at 500 bits and the transmitter and receiver tuning times are varied (but they are assumed to be equal, viz. $T = T_t = T_r$). The maximum throughput for $T = 0$ and $T = 0.1\mu s$ is virtually identical (approximately 9). The maximum throughput for $T = 1\mu s$ is about 6.6, with a low-delay operating range of 0-6. This is an encouraging result for a system with a tuning latency twice that of the packet duration. As tuning time is increased to 5 and 10 μs (the latter is approximately the switching speed of an acoustooptic filter [Brackett, 1990]), performance is seriously degraded, which suggests that either (1) only transmitters and receivers which tune on the order of a few microseconds should be employed, or (2) if slower-tuning ($T > 5\mu s$) transmitters and receivers must be used, then packet length should be increased.

5.2. SYSTEM PERFORMANCE WITH MULTICASTING

In order to study the effect of multicast size on the throughput and delay characteristics of a multicasting network, the multicast limit $1 \leq L < N$ is introduced. L is the maximum number of nodes that may receive a given multicast packet. We assume that the distribution of multicast size for all packet arrivals under a given L is uniform (e.g., given L, the probability that a packet's multicast size is $1 \leq x \leq L$ equals $\frac{1}{L}$).

Fig. 10.6 examines the system's mean delay versus transmitter throughput performance as L is varied. In these examples, a packet size of 500 bits is assumed, transceiver tuning time is negligible and each node has one tunable transmitter and one tunable receiver. As L increases, the transmitter throughput (the mean number of transmitters busy in steady state) reduces. Once L reaches 20, S_T is approximately 1. In other words, once the mean multicast size is 10 or greater, there is a very small probability of having more than one message in the system at any one time.

Fig. 10.7 shows the mean delay versus receiver throughput characteristics for the same system. As expected, the receiver throughput increases as multicast

Table 10.1   Arrays used in the multicast scheduling algorithm.

array	range	description
dest	$0 \leq i < N$	dest[i] = 1 if node i is a recipient of the current packet 0 otherwise
channel	$0 \leq i < Nt$	channel[i] = the channel that transmitter i is tuned to
chanfree	$0 \leq i < W$	chanfree[i] = system time at which channel i will be free
transfree	$0 \leq i < Nt$	transfree[i] = system time at which transmitter i will be free
recfree	$0 \leq i < Nr$	recfree[i] = system time at which receiver i will be free

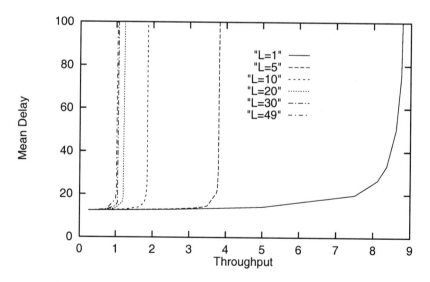

Figure 10.6   Mean delay versus transmitter throughput (S_T), negligible transceiver tuning times.

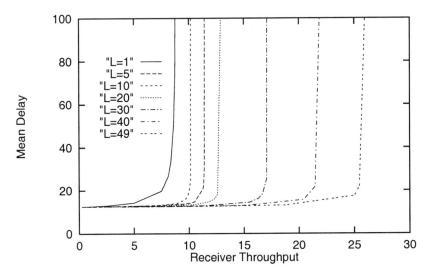

Figure 10.7   Mean delay versus receiver throughput (S_R), negligible transceiver tuning times.

size increases for small values of L, and for systems with a large mean multicast size ($L = 49$), even though only one transmitter (and channel) is in use, 25 receivers are busy. This demonstrates that *the achievable receiver throughput for a WDM single hop network using the MSA protocol is far greater than the maximum throughput of an ideal unicasting system.*

Looking carefully at Fig. 10.7, there appears to be an anomaly – the receiver throughput of the system with $L = 5$ is higher than that of the system with $L = 10$. Since the mean multicast size of the former system is 3, and that of the latter system is 5.5, it would seem to make more sense that latter system should have a higher receiver throughput. This is in fact a somewhat surprising effect of multicasting and can be explained intuitively as follows. Note that, when multicasting is implemented, the bottleneck of this particular architecture is the number of free receivers. When mean multicast size is small, the mean number of receivers busy at steady state is relatively small, as is the mean number of receivers required by a new message entering the system. Thus, the probability that a new message can find the particular receivers it requires for its multicast connection is relatively high. As shown above, once the mean multicast size is greater than approximately 10, the probability of more than one channel being busy is very small. As L is increased, the transmitter throughput drops very rapidly. Thus, at the points where the transmitter throughput is small and L is also small, receiver throughput will actually be less than that of systems with higher transmitter throughputs and that of systems with larger L. Once the transmitter throughput has dropped to approximately 1, the size of the multicast makes very little difference on the amount of time it takes to synchronize a larger number of receivers for a packet transmission. In other words, once $S_T = 1$, S_R increases linearly with L.

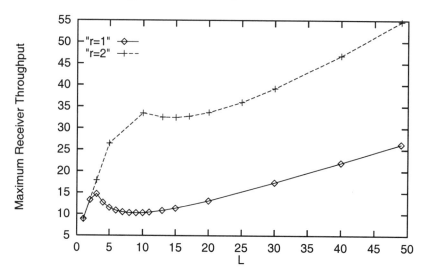

Figure 10.8   Maximum receiver throughput versus multicast limit (L), negligible transceiver tuning times.

This relationship is shown in Fig. 10.8. For small L, S_R is maximized at $L = 3$, for $r = 1$. As L increases from 3 to 25, S_R remains lower than its peak at $L = 3$. Beyond $L = 25$, S_R increases linearly with L. This suggests that, if the mean multicast size is between 2 and 13, it may be more efficient to break large multicasts into groups of smaller multicasts.

Keeping these results in mind, one may ask the following question: if only one channel is being used when the MSA protocol demonstrates its best performance, why do we need a multichannel architecture to begin with? This question motivates us to suggest that WDM-based multichannel multicast algorithms will display the best performance when more than one receiver per node is used. Fig. 10.8 also displays the maximum achievable throughput when the MSA protocol is used on an architecture with 2 receivers per node. Again, we see a range of values of L in which the maximum receiver throughput experiences a local minimum, but this minimum is much less severe than that of the system in which there is one receiver per node. These results agree with the theoretical findings of [Borella and Mukherjee, 1995].

6.   CONCLUSION

This article has presented a multichannel, WDM star-based protocol that efficiently schedules multicast packet transmissions. The algorithm is simple, and its running time is on the order of the total number of nodes and channels in the system. Non-negligible transceiver tuning time and non-zero propagation delays are considered. We have shown that high throughputs can be achieved while requiring a reasonable amount of processing per node. The system's performance displays interesting characteristics as mean multicast size increases, and these results have been justified intuitively. Results agree with previous theoretical analyses and suggest that the multicasting system's performance may be improved by partitioning large multicasts into a number of smaller multicasts, and by using multiple receivers per node. Open problems include an exact analysis of the system's delay versus throughput performance as well as studying the tradeoffs involved in using more elaborate scheduling algorithms.

Notes

1. An exception to these three cases is the LAMBDANET [Goodman et al., 1990] WDM prototype, in which each node is equipped with one fixed transmitter and W fixed receivers, one for each of the W channels.

2. We use the term 'bulk-multicast' to describe bulk arrivals that are destined to a single multicast group. These arrivals are broken into packets, some number of which are transmitted per frame.

3. For simplicity of the algorithm, we assume that all receivers in the multicast group must tune before receiving a packet. In some cases, it is possible to schedule an earlier reception. Let c indicate the channel that the last busy receiver is tuned to when it becomes free. If the data packet is scheduled on channel c, it may be possible to have it arrive at its destinations one tuning latency earlier than it would by using the MSA. However, this scheduling efficiency comes at the cost of extra processing time. To retain algorithmic simplicity and a fast running time for the MSA, we have not considered such enhancements.

4. Due to the large cost of tunable optical transceivers, it is unlikely that a node will be equipped with more than 2 or 3 transmitters or receivers. Thus, t and r in the above equation may be considered constant. This gives us a worst case running time of $O(N + W)$, which reflects the factors (the number of nodes and number of channels, respectively) which are most responsible for the running time of the MSA.

5. For all graphs, mean delay is presented in microseconds.

References

Bogineni, K., Sivalingam, K. M., and Dowd, P. W. (1993). Low-complexity multiple access protocols for wavelength-division multiplexed photonic networks. *IEEE Journal on Selected Areas in Communications*, 11(4):590–604.

Borella, M. S. and Mukherjee, B. (1995). Limits of multicasting in a packet-switched WDM single-hop local lightwave network. *Journal of High Speed Networks*, 4(2):155–167.

Borella, M. S. and Mukherjee, B. (1996). Efficient scheduling of nonuniform packet traffic in a WDM/TDM local lightwave network with arbitrary transceiver tuning latencies. *IEEE Journal on Selected Areas in Communications*, 14(5):923–934.

Brackett, C. A. (1990). Dense wavelength division multiplexing networks: Principle and applications. *IEEE Journal on Selected Areas in Communications*, 8(6):948–964.

Dono, N. R., Green, P. E., Liu, K., Ramaswami, R., and Tong, F. F.-K. (1990). A wavelength division multiple access network for computer communication. *IEEE Journal on Selected Areas in Communications*, 8(6):983–993.

Goodman, M. S., Gimlett, J. L., Kobrinski, H., Vecchi, M. P., and Bulley, R. M. (1990). The LAMBDANET multiwavelength network: Architecture, applications, and demonstrations. *IEEE Journal on Selected Areas in Communications*, 8(6):995–1004.

Gopal, I. and Rom, R. (1994). Multicasting to multiple groups over broadcast channels. *IEEE Transactions on Communications*, 42(7):2423–2431.

Gopal, I. S. and Jaffe, J. M. (1984). Point-to-multipoint communication over broadcast links. *IEEE Transactions on Communications*, 32:1034–1044.

Hammond, J. L. and O'Reilly, P. J. P. (1986). *Performance Analysis of Local Computer Networks*. Addison Wesley.

Jia, F. and Mukherjee, B. (1993). The receiver collision avoidance (RCA) protocol for a single-hop WDM lightwave network. *Journal of Lightwave Technology*, 11(5/6):1053–1065.

Jia, F., Mukherjee, B., and Iness, J. (1995). Scheduling variable-length messages in a single-hop multichannel local lightwave network. *IEEE/ACM Transactions on Networking*, 3(4):477–488.

Lu, J. C. and Kleinrock, L. (1992). A wavelength division multiple access protocol for high-speed local area networks with a passive star topology. *Performance Evaluation Journal*, 16(1-3):223–239.

Mase, K., Takenaka, T., Yamamoto, H., and Shinohara, M. (1983). Go-back-N ARQ schemes for point-to-multipoint satellite communications. *IEEE Transactions on Communications*, 31:583–589.

Mukherjee, B. (1997). *Optical Communication Networks*. McGraw-Hill, New York.

Mukherjee, B. and Jia, F. (1992). Bimodal throughput, nonmonotonic delay, optimal bandwidth dimensioning, and analysis of receiver collisions in a single-hop WDM local lightwave network. In *Proceedings, IEEE Globecom '92*, pages 1896–1900, Orlando, FL.

Rouskas, G. N. and Ammar, M. H. (1994). Multi-destination communication over single-hop lightwave WDM networks. In *Proceedings, IEEE INFOCOM '94*, pages 1520–1527, Toronto, Canada.

Rouskas, G. N. and Sivaraman, V. (1997). Packet scheduling in broadcast WDM networks with arbitrary transceiver tuning latencies. *IEEE/ACM Transactions on Networking*, 5(3):359–370.

Wang, J. L. and Silvester, J. L. (1988). Optimal adaptive ARQ protocols for point-to-multipoint communication. In *Proceedings, IEEE INFOCOM '88*, pages 704–713, New Orleans, LA.

V

PERFORMANCE EVALUATION

Chapter 11

BLOCKING PERFORMANCE OF WAVELENGTH-ROUTING NETWORKS

Suresh Subramaniam

Department of Electrical and Computer Engineering
The George Washington University
Washington, DC 20052
suresh@seas.gwu.edu

Abstract   Performance evaluation of wavelength-routing networks is the topic of this chapter. In particular, an analytical model for the blocking performance under dynamic Poisson traffic is presented. The model has a modest complexity and good `accuracy, and is applicable to arbitrary mesh topologies with arbitrary wavelength converter locations, and arbitrary network state-independent routing, with random wavelength assignment .

1.   INTRODUCTION

Wavelength division multiplexing (WDM) and wavelength routing have emerged as the key multiplexing and switching techniques for future all-optical networks. WDM partitions the fiber bandwidth into a set of (a few tens, and even hundreds of) disjoint channels (wavelengths), each operating at reasonably large bit-rates, e.g., 2.5 Gb/s. Wavelength routing permits the simultaneous use of the same channel in spatially disjoint segments, hence allowing a network to scale easily.

Wavelength-routing networks consist of wavelength-routing nodes interconnected by optical fibers. A wavelength-routing node is one that is capable of switching a signal (possibly dynamically) based on the input port and the wavelength on which the signal arrives [Goodman, 1989]. Such a node can be implemented by a set of wavelength multiplexers and demultiplexers and a set of photonic switches (one per wavelength).

Wavelength-routing networks allow the set-up and teardown of all-optical circuit-switched paths called *lightpaths*. At the present time, the technology to set up static lightpaths in certain special topologies such as linear and ring configurations is advanced enough for commercial use. It is likely that the next few years will see the emergence of all-optical space switches that allow the dynamic configuration of lightpaths. In this chapter, we consider a wavelength-routing network in which on-demand establishment and tear-down of lightpaths is possible.

Two different functionalities of the wavelength-routing nodes are important in the context of lightpath establishment: nodes with no wavelength conversion capability, and nodes which can convert an incoming wavelength to an arbitrary outgoing wavelength. When all nodes have wavelength converters, the situation is analogous to trunk switching in digital telephony [Girard, 1990] as a lightpath on one wavelength can be switched to any outgoing wavelength, if one is available. On the other hand, in a network without any wavelength converting nodes, a lightpath at a certain wavelength on an input fiber has to be switched to an output fiber at the same wavelength. This requirement of *wavelength continuity* affects the network performance; to set up a lightpath request, it is necessary that the *same* wavelength be free on all the links of the route.

The focus of this chapter is the performance evaluation of wavelength-routing networks with given arbitrary topologies and wavelength converter locations. It is appropriate to ask at this point: what is a suitable measure of network performance? The network performance measure is strongly tied to the traffic model to be used. Unfortunately, there does not appear to be a single satisfactory performance criterion or traffic model, not the least because we do not know the traffic characteristics in a network of the future.

In this chapter, the traffic model we consider is a commonly studied statistical model – lightpath requests (also called, *calls*) arrive at a node according to a random point process and an optical circuit is established for the (random) duration of the call. The performance criterion of interest here is the probability that a lightpath request cannot be established due to unavailable network resources, or the lightpath *blocking probability*. Several analytical models have been proposed recently to evaluate the blocking performance of wavelength-routing networks [Barry and Humblet, 1996; Barry and Marquis, 1995; Kovačević and Acampora, 1996; Birman, 1996; Subramaniam et al., 1996; Yates et al., 1996; Sharma and Varvarigos, 1998; Tripathi and Sivarajan, 1999; Zhu et al., 1999]. They differ in their underlying assumptions, and have varying computational complexities and levels of accuracy. The first analytical models to appear in the literature were [Barry and Humblet, 1996; Kovačević and Acampora, 1996; Birman, 1996]. Some important factors affecting blocking probability in wavelength-routing networks were quantified using a simple traffic model in [Barry and Humblet, 1996]. The other two models assumed that calls arrived according to a Poisson process and existed in the network for an exponentially distributed duration, and proceeded to obtain the blocking probabilities for various traffic parameters. Besides the fact that these models were applicable to special topologies, the main drawback of these initial models was that they considered networks with either no wavelength converters at all or networks with converters at every node. Because all-optical wavelength converters are expensive devices, and because of the likely improved blocking performance with wavelength conversion, a more realistic situation is a network with some limited amount of wavelength conversion. Accordingly, subsequent research has addressed this issue [Subramaniam et al., 1996; Yates et al., 1996; Sharma and Varvarigos, 1998; Tripathi and Sivarajan, 1999; Zhu et al., 1999].

A limited amount of wavelength conversion may exist in a network in several forms. The following terminology serves to differentiate the network models used for wavelength conversion. *Limited wavelength conversion* [Ramaswami and Sasaki, 1997] is used to refer to nodes capable of converting a wavelength to some, but not all, wavelengths. *Limited-range wavelength conversion* [Yates et al., 1996] refers to nodes capable of converting a wavelength to some adjacent wavelengths in the spectrum of wavelengths. This is actually a special case of limited wavelength conversion . Finally, *sparse wavelength conversion* [Subramaniam et al., 1996] is used to refer to a network in which some nodes have full wavelength conversion capability, while the rest do not have any. Of course, an actual network's wavelength conversion capability may be realized through any combination of these models.

The analytical model we present here for evaluating the blocking performance is based on the work in [Subramaniam et al., 1996]. The model is applicable to arbitrary network topologies with sparse wavelength conversion, where the converter locations are arbitrary.

2.   MODEL ASSUMPTIONS

The following assumptions are used in the model.

1 *Network Model:*

- The network of N nodes, numbered $1, 2, \ldots, N$, is given. The links are directed, and each link corresponds to a single fiber oriented in the direction of the link. The number of wavelengths is given, and is assumed to be the same on each link. It is denoted by F. The constant wavelength assumption can be easily relaxed, and is used here only for simplicity.

- The locations of the converter nodes are given[1], and the converter configuration is denoted by **a**.

2 *Routing and Wavelength Assignment Model:*

- The route used by a call is independent of the network state. Note that this does not imply that a single fixed route is always used for a lightpath request between a given source and destination. It is possible to have several different routes, but the choice of which route to use is not dependent on the network state. An example of such routing is the random selection of a route from a fixed set of routes for the given source and destination.

- Upon arrival of a lightpath request, a suitable route is chosen as above. Between two consecutive converter nodes on the route, the wavelength[2] used for the call is *randomly* chosen from the set of wavelengths available on all of the links between those two nodes. This assignment scheme is called as *random wavelength assignment* . If no such wavelength is available, the call is not retried on another route, and is considered to be blocked.

3 *Traffic Model:*

- Call requests arrive according to a Poisson process, and each call has a holding time that is exponentially distributed. Because of the routing model used above, we assume that the Erlang loads for a set of routes, rather than for source- destination pairs, are given. Let $\{r_1, r_2, \ldots, r_w\}$ be the set of routes, and let λ_i be the Erlang load on route r_i.

These assumptions are also used to obtain the blocking performance via simulations. Besides these, an additional one is necessary in order to make the analytical computation of the blocking probability tractable. The loads on the various links of the network are correlated because the traffic is offered on entire routes rather than individual links. The degree of correlation depends on a number of factors such as traffic pattern and connectivity of the network. The blocking performance of a circuit-switched network (for example, a telephone network) is generally obtained by decomposing the routes into a set of links, and computing the blocking probability on each link [Girard, 1990]. For analytical tractability, it is usually assumed that the blocking events on different links are statistically independent. In a wavelength-routing network without wavelength converters, not only are the link load values correlated with each other, but also, the usages of individual wavelengths are correlated. It is extremely important to capture this correlation such that the model does not become intractable, while at the same time yields accurate numerical results. Initial model proposals [Kovačević and Acampora, 1996] assumed that the load correlation was negligible, as in the case of circuit-switched networks. Such an assumption led to large differences between blocking probability values obtained via simulations and the model. In the analytical model presented in this chapter, we assume that link loads have Markovian spatial correlation, i.e., the wavelengths used on a link are assumed to depend only on the wavelengths used on the previous link of a route under consideration. This assumption seems to capture the correlation effects to a large degree while keeping the model reasonably tractable. It is possible to extend the correlation effects further at the expense of a more complicated model, but it is not clear whether the increased complexity is worth the (marginally) higher accuracy. We call the model that ignores the correlation of link loads as the *independence model*, and our model as the *correlation model*.

3.   BLOCKING PROBABILITY ANALYSIS

The network-wide blocking probability will be obtained by calculating the individual blocking probabilities on all the routes, and then averaging over them. The route blocking probabilities, in turn, will be calculated based on the distribution of available wavelengths on the route. To elaborate further, let us define a *segment* to be a sub-route with no converters on the intermediate nodes. Given the converter locations, every route can be decomposed into a sequence of segments where there is a converter between consecutive segments. An arriving call on a route must be assigned a wavelength on each of the route's segments. Let $P_b^{(i)}(\mathbf{a})$ be the blocking probability of calls on route r_i for a given converter configuration $\mathbf{a}$. Then, the network-wide average blocking probability is given by

$$\bar{P}_b(\mathbf{a}) = \frac{\sum_{i=1}^{w} \lambda_i P_b^{(i)}(\mathbf{a})}{\sum_{i=1}^{w} \lambda_i}.$$

We next proceed to calculate the route blocking probabilities.

3.1.   COMPUTING ROUTE BLOCKING PROBABILITIES

Let us consider an arbitrary route r_i. Suppose the nodes of route r_i are numbered $v_0, v_1, \ldots, v_{n_i}$, where v_0 and v_{n_i} are the source and destination nodes, respectively, of route r_i, and n_i is the number of hops on r_i. The path is shown in Fig. 11.1. For notational convenience, when computing the blocking probability for route r_i, we assume that converters are present at nodes v_0 and v_{n_i} (This assumption does not affect the blocking probability of route r_i.). Given a converter configuration $\mathbf{a}$, let $\{v_{j_0}, v_{j_1}, v_{j_2}, \ldots, v_{j_{k_i}}, v_{j_{k_i+1}} : 0 = j_0 < j_1 < j_2 < \cdots < j_{k_i} < j_{k_i+1} = n_i\}$, be the set of nodes on r_i with converters. Here, k_i is the number of converters on the intermediate nodes of r_i for the converter configuration $\mathbf{a}$. The placement of k_i converters on r_i yields $k_i + 1$ segments. Fig. 11.1 shows route r_i with $k_i = 3$. A wavelength is said to be *free* on a sub-segment (part of a segment) if it is available (upon call arrival)

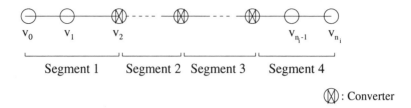

Figure 11.1   Route r_i and corresponding notation.

on all the links of the sub-segment. It is said to be *busy* otherwise. Let $S_l^{(i)}$ be defined as the success probability of a call on route r_i on the last l segments of the route, i.e., the probability of finding a free wavelength (not necessarily the same) on each of the last l segments of route r_i. Then,

$$P_b^{(i)}(\mathbf{a}) = 1 - S_{k_i+1}^{(i)}(\mathbf{a}).$$

Our goal now is to calculate the success probability of a call over the $k_i + 1$ segments of route r_i. We do this in a recursive manner. If the link blocking events were assumed to be statistically independent, this calculation would be fairly straightforward. We now proceed to show how $S_{k_i+1}^{(i)}(\mathbf{a})$ is computed under the Markovian spatial correlation assumption.

Let $Q_v^{(i)}(x)$ be defined as the probability of finding x free wavelengths on the link with source node v on route r_i. Then, by conditioning on the number of free wavelengths on the first hop of route r_i, we may write

$$S_{k_i+1}^{(i)}(\mathbf{a}) = \sum_{x=1}^{F} S_{k_i+1}^{(i)}(\mathbf{a}|x)Q_{v_0}^{(i)}(x),$$

where $S_l^{(i)}(\mathbf{a}|x)$ is the success probability of a call on the sub-route consisting of the last l segments of r_i given that x wavelengths are available on the first hop of the sub-route. Before computing the above probabilities, we develop some more notation.

3.1.1   Notation.   We define the following expressions that will be used in obtaining the blocking probabilities.

- $s_\alpha \overset{\text{def}}{=} j_\alpha - j_{\alpha-1}$ = the number of hops on segment α (counting from the source node) of route r_i, $\alpha = 1, 2, \ldots, k_i + 1$.

- $V_\alpha^{(i)}(s, n, y|x) \overset{\text{def}}{=}$ probability that n wavelengths are free on the first s hops and y wavelengths are free on hop s of segment α of route r_i, given that x wavelengths are free on the first hop of the segment.

- $W_\alpha^{(i)}(y|x) \stackrel{\text{def}}{=}$ probability that y wavelengths are free on the last hop of segment α of route r_i, given that x wavelengths are free on the first hop of the segment.

The probabilities $V(\cdot|\cdot)$ and $W(\cdot|\cdot)$ are associated with a single segment. We compute these by applying recursion on the the number of hops in part of the segment. Because of the assumption of Markovian spatial correlation of the link loads, we construct a two-hop model to compute the above probabilities. The following notation refers to a two-hop sub-route of r_i with intermediate node v.

- $T_v^{(i)}(w|y) \stackrel{\text{def}}{=}$ probability that w wavelengths are free on the second hop of the two-hop sub-route, given that y wavelengths are free on the first hop of the sub-route.

- $U_v^{(i)}(z|y,x) \stackrel{\text{def}}{=}$ probability that z wavelengths continue from the first to the second hop of the sub-route, given that x and y wavelengths are free on the first and second hops of the sub-route, respectively.

- $R(n|x,y,z) \stackrel{\text{def}}{=}$ probability that n wavelengths are free on both hops of the sub-route, given that x and y wavelengths are free on the first and second hop, respectively, and z wavelengths continue from the first hop to the second hop.

The model for computing these probabilities will be discussed a little later. We now return to the computation of the probabilities $V(\cdot|\cdot)$ and $W(\cdot|\cdot)$ using the probabilities $R(\cdot|\cdot)$, $T(\cdot|\cdot)$, and $U(\cdot|\cdot)$.

3.1.2   Blocking Probability on a Route.

Recall that $S_l^{(i)}(\mathbf{a}|x)$ is the success probability on the sub-route consisting of the last l segments of r_i given that x wavelengths are free on the first hop of the sub-route. According to our numbering of segments on route r_i, the first of the last l segments has index $\kappa_l \stackrel{\text{def}}{=} k_i - l + 2$. To further simplify notation, let us define $\xi \stackrel{\text{def}}{=} v_{j_{\kappa_l}}$ as the last node of segment κ_l. With the above notation, we have the following recursive expression for $S_l^{(i)}(\mathbf{a}|x)$.

$$S_l^{(i)}(\mathbf{a}|x) = \sum_{y=1}^{F} \left\{ \left[W_{\kappa_l}^{(i)}(y|x) - V_{\kappa_l}^{(i)}(s_{\kappa_l}, 0, y|x) \right] \left[\sum_{w=1}^{F} T_\xi^{(i)}(w|y) S_{l-1}^{(i)}(\mathbf{a}|w) \right] \right\}.$$

In the above expression, the term in the first pair of square brackets is the joint probability of success on segment κ_l and y free wavelengths on the last hop of the segment. The term inside the second pair of square brackets is the probability of success on the last $l - 1$ segments given that y wavelengths are free on the last hop the segment κ_l. The starting point of this recursion is $S_0^{(i)}(\mathbf{a}|x) = 1$, for $x = 0, \ldots, F$.

Let $\nu \overset{\text{def}}{=} v_{j_{\alpha-1}+s-1}$ be the node that is reached in $s - 1$ hops from the first node of segment α on route r_i. Then, $V_\alpha^{(i)}(s, n, y|x)$ can be expressed recursively by a simple application of the law of total probability as [Subramaniam et al., 1996]

$$V_\alpha^{(i)}(s, n, y|x) =$$

$$\sum_{x_p=0}^{F} \sum_{x_f=0}^{x_p} \sum_{z=0}^{F} R_\nu^{(i)}(n|x_f, y, z)U_\nu^{(i)}(z|x_p, y)T_\nu^{(i)}(y|x_p)V_\alpha^{(i)}(s-1, x_f, x_p|x)$$

Note that for ease of reading, the dependence on s was omitted in the notation for ν. In the above equation, x_f is the number of wavelengths that are free on all of the first $s - 1$ hops of segment α, x_p is the number of free wavelengths on hop $s - 1$, and z is the number of wavelengths occupied by calls continuing from hop $s - 1$ to hop s. The starting point for this recursion is, from the definition of $V(\cdot|\cdot)$,

$$V_\alpha^{(i)}(1, n, y|x) = \begin{cases} 1 & \text{if } n = x = y, \\ 0 & \text{otherwise.} \end{cases}$$

$W_\alpha^{(i)}(y|x)$ is obtained from $V(\cdot|\cdot)$ as

$$W_\alpha^{(i)}(y|x) = \sum_{n=0}^{F} V_\alpha^{(i)}(s_\alpha, n, y|x).$$

The probabilities that remain to be computed are $Q(\cdot)$, $R(\cdot|\cdot)$, $T(\cdot|\cdot)$, and $U(\cdot|\cdot)$. These probabilities are related to a two-hop sub-route, and a model to compute these probabilities approximately were presented in [Subramaniam et al., 1996] for uniform traffic. We now present that model suitably modified for our case. The reader is referred to Fig. 11.2 during the following discussion.

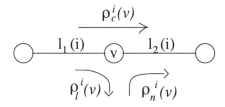

Figure 11.2   Notation for two-hop path.

Let $l_1^{(i)}$ and $l_2^{(i)}$ be the two hops of a sub-route of r_i, and let v be the intermediate node. The *leaving traffic* at node v on route r_i is defined as traffic that uses $l_1^{(i)}$ but does not use $l_2^{(i)}$. Traffic that uses $l_2^{(i)}$ but not $l_1^{(i)}$ is defined as *new traffic* at node v on route r_i, and the traffic that uses both $l_1^{(i)}$ and $l_2^{(i)}$ is defined as *continuing traffic* (or transit traffic).

In order to compute the probabilities $Q(\cdot)$, $T(\cdot|\cdot)$, and $U(\cdot|\cdot)$ for a two-hop segment, we have to obtain the distribution of wavelengths occupied by leaving, continuing, and new traffic on that segment. However, the distribution of wavelengths on the two-hop segment is dependent on the loads offered on that segment by these traffics, which in turn are dependent on the blocking probabilities of calls on all the routes that contribute to the traffic on that segment. It is possible to express this circular dependence between offered traffic on a two-hop segment and the route blocking probabilities in the form of a system of coupled non-linear equations, and solve them numerically in an iterative fashion. To simplify matters, we will not do that; instead, we assume that the effects of route blocking probabilities on the offered traffic on a two-hop segment are negligible.

Let $\rho_l^{(i)}(v)$, $\rho_n^{(i)}(v)$, and $\rho_c^{(i)}(v)$ be the offered loads of leaving, new, and continuing traffic, respectively, at node v on route r_i. Then, given the offered loads on all routes, $\rho_l^{(i)}(v)$, $\rho_n^{(i)}(v)$, and $\rho_c^{(i)}(v)$ can be determined easily as follows.

$$\rho_l^{(i)}(v) = \sum_{\{j: l_1^{(i)} \in r_j,\, l_2^{(i)} \notin r_j\}} \lambda_j,$$

$$\rho_n^{(i)}(v) = \sum_{\{j: l_1^{(i)} \notin r_j,\, l_2^{(i)} \in r_j\}} \lambda_j, \text{ and}$$

$$\rho_c^{(i)}(v) = \sum_{\{j: l_1^{(i)} \in r_j,\, l_2^{(i)} \in r_j\}} \lambda_j,$$

where for a link e and route r, $e \in r$ implies that e lies on r.

Referring to Fig. 11.2, let C_l, C_c, and C_n be the random number of calls belonging to leaving, continuing, and new traffic, respectively, on the two-hop segment at any given time. Therefore, the number of calls that use the first link is $C_l + C_c$ and the number of calls that use the second link is $C_c + C_n$, and both of these are no more than F at any time.

In [Subramaniam et al., 1996], approximate expressions for $Q(\cdot), T(\cdot|\cdot)$, and $U(\cdot|\cdot)$ were obtained by assuming that C_l, C_c, and C_n form a truncated discrete-time Markov chain[3]. This assumption allows us to write the stationary distribution of the number of calls belonging to the three traffic types in product form. The model is approximate because the system is, in fact, not a Markov chain [Zhu et al., 1999]. The number of state variables increases if the system is to be made a Markov chain [Zhu et al., 1999]. In the interest of keeping the model tractable, we employ the said approximation, and as we will see in

Section 4, the approximation does yield reasonably accurate numerical results. The resulting truncated[4] three-dimensional Markov chain is shown in Fig. 11.3.

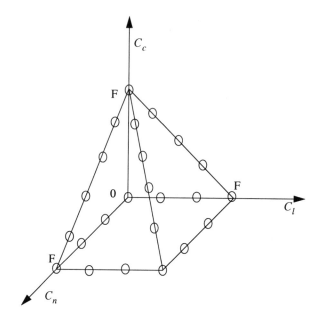

Figure 11.3 The Markov chain for computing the wavelength occupancy distribution on a two-hop segment.

The joint distribution of finding a wavelengths occupied by leaving traffic, b wavelengths occupied by continuing traffic, and c wavelengths occupied by new traffic at node v on route r_i is given by [Bertsekas and Gallager, 1992]

$$\pi_v^{(i)}(a,b,c) = \frac{\dfrac{\left[\rho_l^{(i)}(v)\right]^a}{a!}\dfrac{\left[\rho_c^{(i)}(v)\right]^b}{b!}\dfrac{\left[\rho_n^{(i)}(v)\right]^c}{c!}}{\displaystyle\sum_{z=0}^{F}\sum_{x=0}^{F-z}\sum_{y=0}^{F-z}\dfrac{\left[\rho_l^{(i)}(v)\right]^x}{x!}\dfrac{\left[\rho_c^{(i)}(v)\right]^z}{z!}\dfrac{\left[\rho_n^{(i)}(v)\right]^y}{y!}},$$

for $0 \le a+b \le F$, $0 \le b+c \le F$.

Now, the probabilities $U(\cdot|\cdot)$, $T(\cdot|\cdot)$, and $Q(\cdot)$, can be computed using the above steady-state distribution. Before we do that, we give an expression for $R(n|x,y,z)$, the probability of finding n free wavelengths on both hops of a two-hop segment given that x wavelengths are free on the first, y are free on the second hop, and z are occupied by calls continuing from the first hop to the second. The combinatorial expression is based on the assumption that the random wavelength assignment scheme was used.

$$R(n|x,y,z) = \frac{\binom{x}{n}\binom{F-x-z}{y-n}}{\binom{F-z}{y}}$$

for $\min(x,y) \geq n \geq \max(0, x+y+z-F)$, and is 0 otherwise.

Recalling our definitions of $U(\cdot|\cdot)$, $T(\cdot|\cdot)$, and $Q(\cdot)$, those probabilities are given by the following expressions. The expressions directly follow from the definitions of the probabilities.

$$U_v^{(i)}(z|y,x) = \frac{\pi_v^{(i)}(F-x-z,z,F-y-z)}{\sum\limits_{u=0}^{\min(F-x,F-y)} \pi_v^{(i)}(F-x-u,u,F-y-u)},$$

$$T_v^{(i)}(y|x) = \frac{\sum\limits_{z=0}^{\min(F-x,F-y)} \pi_v^{(i)}(F-x-z,z,F-y-z)}{\sum\limits_{z=0}^{F-x}\sum\limits_{u=0}^{F-z} \pi_v^{(i)}(F-x-z,z,u)},$$

and

$$Q_v^{(i)}(w) = \sum\limits_{z=0}^{F-w}\sum\limits_{y=0}^{F-z} \pi(F-w-z,z,y).$$

This completes the description of the analytical model for computing the route blocking probabilities, and hence, the network-wide blocking probability.

4. NUMERICAL RESULTS

We now present some sample numerical results that demonstrate the accuracy of the model, as well as the effects of various parameters on the blocking performance of wavelength-routing networks. We first start with regular network topologies and uniform traffic (in which an arriving call at a node is equally likely to have any other node as the destination) to see the improvement in blocking performance that wavelength conversion yields. In the graphs, each simulation data point was obtained using 10^6 call arrivals.

First, we plot the call blocking probability against the load per station for a 100-node network when the number of wavelengths per fiber are 5 and 20 in Fig. 11.4(a) and 11.4(b), respectively. Analytical and simulation results are plotted for the case in which there are no converter nodes, and for the case in which there is a converter at every node. In both cases, the converter case curves lie below the corresponding no-converter case curves. The close match between the analytical and simulation results indicates that the model is adequate in analytically predicting the performance of even very sparse networks, in which there is significant correlation between link loads. For comparison, we also plot the analytical results when the link load correlation is not taken into account (the independence model). It is seen that the independence model severely overestimates the blocking probability. We observe that wavelength converters are more useful when the number of wavelengths per fiber is larger and the load is lower. An explanation for this is the following. When the number of wavelengths is larger, blocking occurs primarily not due to a lack of bandwidth (wavelengths) but due to the inability to use the bandwidth efficiently in the absence of conversion. Thus, converters are more useful when the number of wavelengths is larger. Under heavy loads, blocking occurs primarily due to a lack of sufficient number of wavelengths and the presence of converters does not have as much effect as at lighter loads.

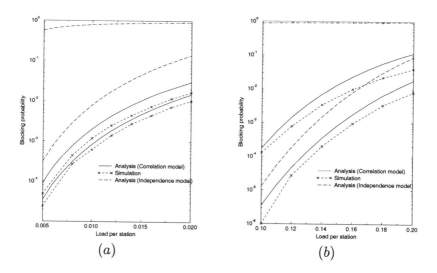

Figure 11.4   The blocking probability vs. the load per station for a 100-node ring network. (a) $F = 5$, and (b) $F = 20$.

Results for a network topology that is richer in connectivity than the ring, namely, the mesh-torus are shown next. In Fig. 11.5, we plot the simulation and analytical (independence and correlation models) results for a 11×11 mesh-torus network with 5 wavelengths per fiber. The results are plotted for a network in which no node has a converter and one in which all nodes have converters. A shortest path was randomly chosen to set up the connection for each arriving lightpath request. The results of our analytical model and the simulation results match very closely. The independence model is less accurate than the correlation model but not significantly so, indicating that the load correlation between successive links is very high in sparse networks and decreases as the network becomes more connected. We observe from the figure the large improvement in performance with wavelength conversion , unlike in the case of the ring. In a mesh-torus with uniform traffic, wavelength usages on the various links are not strongly correlated, despite the fact that path lengths are reasonably large. This is due to the fact that there is more "mixing" of traffic [Barry and Humblet, 1996] at each node in a mesh-torus , when compared to a ring. This leads to the significant performance improvement with wavelength conversion .

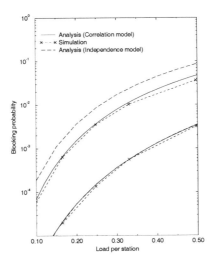

Figure 11.5   The blocking probability vs. the load per station for a 11×11 mesh-torus network with 5 wavelengths per fiber.

It can be shown that the performance improvement with wavelength conversion does not keep increasing with network connectivity. As the network becomes more and more connected, the path lengths will decrease, thereby decreasing the benefits of wavelength conversion . For example, the blocking performance of a fully connected network that uses the direct hop for every call is clearly unaffected by wavelength conversion .

To demonstrate the applicability of the model to irregular topologies with arbitrary given converter locations and non-uniform traffic, we consider the NSFNET T-1 backbone network topology shown in Fig. 11.6. The NSFNET topology has 14 nodes, and hence there are 182 possible ordered node-pairs. We choose a fixed shortest path for each node-pair, and the offered load for each node-pair is made proportional to actual measured packet traffic during a 15-minute time interval [Mukherjee et al., 1996][5]. A scaling factor is used to multiply those values to obtain the actual offered loads, so that the blocking probability values are in the neighborhood of 10^{-3}. We do not present the actual numbers here, and the interested reader is referred to [Mukherjee et al., 1996]. The number of wavelengths on each link is assumed to be 10.

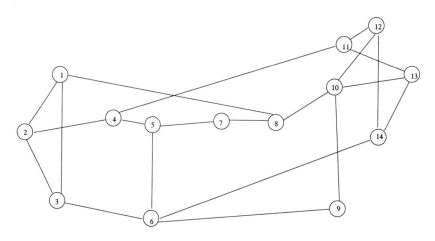

Figure 11.6   The T-1 NSFNET backbone network topology.

In Fig. 11.7, the network-wide blocking probability obtained using the correlation model and simulations is plotted against the scaling factor (and, indirectly, the network load) for two cases: (i) no converters are present; (ii) five converters are present at arbitrarily fixed locations. The close match between the the analytical and simulation results can be observed. The accuracy decreases only for large blocking probability values ($> 10^{-2}$).

While the network-wide average blocking probabilities are reasonably accurately predicted by the model, one is also interested in knowing if the model is good enough to predict individual route blocking probabilities accurately. The blocking probability is plotted for each path in Fig. 11.8 for a constant scaling factor (4.5×10^{-7}). The paths are numbered in ascending order of their blocking probability values obtained using simulations. Note that there are actually 182 paths in the NSFNET and only those paths which yielded a blocking probability of at least 10^{-4} in the simulations are shown in the figure[6].

5.   CONCLUDING REMARKS

We have presented an analytical model for obtaining the blocking performance of wavelength-routing networks with sparse wavelength conversion . The model is efficient enough to be applicable to large network topologies, and

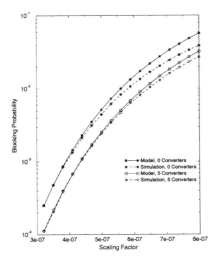

Figure 11.7   Blocking probability vs. the scaling factor in the NSFNET.

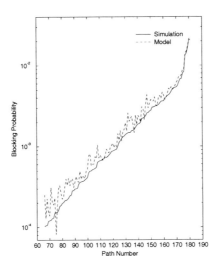

Figure 11.8   The blocking probabilities for various paths in the NSFNET.

at the same time, sophisticated enough to be accurate for sparsely connected networks which tend to have non-negligible link load correlations . The application of the model to several topologies indicates that wavelength conversion plays a significant role in determining blocking performance in some reasonably well-connected topologies (e.g., mesh-torus), whereas it does not affect the performance much in other topologies such as ring and NSFNET.

The following chapter proposes a new dynamic routing scheme, and extends this model to analyze that scheme. For other analytical models, the reader is referred to the references at the end of the chapter. A summary of several analytical models is presented in [Ramamurthy and Mukherjee, 1998]. There is also some work on evaluating the blocking performance under non-Poisson traffic [Subramaniam et al., 1997]. Finally, we reiterate that blocking performance is just one possible performance measure. It is not clear if this is the most relevant measure. Perhaps, in the near-term, the ability of a network to successfully provision lightpaths may be a more important measure. The reader is referred to [Ramaswami and Sivarajan, 1998] and the references therein for other performance criteria and models.

Acknowledgments

The author would like to thank graduate student Amrinder Arora for generating some of the numerical results. This work was partially supported by the NSF under grant ANI-9973111.

Notes

1. In [Subramaniam et al., 1996], the locations of converters were not assumed to be given, and sparse wavelength conversion was modeled by a single parameter called the conversion density of the network. The conversion density is the fraction of network nodes that have conversion capability. The model was a convenient tool to obtain the average blocking performance, without the knowledge of the actual converter locations.

2. Recall that the same wavelength must be used on all the links between two converter nodes, and that a wavelength may be converted to any wavelength at a converter node.

3. The notation used in [Subramaniam et al., 1996] was slightly different.

4. The truncation is due to the fact that $C_l + C_c \leq F$ and $C_c + C_n \leq F$.

5. This is not necessarily indicative of loads in a future wavelength-routing network; it is used here for lack of any projected traffic data and for demonstration purposes. Also, the data in [Mukherjee et al., 1996] had some self-traffic which we have ignored.

6. These simulation values were obtained using 10^8 total call arrivals, and therefore, blocking probability values under 10^{-4} may not be accurate.

References

Barry, R. A. and Humblet, P. A. (1996). Models of blocking probability in all-optical networks with and without wavelength changers. *IEEE J. Sel. Areas Comm.*, 14(5):858–867.

Barry, R. A. and Marquis, D. (1995). Evaluation of a model of blocking probability in all-optical mesh networks without wavelength changers. In *All-Optical Communication Systems: Architecture, Control, and Network Issues, Proc. SPIE*, volume 2614, pages 154–167.

Bertsekas, D. and Gallager, R. (1992). *Data Networks*. Prentice Hall.

Birman, A. (1996). Computing approximate blocking probabilities for a class of all-optical networks. *IEEE J. Sel. Areas Comm.*, 14(5):852–857.

Girard, A. (1990). *Routing and Dimensioning in Circuit-Switched Networks*. Addison-Wesley.

Goodman, M. S. (1989). Multiwavelength networks and new approaches to packet switching. *IEEE Communications Magazine*, 27(10):27–35.

Kovačević, M. and Acampora, A. S. (1996). Benefits of wavelength translation in all-optical clear-channel networks. *IEEE J. Sel. Areas Comm.*, 14(5):868–880.

Mukherjee, B., Banerjee, D., Ramamurthy, S., and Mukherjee, A. (1996). Some principles for designing a wide-area WDM optical network. *IEEE/ACM Trans. Networking*, 4(5):684–696.

Ramamurthy, B. and Mukherjee, B. (1998). Wavelength conversion in WDM networking. *IEEE J. Sel. Areas Comm.*, 16(7):1061–1073.

Ramaswami, R. and Sasaki, G. H. (1997). Multiwavelength optical networks with limited wavelength conversion. In *Proc. INFOCOM '97*, pages 489–498.

Ramaswami, R. and Sivarajan, K. N. (1998). *Optical Networks: A Practical Perspective*. Morgan Kaufmann.

Sharma, V. and Varvarigos, E. A. (1998). Limited wavelength translation in all-optical WDM mesh networks. In *Proc. INFOCOM*, pages 893–901.

Subramaniam, S., Azizoğlu, M., and Somani, A. K. (1996). All-optical networks with sparse wavelength conversion. *IEEE/ACM Trans. Networking*, 4(4):544–557.

Subramaniam, S., Somani, A. K., Azizoğlu, M., and Barry, R. A. (1997). A performance model for wavelength conversion with non-Poisson traffic. In *Proc. INFOCOM*, pages 499–506.

Tripathi, T. and Sivarajan, K. N. (1999). Computing approximate blocking probabilities in wavelength routed all-optical networks with limited-range wavelength conversion. In *Proc. INFOCOM*, pages 329–336.

Yates, J., Lacey, J., Everitt, D., and Summerfield, M. (1996). Limited-range wavelength translation in all-optical networks. In *Proc. INFOCOM*, pages 954–961.

Zhu, Y., Rouskas, G. N., and Perros, H. (1999). Blocking in wavelength routing networks, part I: The single path case. In *Proc. INFOCOM*, pages 321–328.

Chapter 12

DYNAMIC WAVELENGTH ROUTING TECHNIQUES AND THEIR PERFORMANCE ANALYSES

Ling Li
Department of Electrical and Computer Engineering
Iowa State University, Ames, IA 50011
lingli@iastate.edu

Arun K. Somani
Department of Electrical and Computer Engineering
Iowa State University, Ames, IA 50011
arun@iastate.edu

Abstract   The performances of fixed-path routing in WDM networks have been studied in the previous chapter. We present two dynamic routing algorithms based on path and neighborhood link congestion in all-optical networks in this chapter. We first introduce fixed-paths least-congestion (FPLC) routing in which the shortest path may not be preferred to use. We then extend the algorithm to develop a new routing method: dynamic routing using neighborhood information. It is shown by using both analysis and simulation methods that FPLC routing with the first-fit wavelength assignment method performs much better than fixed-path and alternate-path routing methods in mesh-torus networks (regular topology) and in the NSFnet T1 backbone network (irregular topology). Routing using neighborhood information also achieves good performance when compared to alternate-path routing.

1.   INTRODUCTION

One of the main requirements in all-optical WDM networks is that the same wavelength must be assigned to a connection on every link on a path if wavelength converters are not available at the switching nodes. A connection request encounters higher blocking probability than it does in circuit-switched networks because of the wavelength continuity constraint. Therefore, routing and wavelength assignmentalgorithms play a key role in improving the performance of WDM networks [Mukherjee, 1997; Ramaswami and Sivarajan, 1998]. The performances of fixed-path routing in WDM networks have been studied in the previous chapter and in [Li and Somani, 1999d; Kovačević and Acampora, 1996; Barry and Humblet, 1996]. Many researchers have also proposed the use of alternate-path routing algorithms [Harai et al., 1997; Karasan and Ayanoglu, 1998; Ramamurthy and Mukherjee, 1998; Li and Somani, 1999c]. Since the paths are statically computed and an attempt is made to set up a connection request on fixed paths without acquiring the information of current network status, it is not possible to further improve the network performance in terms of blocking probability by using these routing approaches.

Dynamic routing approaches are more efficient than static routing methods [Harai et al., 1997; Karasan and Ayanoglu, 1998; Ramamurthy and Mukherjee, 1998; Banerjee and Mukherjee, 1996; Lowe and Hunter, 1997; Li and Somani, 1999a; Li and Somani, 1999b]. The main problems of dynamic routing methods are longer setup delays and higher control overheads including, in some cases, the introduction of a central control node that keeps track of the network's global state. However, simulation results show that the dynamic routing method can significantly improve the network performance compared to the fixed-path routing and the alternate-path routing. Routing and wavelength assignments are considered jointly and adaptively in [Banerjee and Mukherjee, 1996; Mokhtar and Azizoğlu, 1998]. All feasible paths between a source-destination pair are computed and one of them is selected according to a specific criterion to set up a request. Least loaded routing (LLR) algorithms as dynamic routing methods are introduced in [Birman, 1996; Chan and T.P.Yum, 1994], and their performances for fully connected networks are investigated.

In this chapter, we consider alternate dynamic routing algorithms in all-optical networks. We first introduce a dynamic routing algorithm, called fixed-paths least-congestion routing (FPLC), and compare its performance to that of the fixed-path routing and the alternate-path routing. This algorithm routes a connection request on the least congested path out of a set of predetermined paths. A set of routes[1] connecting the source-destination pair are searched in parallel and the route with the maximum number of idle wavelengths is selected to set up the connection. If a request cannot be accommodated by any of the routes, it is blocked.

The FPLC still has higher setup delay and higher control overhead. To overcome these shortcomings, a new routing method using neighborhood information is also investigated in this chapter. In this method, for each source-destination pair, a set of preferred paths are pre-computed and stored at the source. Moreover, instead of searching all the links on the preferred routes for availability of free wavelengths, only the first k links on each path are searched. A route is selected based on the availability of free wavelengths on the first k links on the preferred paths. If several free wavelengths are available on the selected route, a wavelength is selected according to a pre-specified wavelength assignment algorithm. If no free wavelengths are available on the first k links of all the preferred routes, the request is blocked. An essential observation here is that the parameter k depends on the diameter and topology of the network and the network performance requirement. It is shown that a value of $k = 2$ is generally enough to ensure good network performance in a 4×4 mesh-torus network and in the NSFnet T1 backbone network.

Wavelength assignment is another unique problem of optical networks. Algorithms for improving network performance using different information are proposed in [Karasan and Ayanoglu, 1998; Zhang and Acampora, 1995; Subramaniam and Barry, 1997]. In [Subramaniam and Barry, 1997], a dynamic wavelength assignment algorithm, the Max_Sum ($M\Sigma$), is proposed and compared to other algorithms, i.e., the First-Fit (FF), the Most-Used (MU), the Min-Product (MP), the Least Loaded (LL) and the Random (R). However, global information is required in most of these algorithms. Since distributed algorithms are considered in this chapter, two general wavelength assignment methods are examined:

1 Random Assignment: The assigned wavelength is selected uniformly randomly from the available wavelengths. The approximate analyses and the corresponding simulations in this chapter are based on using this method.

2 First-fit Assignment: The wavelength is assigned according to a predefined order [Kovačević and Acampora, 1996]. We assume that the wavelengths are indexed and the free wavelength with the smallest index is selected.

Several approximate analytical methods on the blocking probabilities of networks are proposed in the literature. In [Kovačević and Acampora, 1996; Barry and Humblet, 1996], two models to compute the approximate blocking probability are presented. However, these models are inappropriate for networks with sparse topologies because they do not consider the correlation among the use of wavelengths between successive links of a path. An improved model with the consideration of this dependence is proposed in [Barry and Marquis, 1995]. A Markov chain based reduced load model with state-dependent arrival rate is presented in [Birman, 1996]. A more accurate model with modest complexity in [Subramaniam et al., 1996] accounts for link load correlation. As pointed out in [Zhu et al., 1999], the MC model is an approximate model because the arrival rates vary with the state of the Markov chain. We use and extend the link load correlation model to analyze the performances of the FPLC and the neighborhood-information-based routing.

2.   FIXED-PATHS LEAST-CONGESTION ROUTING

In the fixed-paths least-congestion routing, a set of routes to be used for each source-destination pair in a network is first statically computed and stored at each source node.   Two edge-disjoint shortest paths for each source-destination pair are used in the analysis and simulation. We refer to these routes as the first and second route. The length of the first route is less than or equal to that of the second route. If two routes have the same number of idle wavelengths, the first route is selected to set up the request. The number of preferred routes are restricted to two because network resources cannot be used efficiently if many longer routes are allowed in the network. One reason that the two routes are required to be edge-disjoint is because we search two paths in parallel. Another consideration is fault tolerance. If one path fails, the connection can be re-routed to another path [Ramamurthy and Mukherjee, 1999a; Ramamurthy and Mukherjee, 1999b; Mohan and Somani, 2000; Li, 1999]. We also notice that using more than two paths does not significantly improve performance [Karasan and Ayanoglu, 1998; Ramamurthy and Mukherjee, 1998].

Distributed network control for WDM networks is discussed in [Ramaswami and Segall, 1996; Mei and Qiao, 1997]. Two reservation protocols, source initiated reservation (SIR) and destination initiated reservation (DIR) are introduced and compared in [Mei and Qiao, 1997]. As the names of the protocols suggest, the SIR reserves resources when a setup request is sent from the source to the destination. The DIR reserves resources when a setup request is accepted by the destination and an ack-packet is sent back to the source. The results show that DIR performs better than SIR when the propagation delay is small. However, the DIR cannot improve throughput when the propagation delay is large and no wavelength converter is available at the network. The control protocol for our FPLC routing algorithm is similar to DIR. However, to help the destination to make a correct decision in the absence of wavelength converters, we also include information in the request message going towards the destination regarding how many different requests may be competing for a wavelength on the same link. For this purpose, we assume that each node keeps a counter for each wavelength on the links connected to this node. When a wavelength is requested by a wavelength-searching message from a source node, the corresponding counter is increased by one before the request is forwarded. This counter information is added to the searching message as a hint to indicate the probability of other paths using this wavelength. This hint can be used by the wavelength assignment algorithm to avoid a possible duplicate use of a wavelength.

The FPLC routing algorithm consists of two steps:

- Step 1. Upon arrival of a connection setup request destined to node d, node i performs the following steps:

  1 If i is the source node, search the available number of wavelengths on two routes in parallel by sending *needle packets* [2] requesting a path setup towards the next nodes on the two routes. The counters of the available wavelengths on the outgoing link, S_λ, are increased by one, and the path information is also included in the *needle packet*.

  2 If i is an intermediate node, compute $S_\lambda = S_\lambda \cap S_{available}$. If $S_\lambda \neq \phi$, increase the counters for these wavelengths, add the counter information to the packet, and forward it to the next node. If $S_\lambda = \phi$, node i responds to the source node with a nack-packet, but also continues sending a nack-packet indicating no free wavelength available on the route to the destination node for a resolution.

  3 If $i = d$, this request has found the route to the destination. Wait for the second request to arrive. Select the route that has the maximum number of free wavelengths available and one wavelength from the available wavelengths on the route according to the wavelength assignment method to set up the request. If two routes have the same number of idle wavelengths, the first route is selected. Reserve the selected wavelength for the request, then send an ack-packet to the source node on the selected route and a nack-packet to the source node on other routes (it is not necessary to send back any packet on the route on which a nack-packet is received, as a nack-packet has already been sent back by an intermediate node).

- Step 2. Upon arrival of a responding packet on the return path, node i performs the following steps:

  1 If i is an intermediate node, reserve the selected wavelength for the request, decrease the wavelengths' counters related to the request, and forward the packet towards the source node. If the selected wavelength has been reserved by another request, send a nack-packet to the source node. The reserved wavelength on previous links from the destination node is released after a short period of time if no request is set up on it.

  2 If $i = s$, the source node is reached. The connection is set up on the selected route using the selected wavelength. If no wavelength is available on any of the two routes, the connection request is blocked

2.1.   ANALYSIS OF THE FPLC ROUTING

We assume that call requests arrive at each source-destination pair according to a Poisson process with rate λ. The destination of a call is uniformly distributed to other nodes. Call holding time is exponentially distributed with mean $(1/\mu)$. The number of wavelengths, F, is the same on all links. Wavelengths are randomly assigned to a session from the set of free wavelengths on the associated path. To make the analysis tractable, we also assume that the wavelength searching and reservation time is negligibly small.

2.1.1   Link Load Correlation Model.   The link load correlation model has been introduced in the previous chapter. The basic idea of the model is that the blocking probability on a two-hop path can be computed with the consideration of link load correlation. Then the blocking probability on a l-hop path can be computed recursively by viewing the first $l-1$ hops as the first hop and the lth hop as the second hop of a two-hop path. The following steady-state probabilities are defined:

- $S(y_f|x_{pf}) = \Pr\{y_f$ wavelengths are free on a link of a path $\mid x_{pf}$ wavelengths are free on the previous link of a path$\}$.

- $U(z_c|y_f, x_{pf}) = \Pr\{z_c$ calls (wavelengths) continue to the current link from the previous link $\mid y_f$ wavelengths are free on the current link, and x_{pf} wavelengths are free on the previous link$\}$.

- $R(n_f|x_{ff}, y_f, z_c) = \Pr\{n_f$ wavelengths are free on a two-hop path $\mid x_{ff}$ wavelengths are free on the first hop, y_f wavelengths are free on the second hop, and z_c calls use both hops$\}$.

- $T_P^{(l)}(n_f, y_f) = \Pr\{n_f$ wavelengths are free on an l-hop path P and y_f wavelengths are free on hop $l\}$.

$T_P^{(l)}(n_f, y_f)$ is given by [Subramaniam et al., 1996]

$$T_P^{(l)}(n_f, y_f) = \sum_{x_{pf}=0}^{F} \sum_{x_{ff}=0}^{F} \sum_{z_c=0}^{\min(F-x_{pf}, F-y_f)} R(n_f|x_{ff}, z_c, y_f)$$

$$U(z_c|y_f, x_{pf})S(y_f|x_{pf})T_P^{(l-1)}(x_{ff}, x_{pf}). \tag{12.1}$$

Let $Q_P(w_f)$ be the probability that w_f wavelengths are free on a path P with length $l(P)$. $Q_P(w_f)$ becomes

$$Q_P(w_f) = \sum_{y_f=0}^{F} T_P^{(l(P))}(w_f, y_f). \tag{12.2}$$

A fundamental assumption made in the correlation model is that the path used by a call does not depend on the state of the links on the path. For a fixed shortest-path routing on regular networks, it is possible to assume that the effect of blocking probability on the carried load can be neglected, and the arrival rate on each link is the same for keeping the analysis simple. However, these assumptions become invalid when the FPLC is used. In this case, a path for a request is selected using the current network status. Thus the arrival rate on each link is continuously changing. No steady state may be reached in the strict sense when the FPLC is used. We propose to use a technique based on the Erlang Fixed-Point method for Alternate routing [Girard, 1990] to solve this problem. We need the following further notations:

- Let $R_j^{(1)}$ be the set of first shortest routes that employ link j, and $R_j^{(2)}$ be the set of second shortest routes that employ link j.

- Let $R_{i,j}^{(1)}$ be the set of first shortest routes that have a subset of route from link i to j. $R_{i,j}^{(2)}$ is similarly defined for the set of second shortest routes.

- Let $Pr(P_\alpha^1)$ and $Pr(P_\alpha^2)$ be the probabilities that a call for a source-destination pair α is set up on the first and second path, respectively.

In the FPLC, a call request is set up on the first shortest path if the number of free wavelengths on the second shortest path is less than the number of free wavelengths on the first shortest path. Otherwise, it is set up on the second shortest path assuming that the path has at least one free wavelength. Therefore,

$$Pr(P_\alpha^1) = \sum_{i=1}^{F} Q_{P_\alpha^1}(i)(\sum_{n=0}^{i} Q_{P_\alpha^2}(n)), \qquad (12.3)$$

$$Pr(P_\alpha^2) = \sum_{i=1}^{F} Q_{P_\alpha^2}(i)(\sum_{n=0}^{i-1} Q_{P_\alpha^1}(n)). \qquad (12.4)$$

Recall that λ is the call arrival rate at each node. The arrival rate of calls that enter at link i and continue to link j, $\rho_c(i,j)$, becomes

$$\rho_c(i,j) = \sum_{P_j \in R_{i,j}^1} \lambda Pr(P_j) + \sum_{P_j \in R_{i,j}^2} \lambda Pr(P_j). \qquad (12.5)$$

The arrival rate of calls that leave from link i, $\rho_l(i)$, includes calls that use link i as the first or second route, but do not continue to link j,

$$\rho_l(i) = \sum_{P_i \in R_i^1} \lambda Pr(P_i) + \sum_{P_i \in R_i^2} \lambda Pr(P_i) - \rho_c(i,j). \qquad (12.6)$$

The arrival rate of calls that enter at link j, $\rho_e(j)$, includes calls that use link j as the first or second route, but do not include calls that continue from link i to link j. $\rho_e(j)$ is obtained by

$$\rho_e(j) = \sum_{P_j \in R_j^1} \lambda Pr(P_j) + \sum_{P_j \in R_j^2} \lambda Pr(P_j) - \rho_c(i,j). \qquad (12.7)$$

Given the arrival rates to each link, the conditional probabilities $S(y_f|x_{pf})$, $U(z_c|y_f, x_{pf})$, and $R(n_f|x_{ff}, y_f, z_c)$ can be derived (see [Subramaniam et al., 1996] for details). Let L_α be the blocking probability for source-destination pair α. L_α can be derived from Eqs. (12.1) and (12.2) as

$$L_\alpha = Q_{P_\alpha^1}(0) \times Q_{P_\alpha^2}(0). \qquad (12.8)$$

Let J be the number of links in a network. Let ϵ be a small positive number that is used as convergence criterion. The algorithm given below iteratively computes the approximate blocking probabilities for the traffic on all the routes.

1 Initialization. For each source-destination pair α let $\bar{L}_\alpha = 0$. Choose $\rho_e(i)$, $\rho_c(i,j)$, and $\rho_l(i)$, $i,j = 1, \ldots, J$ arbitrarily for all links.

2 Calculate $Q_P(w_f)$ for every path of each source-destination pair using Eqs. (12.1) and (12.2).

3 Calculate the blocking probability L_α for every source-destination pair α using Eq. (12.8). If $\max_\alpha |L_\alpha - \bar{L}_\alpha| < \epsilon$ then terminate, otherwise let $\bar{L}_\alpha = L_\alpha$, go to next step.

4 Calculate ρ_e, ρ_l, and ρ_c for each link using Eqs. (12.5), (12.6) and (12.7), then go back to step 2.

Since the arrival rate for each link can be computed individually, this method is also suitable for analysis of irregular networks. The method is applicable to alternate routing approaches with small modifications of Eqs. (12.3) and (12.4).

2.1.2   Reduced Load Model.   The reduced load model proposed in [Harai et al., 1997; Birman, 1996] is also extended to the FPLC. We present the model here for comparison with the link load correlation model. The results are compared and discussed in the next subsection.

The arrival rate on each link for different link status is considered in the reduced load model. We assume that the inter-arrival time on link j be exponentially distributed with parameter $\alpha_j(m_j)$ when link j has m_j free wavelengths.

Let X_j be a random variable representing the number of wavelengths available on link j in a steady state. The idle capacity distribution on link j is denoted by

$$q_j(m_j) = Pr[X_j = m_j], m_j = 0, \ldots, F .$$

Following [Harai et al., 1997; Birman, 1996; Kovačević and Acampora, 1996; A.Kashper and Ross, 1993], we also assume that X_j's are independent. Thus X_j can be viewed as a birth-and-death process. Using the result of [Birman, 1996], the idle capacity distribution of X_j is given by

$$q_j(m_j) = \frac{F(F-1)\ldots(F-m_j+1)}{\alpha_j(1)\alpha_j(2)\ldots\alpha_j(m_j)} \qquad (12.9)$$

where

$$q_j(0) = \left[1 + \sum_{m_j=1}^{F} \frac{F(F-1)\ldots(F-m_j+1)}{\alpha_j(1)\alpha_j(2)\ldots\alpha_j(m_j)}\right]^{-1} . \qquad (12.10)$$

Let the link set of P_j be $\{j, j_1, j_2, \ldots, j_{h(P_j)-1}\}$. Link numbers in the set are not in order. Let $p_i(.)$ be the probability that i wavelengths are available on route P_j given that each link has $m_j, m_{j_1}, \ldots, m_{j_{h(P_j)-1}}$ free wavelengths, respectively. Then,

$$p_i(m_{j_1}, \ldots, m_{j_n}) = \sum_{k=0}^{F} p_i(k, m_{j_n})p_k(m_{j_1}, \ldots, m_{j_{n-1}}) \qquad (12.11)$$

where $p_i(x, y)$ is the probability that there are i free wavelengths on two-hop path on which each link has x and y wavelengths available, respectively. From [Kovačević and Acampora, 1996], this conditional probability is

$$p_i(x, y) = \frac{\dbinom{x}{i}\dbinom{F-x}{y-i}}{\dbinom{F}{y}} .$$

$u_i(m_j; P_j)$, the probability of i free wavelengths available on route P_j given that m_j wavelengths are idle on link j, is given by

$$u_i(m_j; P_j) = \sum_{m_{j_1}=i}^{F}\sum_{m_{j_2}=i}^{F} \cdots \sum_{m_{j_{h(P_j)-1}}=i}^{F} q_{j_1}(m_{j_1})q_{j_2}(m_{j_2}), \ldots,$$

$$\times q_{j_{h(P_j)-1}}(m_{j_{h(P_j)-1}})p_i(m_j, m_{j_1}, m_{j_2}, \ldots, m_{j_{h(P_j)-1}}).$$

The probability of i wavelengths available on route P_j is given by

$$v_i(P_j) = \sum_{m_j=i}^{F} q_j(m_j)u_i(m_j; P_j).$$

Let $X_{P_j^1}$ and $X_{P_j^2}$ be the random variables representing the number of idle wavelengths on routes P_j^1 and P_j^2, respectively. The inter-arrival time distribution parameter on link j, $\alpha_j(m_j)$, is given by

$$\alpha_j(m_j) = \sum_{P_j^1 \in R_j^1} \lambda(\sum_{i=1}^{m_j} u_i(m_j; P_j^1)Pr[X_{P_j^2} \leq i])$$

$$+ \sum_{P_j^2 \in R_j^2} \lambda(\sum_{i=1}^{m_j} u_i(m_j; P_j^2)Pr[X_{P_j^1} < i]) \qquad (12.12)$$

where $Pr[X_{P_j^1} \leq i])$ denotes the probability that the number of idle wavelengths on route P_j^1 is less than or equal to i and it is given by

$$Pr[X_{P_j^1} \leq i] = 1 - \sum_{k=i+1}^{F} v_k(P_j^1).$$

Similarly,

$$Pr[X_{P_j^2} < i] = 1 - \sum_{k=i}^{F} v_k(P_j^2).$$

In the formulation of Eq. (12.12), the first term is for selecting the first route from set R_j^1 that has i idle wavelengths and P_j^2 (the corresponding alternate path) has less than or equal to i free wavelengths; second term is for selecting the route from set R_j^2 where the corresponding alternate path has less than i free wavelengths.

The blocking probability L_a becomes

$$L_a = Pr[X_{P_a^1} = 0]Pr[X_{P_a^2} = 0] \qquad (12.13)$$

where P_a^1 and P_a^2 denote the first and second shortest routes connecting the source-destination pair a.

Due to a circular relationship between $q_j(m_j)$ and $\alpha_j(m_j)$, it is not easy to compute L_a. Therefore, we again fall back on an iterative algorithm to compute the approximate blocking probabilities.

1 Initialization. For all source-destination pair a let $\bar{L}_a = 0$. Choose $\alpha_j(.), j = 1, \ldots, J$ arbitrarily for all links.

2 Determine $q_j(.), j = 1, \ldots, J$, using Eqs.(12.9) and (12.10).

3 Calculate $\alpha_j(.), j = 1, \ldots, J$, using Eq.(12.12).

4 Calculate the blocking probability L_a for all source-destination pair a using Eq.(12.13). If $\max_a |L_a - \bar{L}_a| < \epsilon$ then terminate, otherwise let $\bar{L}_a = L_a$, go to step 2.

2.2. NUMERICAL RESULTS AND DISCUSSION

The analytical methods of the previous subsection are used here to calculate blocking probabilities and are compared to simulation results. The physical network, a mesh-torus network with 16 nodes, is depicted in Fig. 12.1. All links are bi-directional. The analytical results are then compared to blocking probabilities obtained by simulation.

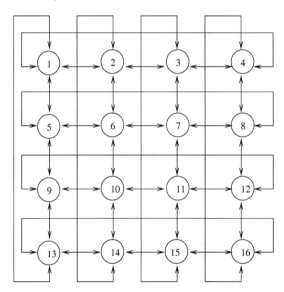

Figure 12.1 A mesh-torus network with 16 nodes.

Two edge-disjoint shortest paths are required for each source-destination pair in the network. Different route selection methods will yield different network performance because some links maybe over-used and others maybe under-used if route selection is not symmetric. Since uniform traffic is assumed and the arrival rate to all source-destination pairs is assumed to be identical, we try to distribute traffic evenly on each link by using x-y and y-x routing. In x-y and y-x routing, each node is denoted by an integer pair (x, y). The rules to determine the first path and the second path for a connection setup request from (x_1, y_1) to (x_2, y_2) is as follows:

1 The first path first uses x-links connecting (x_1, y_1) to (x_2, y_1) and then uses y-links connecting (x_2, y_1) to (x_2, y_2).

2 The second path first uses y-links connecting (x_1, y_1) to (x_1, y_2) and then uses x-links connecting (x_1, y_2) to (x_2, y_2).

3 If two nodes are on the same row (or column), the shortest path on the row (or column) is selected as the first path. The second path is the shortest of the following two possible paths. If two possible paths have the same length, one of them is randomly selected as the second path.

   (a) The path in reverse direction of the first path, i.e., $(x_1, y_1) \to ((x_1 - 1) \bmod N, y_1) \to \ldots (x_2, y_2)$.

   (b) Go to the next column (or row) first, then go to the destination using shortest path, i.e., $(x_1, y_1) \to (x_1, (y_1 + 1) \bmod N) \to \ldots (x_2, y_2)$.

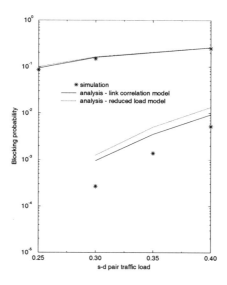

Figure 12.2   Blocking probability versus traffic load in a 4×4 mesh-torus network.

Fig. 12.2 shows the network blocking probability versus the traffic load per source-destination pair for a given number of wavelengths on each link, $F = 4$ and $F = 8$. In this figure, curves are for analytical results and points are for simulation results. In the approximate analysis, iteration is required and the convergence criteria is set to be 10^{-6} for the blocking probabilities. Each data point in the simulations is obtained using 10^7 call arrivals. The same criteria are set for all the analysis and simulation results in this chapter. Note that for the same source-destination pair traffic load, if F is small then the traffic load per wavelength is heavy and vice-versa. From the figure, we observe that analytical results are in good agreement with simulation results for heavy to moderate traffic ($F = 4$) for both of the models. However, the blocking probabilities are slightly overestimated for light traffic ($F = 8$). The analytical results are not very accurate when traffic load is extremely light ($F = 8$ and source-destination pair traffic load ≤ 0.30). In any case, the link load correlation model always gives more accurate results compared to the reduced load model. It is also faster to compute as the results can be obtained in a few minutes, whereas the reduced load model took several hours to analyze a 4×4 mesh-torus network on a sun Ultra-1 workstation. Therefore we use only the link correlation model in the following sections.

2.3. PERFORMANCE COMPARISON

In this subsection, we compare three different routing methods based on our analytical models and simulation results. The 4×4 mesh-torus network topology depicted in Fig. 12.1 is used. The following routing methods are employed for comparison purposes.

1 Shortest Path routing (SP): This method selects any of the free wavelengths available on the path according to the specified wavelength assignment algorithm. A connection request is blocked if there is no idle wavelength available on the path.

2 Alternate Shortest Path routing (ASP): This method first tries to set up a connection request on the first shortest path. If blocked on the first shortest path, the request is overflowed to the second path. The request is blocked if it cannot be set up on either of the two paths.

3 Fixed-Paths Least-Congestion routing (FPLC): Two preselected shortest paths are searched in parallel and the path with the maximum number of free wavelengths is employed to set up the connection. If no wavelength is idle on the two paths, the request is blocked.

Fig. 12.3, 12.4 and 12.5 show the performance results for the three different routing methods. We plot average link utilization for each link, average wavelength utilization, and average blocking probability in these figures, respectively. The average link utilization shown in Fig. 12.3 is computed using the following equation,

$$\overline{L_u}(j) = \sum_{w=0}^{F} (F - w) q_j(w).$$

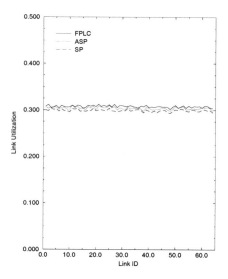

Figure 12.3   Average link utilization for different routing methods.

The average wavelength utilization illustrated in Fig. 12.4 is defined as the average probability that a link has w busy wavelengths,

$$\overline{W_u}(w) = \frac{\sum_{j=1}^{J}(1 - q_j(w))}{J} .$$

For Fig. 12.3 and Fig. 12.4, the corresponding traffic load for each source-destination pair is 0.3 and the number of wavelengths is $F = 8$. We observe from Fig. 12.3 that the FPLC approach slightly increases the average link utilization compared to the ASP and the SP. The reason for this increase is that this algorithm does not use the shortest paths all the time. In the FPLC, longer paths are allowed to be used even when there are idle wavelengths on the shortest path. Another, and more important, reason is that more connection requests are accommodated by using the FPLC than the ASP and the SP as the blocking probability is small. This can be observed from Fig. 12.5 (F=8) where the FPLC improves the blocking probability when the traffic load is low. As the traffic load increases, the performance of the FPLC and ASP are close.

One of the reasons that the FPLC performs better than the ASP and the SP can be seen from Fig. 12.4. The horizontal axis in the figure shows the number of busy wavelengths on a link. The vertical axis shows the probability that these many wavelengths on an average are busy. An ideal routing method should produce a curve with a sharp peak depicting better distribution of traffic load. Fig. 12.4 shows that the FPLC tends to get a sharper peak compared to the ASP and the SP.

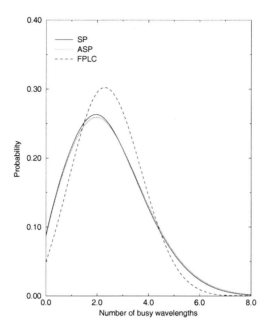

Figure 12.4   Average probability of a link with w wavelengths being busy versus w for different routing methods.

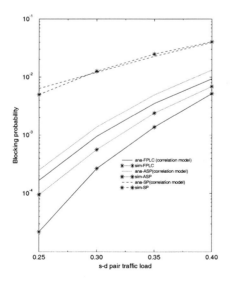

Figure 12.5   Performance comparison for three routing methods.

We also notice that in the FPLC, the network performance can be significantly improved by using the first-fit wavelength assignment compared to the random wavelength assignment. Fig. 12.6 shows the difference of the ASP and the FPLC using different wavelength assignment methods through simulation.

As reported in [Karasan and Ayanoglu, 1998; Somani and Azizoğlu, 1997], compact wavelength assignment methods (i.e. most-used) perform better than spread wavelength assignments. The first-fit wavelength assignment strategy can slightly improve the blocking probability in the ASP [Harai et al., 1997]. While using the FPLC, the first-fit wavelength assignment performs much better than the random strategy under light load condition. The reason can be explained intuitively as follows: In the FPLC, a request is set up on the least congested route. The wavelengths with lower indexes have more chances to be employed when compared to the ASP. Using the first-fit wavelength assignment , this compact wavelength assignment effect is enhanced. For example, two wavelengths (λ_1 and λ_2) are used on the first shortest path; λ_3 may be used to set up a request on the first shortest path for ASP. However, in the FPLC using first-fit wavelength assignment, λ_1 and λ_2 still have higher chances to be used to set up a request on the second shortest path. This compact wavelength assignment effect leaves more free wavelengths in the network. Thus the performance is improved. In mesh-torus networks, it seems that the second shortest path for a source-destination pair may not be longer than the first shortest path. However, among the 240 first shortest paths in a 4×4 mesh-torus network, 64 of them are 1-hop long. The corresponding second shortest paths are 3-hop long. We also show in Section 4 that the FPLC achieves better performance than the SP and the ASP in the NFSNET T1 backbone in which the second shortest path could be 4-hops longer than the first shortest paths.

The simulation results also show that more sophisticated wavelength assignment algorithms, such as using the most-used wavelength assignment, do not significantly improve the network performance in FPLC. Since accurate global state of a large network is hard to obtain, the first-fit wavelength assignment is a good candidate to assign wavelengths in a distributed routing network.

3. ROUTING USING NEIGHBORHOOD INFORMATION

In the previous section we notice that the fixed-paths least-congestion routing method improves network performance compared to static routing approaches. However, problems still exist when using this dynamic routing method in a large network. Although we use a distributed parallel algorithm to overcome some of the problems, the main difficulties remaining are the control overhead, setup delay, and possible conflicts in wavelength usage when multiple paths are being set up simultaneously on one link. In this section, we introduce

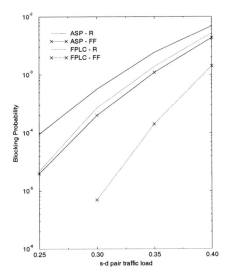

Figure 12.6 Performance comparison for the ASP and the FPLC using different wavelength assignment methods.

a new routing algorithm using neighborhood information in which the setup efficiency and lower control overhead of static routing versus the low blocking probability trade-off is introduced. Using analytical and simulation methods, it is shown that routing using neighborhood information can achieve good performance compared to static routing approaches which overcome some of the difficulties. Using this approach also reduces the possible conflicts in wavelength usage.

In the neighborhood-information-based routing, similar to the FPLC, a set of edge-disjoint shortest paths are statically computed and stored at each node. Upon the arrival of a connection request, the source node first decides on which route to set up the request. This route selection process is similar to that in the FPLC. However, instead of searching all the links on the preferred paths, the source node s collects and analyzes the neighborhood information on two edge-disjoint routes for the first k links only. There are two possibilities for collecting the neighborhood information: it can be either collected when needed or exchanged periodically. The number of free wavelengths on the two paths up to distance k are compared, and a path that has the maximum number of idle wavelengths is selected. If no common wavelength is free on any of the two routes up to distance k, the request is blocked.

After a route is selected, the source node s then initiates a wavelength assignment process by sending a *needle packet* to the next node on the selected path. A set of wavelengths, S_λ, which are available on the outgoing link, and the path information are included in the *needle packet*. Each intermediate node i computes $S_\lambda = S_\lambda \cap S_{available}$. If $S_\lambda \neq \phi$, it increases the requested-wavelength counter and forwards the packet to the next node on the route. The destination node selects a wavelength from the set of available wavelengths according to the pre-specified wavelength assignment methods and assigns it to the request. It responds to the source node with an ack-packet and the selected wavelength. The counters are decreased when a response packet on the path is returned, as is the case in the FPLC routing. If $S_\lambda = \phi$, the intermediate node responds to the source node with a nack-packet.

If no wavelength is available on the selected route, the connection request is blocked. The counters related with the request are decreased by one when a response packet is received by an intermediate node. This routing algorithm is called $FPLC - N(k)$.

3.1.   ANALYSIS OF THE FPLC-N(K)

We extend the approximate analytical models discussed in the previous section to the FPLC-N(k). We use the same terminology as before.

Let $\widehat{Q}_{P_j}^k(w)$ present the probability that there are w free wavelengths available on route P_j up to neighborhood distance k. $\widehat{Q}_{P_j}^k(w)$ is given by

$$\widehat{Q}_{P_j}^k(w) = \sum_{y_f=0}^{F} T_{P_j}^{(k)}(w, y_f).$$

Let $V_P^l(n_f, x_{pf})$ be the probability that n_f wavelengths are free on an l-hop path P, and x_{pf} wavelengths are free on the first link. Then,

$$V_P^l(n_f, x_{pf}) = \sum_{y_f=0}^{F} \sum_{z_c=0}^{\min(F-x_{pf}, F-y_f)} \sum_{x_{ff}=0}^{y_f} R(n_f | x_{pf}, z_c, x_{ff})$$
$$U(z_c | y_f, x_{pf}) S(y_f | x_{pf}) V_P^{(l-1)}(x_{ff}, y_f). \tag{12.14}$$

Considering the first k hops as the first hop, as we do in the previous section, $V^{(l-k)}(0, x_{pf})$ is the probability that x_{pf} wavelengths are available on the first k hops of an l-hop path, but none of the x_{pf} wavelengths is available on the following $l - k$ hops.

Let $\widehat{Pr}(P_\alpha^1)$ be the probability that a call for a source-destination pair α is set up on the first shortest path P_α^1 and $\widehat{Pr}(P_\alpha^2)$ be the probability that a request is set up on the second shortest path P_α^2 using neighborhood information. In the FPLC-N(k), a call request is attempted to be set up on the path that has more free wavelengths on the first k links. The request is successfully set up if at least one wavelength is free on the first k links, and at least one of the free wavelengths can continue on the following links of the path. Therefore,

$$\widehat{Pr}(P_\alpha^1) = \sum_{i=1}^{F} \widehat{Q}_{P1}^k(i)(\sum_{n=0}^{i} \widehat{Q}_{P2}^k(n))(1 - V_{P1}^{(l-k)}(0,i)), \qquad (12.15)$$

$$\widehat{Pr}(P_\alpha^2) = \sum_{i=1}^{F} \widehat{Q}_{P2}^k(i)(\sum_{n=0}^{i-1} \widehat{Q}_{P1}^k(n))(1 - V_{P2}^{(l-k)}(0,i)). \qquad (12.16)$$

The arrival rate of calls that enter at link i and continue to link j becomes

$$\rho_c(i,j) = \sum_{P_j \in R_{i,j}^1} \lambda \widehat{Pr}(P_j) + \sum_{P_j \in R_{i,j}^2} \lambda \widehat{Pr}(P_j). \qquad (12.17)$$

The arrival rate of calls that leave from link i becomes

$$\rho_l(i) = \sum_{P_i \in R_i^1} \lambda \widehat{Pr}(P_i) + \sum_{P_i \in R_i^2} \lambda \widehat{Pr}(P_i) - \rho_c(i,j). \qquad (12.18)$$

The arrival rate of calls that enter at link j becomes

$$\rho_e(j) = \sum_{P_j \in R_j^1} \lambda \widehat{Pr}(P_j) + \sum_{P_j \in R_j^2} \lambda \widehat{Pr}(P_j) - \rho_c(i,j). \qquad (12.19)$$

Blocking probability L_α for the FPLC-N(k) consists of three terms: (1) the probability that the first route is selected using neighborhood information (there are i free wavelengths on the first route up to neighborhood distance k and the corresponding alternate route has less than or equal to i idle wavelengths up to neighborhood distance k) but blocked at other links on the first route after distance k; (2) the probability that the second route is selected using neighborhood information (there are i free wavelengths on the second route up to neighborhood distance k, and the corresponding alternate route has less than i idle wavelengths up to neighborhood distance k) but blocked at other links on the second route after distance k; and (3) the probability that no wavelength is available on any of the two routes up to distance k. Thus,

$$\begin{aligned} L_\alpha &= \sum_{i=1}^{F} \widehat{Q}_{P1}^k(i)(\sum_{n=0}^{i} \widehat{Q}_{P2}^k(n))V_{P1}^{(l-k)}(0,i) \\ &+ \sum_{i=1}^{F} \widehat{Q}_{P2}^k(i)(\sum_{n=0}^{i-1} \widehat{Q}_{P1}^k(n))V_{P2}^{(l-k)}(0,i) \\ &+ \widehat{Q}_{P1}^k(0)\widehat{Q}_{P2}^k(0). \end{aligned} \qquad (12.20)$$

Using the same algorithm given in Section 2.1, we can compute the blocking probabilities for the traffic on all the routes.

3.2.   NUMERICAL RESULTS AND DISCUSSION

We assess the accuracy of our approximation in this section. The physical network, a 4×4 mesh-torus network depicted in Fig. 12.1, is employed. The routes used for each source-destination pair are similar to those used in the FPLC as described in Section 2.2.

Fig. 12.7 shows the blocking probability versus the traffic load for a given number of wavelengths on each link ($F = 8$) using the FPLC-N(1). The route to set up a connection request is determined by 1-neighborhood information. Similar to the original observation for the FPLC, we again notice that the analytical and the simulation results match when the traffic load is heavy to moderate, but analytical results are less accurate for light traffic.

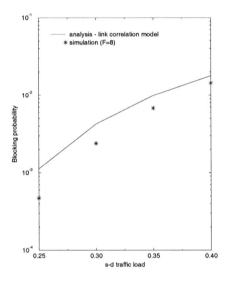

Figure 12.7   Blocking probability versus traffic load using the FPLC-N(1).

The performance of the FPLC-N(k) with different values of k is compared using analytical results shown in Fig. 12.8 (F=8). The diameter of our network topology is 4, so the FPLC-N(3) is the same routing method as the FPLC. Thus, the FPLC is the lower bound on the FPLC-N(k). From the figure it is observed that the blocking probability is not linearly decreasing with the increase of neighborhood distance. The difference of the blocking probabilities between $k = 0$ and $k = 1$ is much higher than the difference between $k = 1$ and $k = 2$ and the difference between $k = 2$ and $k = 3$ under light traffic conditions.

From Fig. 12.6 we know that for the FPLC, the first-fit wavelength assignment method performs much better than random strategy under light traffic. The same is true for the FPLC-N(k) as shown by our simulation results de-

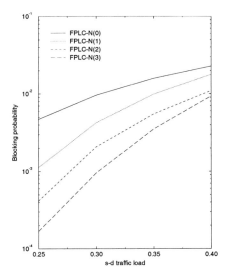

Figure 12.8   Performance comparison using different neighborhood information.

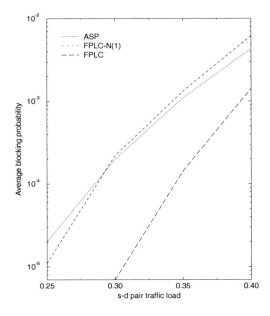

Figure 12.9   Performance comparison of using the ASP, the FPLC-N(1) and the FPLC.

picted in Fig. 12.9 (F=8). In comparison to the results of Fig. 12.7, we note from Fig. 12.9 that the blocking probability is significantly improved when the first-fit wavelength assignment is used. The FPLC-N(1) can achieve the performance similar to the ASP. Thus, we conclude that 1-neighborhood information is sufficient to ensure good network performance in terms of blocking probability in a 4×4 mesh-torus network.

4.   PERFORMANCE ANALYSIS OF THE NSFNET

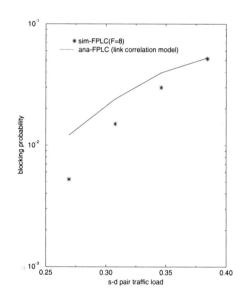

Figure 12.10   Blocking probability versus traffic load in the NSFnet.

Since we analyze the blocking performance individually for each s-d pair, the analytical model is also applicable for irregular networks. The NSFnet T1 backbone network shown in Fig. 12.11 with 14 nodes and 21 bi-directional links is studied in this section. We compare the performance of the FPLC to the fixed-path routing and alternate-path routing. We show using analytical and simulation results that one neighborhood information is sufficient to guarantee high performance in the FPLC-N(k) compared to that in the alternate-path routing.

The analytical results using the link correlation model and simulation results for the NSFnet are shown in Fig. 12.10. The figure shows that the link correlation model performs well on the NSFnet. The analytical results are accurate under heavy and moderate traffic loads.

Simulation results for the alternate-path routing, the FPLC, the FPLC-N(1) and the FPLC-N(2) are shown in Fig. 12.12. The performance of the FPLC

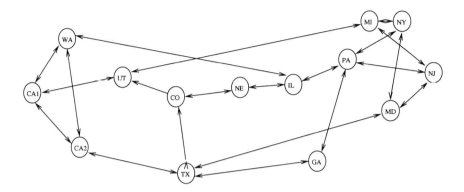

Figure 12.11   The NSFnet T1 backbone network.

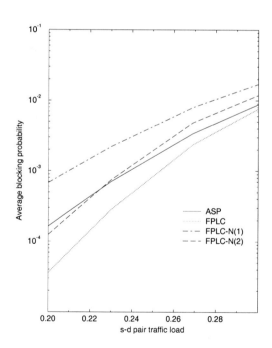

Figure 12.12   Simulation results of the average blocking probability in the NSFnet.

is better than the alternate-path routing in the NSFnet when the traffic load is light and it gets closer to the alternate-path routing as the traffic load increases. Routing using neighborhood information does not perform as well in the NSFnet as it does in mesh-torus network. However, the performance of the FPLC-N(1), which employs 1-neighborhood information, is close to the alternate-path routing. By using 2-neighborhood information, the FPLC-N(2) can achieve similar performance to the alternate-path routing. Thus one can use the FPLC-N(1) or the FPLC-N(2) to achieve similar performance as the alternate-path routing and keep the setup time and control overhead low.

5.   CONCLUSIONS

We developed two new dynamic routing methods in all-optical wavelength-routed networks. An approximate analytical approach is developed for the fixed-paths least-congestion routing and the routing using neighborhood information algorithms. Numerical results show that the fixed-paths least-congestion routing with the first-fit wavelength assignment method significantly improves network performance compared to the alternate paths routing algorithms. The reason is that more wavelengths are left free on a network when the FPLC with the first-fit wavelength assignment method is used.

The routing using neighborhood information algorithm is employed as a trade-off between network performance in terms of blocking probability versus setup delay and control overhead when using dynamic routing algorithms. It is shown that the routing using neighborhood information method achieves good performance when compared to static routing approaches. 1-neighborhood information is sufficient to ensure network performance in a 4×4 mesh-torus network and in the NSFnet T1 backbone network.

Notes

1. The words "path" and "route" are used interchangeably throughout this chapter

2. The *needle packets* use a control network, either out-of-band or in-band, as proposed in [Ramaswami and Segall, 1996]

References

Chung, S., Kashper, A., and Ross, K. (1993). Computing approximate blocking probabilities for large loss networks with state-dependent routing. *IEEE / ACM Transactions on Networking*, 1:105–115.

Banerjee, D. and Mukherjee, B. (1996). A practical approach for routing and wavelength assignment in large wavelength-routed optical networks. *IEEE J. Sel. Areas Comm.*, 15(5):903–908.

Barry, R. A. and Humblet, P. A. (1996). Models of blocking probability in all-optical networks with and without wavelength changers. *IEEE J. Sel. Areas Comm.*, 14(5):858–867.

Barry, R. A. and Marquis, D. (1995). An improved model of blocking probability in all-optical networks. In *LEOS 1995 Summer Topical Meeting*.

Birman, A. (1996). Computing approximate blocking probabilities for a class of all-optical networks. *IEEE J. Sel. Areas Comm.*, 14(5):852–857.

Chan, K. and T.P.Yum (1994). Analysis of least congested path routing in WDM lightwave networks. In *Proc. INFOCOM '94*, pages 962–969.

Girard, A. (1990). *Routing and Dimensioning in Circuit-Switched Networks*. Addison-Wesley.

Harai, H., Murata, M., and Miyahara, H. (1997). Performance of alternate routing methods in all-optical switching networks. In *Proc. INFOCOM '97*.

Karasan, E. and Ayanoglu, E. (1998). Effects of wavelength routing and selection algorithms on wavelength conversion gain in WDM optical networks. *IEEE/ACM Trans. Networking*, 6(2):186–196.

Kovačević, M. and Acampora, A. S. (1996). Benefits of wavelength translation in all-optical clear-channel networks. *IEEE J. Sel. Areas Comm.*, 14(5):868–880.

Li, L. (1999). Optimal resource placement in multifiber mesh-survivable WDM networks. In *Proc. On-going Student Research, FTCS'99*.

Li, L. and Somani, A. K. (1999a). Blocking performance analysis of fixed-paths least-congestion routing in multifiber WDM networks. In *All-Optical Networking 1999: Architecture, Control, and Management Issues, Proc. SPIE*.

Li, L. and Somani, A. K. (1999b). Dynamic wavelength routing using congestion and neighborhood information. *IEEE/ACM Trans. Networking*, 7(5):779 – 786.

Li, L. and Somani, A. K. (1999c). Fiber requirement in multifiber WDM networks with alternate-path routing. In *Proc. ICCCN '99*.

Li, L. and Somani, A. K. (1999d). A new analytical model for multifiber WDM networks. In *Proc. GLOBECOM*, pages 1007–1011.

Lowe, E. and Hunter, D. (1997). Performance of dynamic path optical networks. *IEE Proceedings-Optoelectronics*, 144(4):235–2397.

Mei, Y. and Qiao, C. (1997). Efficient distributed control protocol for WDM all-optical networks. In *Proc. ICCCN*, pages 150–153.

Mohan, G. and Somani, A. K. (2000). Routing dependable connections with specified failure restoration guarantees in WDM networks. In *Proc. INFOCOM '2000*.

Mokhtar, A. and Azizoğlu, M. (1998). Adaptive wavelength routing in all-optical networks. *IEEE/ACM Trans. Networking*, 6(2):197–206.

Mukherjee, B. (1997). *Optical Communication Networks*. McGraw Hill.

Ramamurthy, S. and Mukherjee, B. (1998). Fixed-alternate routing and wavelength conversion in wavelength-routed optical networks. In *Proc. GLOBECOM '99*, volume 4, pages 2295–2303.

Ramamurthy, S. and Mukherjee, B. (1999a). Survivable WDM mesh networks, Part I – protection. In *Proc. INFOCOM '99*, pages 744–751.

Ramamurthy, S. and Mukherjee, B. (1999b). Survivable WDM mesh networks, Part II – restoration. In *Proc. ICC '99*, pages 2023–2030.

Ramaswami, R. and Segall, A. (1996). Distributed network control for wavelength routed optical networks. In *Proc. INFOCOM '96*, pages 138–147.

Ramaswami, R. and Sivarajan, K. N. (1998). *Optical Networks: A Practical Perspective*. Morgan Kaufmann.

Somani, A. K. and Azizoğlu, M. (1997). All-optical LAN interconnection with a wavelength selective router. In *Proc. INFOCOM '97*, pages 1278–1285.

Subramaniam, S., Azizoğlu, M., and Somani, A. K. (1996). All-optical networks with sparse wavelength conversion. *IEEE/ACM Trans. Networking*, 4(4):544–557.

Subramaniam, S. and Barry, R. A. (1997). Wavelength assignment in fixed routing WDM networks. In *Proc. ICC*, pages 406–410.

Zhang, Z. and Acampora, A. S. (1995). A heuristic wavelength assignment algorithm for multihop WDM networks with wavelength routing and wavelength re-use. *IEEE/ACM Trans. Networking*, 3(3):281–288.

Zhu, Y., Rouskas, G. N., and Perros, H. (1999). Blocking in wavelength routing networks, part I: The single path case. In *Proc. INFOCOM*, pages 321–328.

VI

MISCELLANEOUS TOPICS

Chapter 13

Optical Access Networks for the Next Generation Internet

Eytan Modiano
Massachusetts Institute of Technology

Abstract: We describe a WDM-based optical access network architecture for providing broadband Internet services. The architecture uses a passive collection and distribution network and a configurable Feeder network. Unlike earlier papers that concentrate on the physical layer design of the network, we focus on higher layer architectural considerations. In particular we discuss the joint design of the electronic and optical layers including: the choice of electronic multiplexing and switching between the IP and WDM layers; joint optical and electronic protection mechanisms; network reconfiguration algorithms that alter the logical topology of the network in response to changes in traffic; traffic grooming algorithms to minimize the cost of electronic multiplexing; and WDM Medium Access Control protocols.

1.   INTRODUCTION

Over the past decade the growth in the use and capabilities of communication networks has transformed the way we live and work. As we progress further into the information age, the reliance on networking will increase. While today's network traffic is still dominated by voice, there is an increasing demand for data services with broader bandwidth needs and a wide range of Quality of Service (QoS) requirements. These emerging demands offer new challenges and opportunities for the design of *access networks*. The access network refers to the portion of the communication infrastructure responsible for reaching the customer premises. Because of the proximity to the end-user, an access network is quite different from a backbone network and hence offers additional technological and economic challenges. While backbone networks have been able to take advantage of developments in high speed transmission and switching systems to tremendously increase their transmission capacity, access networks have not advanced accordingly, thus the transmission rates available to subscribers are still rather limited.

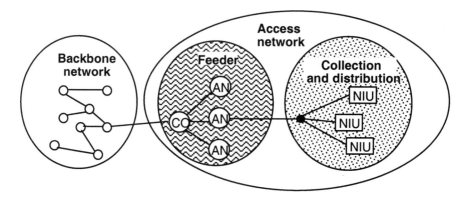

Figure 1. Generic access network architecture

This paper describes an architecture for a WDM based optical access network and examines critical issues in its design. Our discussion will follow the generic access network architecture shown in Figure 1. As shown in the figure the access network consists of two parts: *the collection and distribution* network (C/D) responsible for connectivity between the Network Interface Unit (NIU) at the customer premise, and a remote Access Node (AN), and the *Feeder* network that connects the various ANs to a Central Office (CO). Central offices are connected via a backbone network that is generally not considered part of the access network. Previous efforts in the design of optical access networks focused, almost exclusively, on the design of Passive Optical Networks (PONs) for the *C/D* part of the access network. The PON architecture was motivated by the need for low cost, simple maintenance and powering considerations. Hence, most of that effort was concentrated on the physical design of the PON. A number of PON architectures were developed and systems were demonstrated [FR96, LS89, MC89, FPS89, WL89]. However, little attention has been paid to the overall access network architecture. The focus of this article is on an overall access network architecture that includes the C/D network as well as the *Feeder* network and discusses architectural considerations in their design, paying particular attention to network layer issues.

Despite some successful experiments, optical access has failed to materialize, primarily because of the relatively high cost of optical equipment (e.g., lasers). In recent years efforts have been geared toward the development of low cost PON architectures aimed at reducing system costs. Recent progress toward the development of low cost optical technology for use in the local loop have brought the cost of optical local loop to the point that it is becoming competitive with electronic alternatives, especially in new locations where no access infrastructure exists [LU98]. Nonetheless, since in most locations a copper and coax infrastructure already exists, hybrid architectures that utilize the existing copper or coax have emerged and appear to offer a lower cost alternative to an all-fiber solution. A Fiber To The Curb (FTTC) architecture uses a curb-side Optical Network Unit (ONU) to serve several subscribers. The ONU is connected to an AN using a PON and connectivity between the ONU and the subscriber is provided over existing twisted pair [PB95]. Similarly, the cable TV industry is adopting a Hybrid Fiber Coax (HFC) architecture where the curb side ONU (typically referred to as the Fiber Node) is connected to the subscribers over existing coax [LU96].

These approaches appear attractive for meeting present and near-term demands of most residential customers. However, they are limited to a transmission capacity of a few tens of Mega-bits-per-second (Mbps). It is widely believed that in the future, applications such as video on demand may require transmission rates in the 100's of Mbps or even Giga-bits-per-second (Gbps) [RS92]. Furthermore, certain high-end businesses already have needs for these kinds of transmission rates. It is for these applications and users that an optical access solution will be necessary. The challenge for optical access networks is to provide a cost effective interface between end-users and very high capacity WDM-based backbone networks.

This challenge is the focus of the DARPA-sponsored Next Generation Internet (NGI) ONRAMP consortium, a pre-competitive consortium including MIT, JDS Fitel, Nortel Networks, Cabletron systems, AT&T and MIT Lincoln Laboratory. The goal of the consortium is to design and build an optical access network which exploits WDM and other emerging technologies to support next generation Internet services. At least two types of services will be offered: an all-optical service, which establishes an uninterrupted lightpath from source to destination, and an IP service, which connects the source and destination via IP routers. In support of IP and other types of traffic, the network will feature optical flow switching, MAC protocols to share wavelengths among bursty users, dynamic provisioning and reconfiguration, automatic protection switching, and a robust and responsive network control and management system. The ONRAMP test-bed will be described in more detail in section IV of this paper.

2.   ACCESS NETWORK SERVICES, SIZE AND CAPACITY

Before designing a high-speed optical access network, it is necessary to know what services the network will provide, what geographical span the access network will cover and what traffic capacity it will support. We start by discussing basic access network services and applications. While it may be difficult, at present, to foresee specific applications that require giga-bit per second transmission rates, it is important to note that past forecasts of new applications have not been terribly accurate. In fact, the World-Wide-Web application was not in anyone's predictions much before its appearance. Hence, we will not attempt to foretell any specific future applications but rather try to provide an abstraction of the type of services that can be offered in the future. We believe that future access networks should be able to provide the following two basic services:

1. *Conventional electronic network services*: - e.g., SONET, ATM, Frame Relay, and IP services.
2. *Switched optical lightpaths* – e.g., point to point optical connections that take up a full wavelength. These lightpaths connections can be used to support a number of applications, such as:
   a) *Large point-to-point circuit-switched trunks on demand* – e.g., OC-48, OC-192 and above to deal with stream traffics but with setup time on the order of tens of milli-seconds.
   b) *Optical flow switching* - for bursty, unscheduled large file transfer (100 Mbyte to 10 Gbyte) at high access rates (> 2 Gbps). This service can be provided by dynamically setting up an end-to-end optical lightpath for the duration of the transaction.
   c) *Analog services* - narrow and broadband analog services with high amplitude, phase and timing fidelity preserving features.

Access networks must be designed for urban, suburban and rural locations. Depending on the type of location that the access network is being designed for, it will a have different size and capacity. For example, an access network for the New York city can be as small as a few miles in diameter, while an access network in Minnesota is likely to have 100's of miles in diameter. These differences will inevitably impact the physical design of a network. In this paper we will consider an access network that is designed to support a metropolitan area with approximately 500,000 people in population and spanning between 100 and 1000 square miles. Clearly, very dense metropolitan areas can be supported by multiple access networks. Using simple arithmetic one can see that approximately 500 such access networks may exist in the entire U.S. While such an access network covers an area with 500,000 people in population, only a relatively small number of users will require the services of a high-speed optical access network. We assume, that at least initially, only high-end businesses will require optical access and that each access network will support somewhere between 100 and 1000 businesses. We will show later, that these numbers have a significant impact on the design of the network.

Another important design criteria is the traffic capacity that such an access network must support. While it is rather difficult to predict future traffic requirements, the U.S. has been experiencing significant growth in internet traffic over the past few years. Furthermore, since optical access networks will not be needed <u>unless</u> such growth materializes, it is reasonable to design a future optical access network that is based on significant traffic growth projections. While today's backbone network traffic in the entire U.S. is less than 1 tera-bits per second, traffic increase at a factor of 2 per year will result in backbone traffic on the order of 1000 tera-bits per second within 10 years. Even, if a factor of two rate of increase is not sustained, it is reasonable to assume that in the foreseeable future a backbone capacity of this order will be required. With 500 access networks in the U.S., each access network will have to support approximately a tera-bit of capacity.

We will focus on the design of an access network that can accommodate these services, geographical span and capacity. In section III we will discuss the physical network architecture and in section IV we will discuss the ONRAMP test-bed.

3.   ACCESS NETWORK ARCHITECTURE

The proposed architecture consists of a configurable *Feeder* network and passive *Collection and Distribution (C/D)* network, as shown in Figure 2. The choice of a passive C/D network is driven by the need for low equipment and maintenance costs at or near the customer premise where equipment is shared among a small number of users. However, in the *Feeder* network the cost of configurable components can be justified because equipment is shared among many more users. Furthermore, in the relatively long distance *Feeder* network, the cost of the fiber can be substantial and hence both electronic and optical multiplexing can be used to make efficient use of the fiber. Lastly, configurability in the *Feeder* network is needed to provide rapid and efficient restoration.

As shown in Figure 2, the proposed *Feeder* network is a configurable WDM ring and the C/D network is a WDM PON. Subscribers communicate with an Access Node (AN) over the passive C/D network. At the AN their communication is switched, either optically or electronically, over the feeder network and onto another AN or to the Central Office (CO). The Feeder network uses a combination of optical and electronic techniques, where the electronic layer is aware of the optical layer and vice versa, to make full use of the WDM layer. This is in contrast to existing approaches that separate the optical and electronic layers.

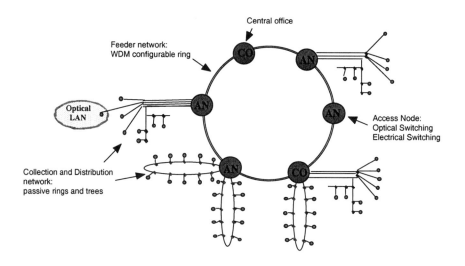

Figure 2. A WDM-based architecture that includes a configurable feeder ring network and passive C/D rings and trees

3.1   Feeder network architecture

The *Feeder* network has a configurable WDM ring architecture. In order to make efficient use of the fiber in the Feeder network, each node contains a combination of electronic and optical switching equipment. Electronic switches are needed to provide necessary electronic services, such as IP, ATM, SONET or Frame Relay. In addition, electronic switches and multiplexers can also be used to provide efficient statistical multiplexing and fast protection mechanisms. While optical switching is not strictly required in the network, it can significantly simplify the electronic layer by providing optical layer services such as dynamic reconfiguration of the network topology, optical protection and restoration, and traffic grooming.

In this section we will discuss issues in the design of both the optical and electronic layers. In particular we will discuss optical layer services that can improve network performance and reduce the cost and complexity of the electronic layer. We will also describe the joint design of the electronic and optical layers of the network so that optical services are used for functions best provided optically and electronic services for functions best provided electronically.

3.1.1   Use of electronic multiplexing

One important issue in the design of the feeder network is the form of electronic multiplexing offered at the ANs. Customers may require a variety of electronic services such as SONET, ATM, Frame Relay or IP. One possible solution is to have the network provide all of these services directly at the AN; however, this approach would require a significant amount of electronic equipment at an AN. The other extreme would be to only offer optical services (e.g., lightpaths) at the ANs and to provide all of the electronic services only at the central office location, which would be accessed by customers optically. This latter option, however, would inefficiently use the fiber, since little statistical multiplexing would be done until reaching the central office.

A compromise approach is to provide some electronic services (and hence statistical multiplexing) at the ANs and back-haul all of the traffic to the Central Office where the other electronic services would be available. For example, some electronic multiplexer (e.g.,IP router) can be provided at the AN and alternative services could be carried over IP to the Central Office where those services would be available. Of course, with this option a number of problems might arise such as protocol compatibility issues and inefficiencies due to a multi-layered protocol stack. For example, currently a typical high-end customer's Internet connection involves a multi-layered protocol stack as shown in Figure 3a. Such customers typically gain access to service provider networks via Frame Relay where the IP packets are encapsulated in frames. In the backbone network, the frames are sometimes mapped into ATM cells, which, in turn, are carried over Synchronous

Optical Networking (SONET) transport frames.[1] The multitude of layers produces gross bandwidth inefficiencies, adds to the latencies of connections, and inhibits providing quality of service assurances. Worse, the layers are largely unaware of each other causing duplication of network services.

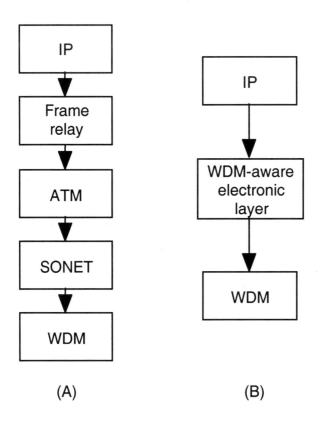

Figure 3. a) Typical protocol stack, b) Simplified protocol stack

One solution is the simplified protocol stack shown in Figure 3b, where the IP traffic is carried directly by a simplified electronic layer. Such an arrangement would not only reduce the overhead associated with the different layers but would also allow the electronic layer to be "WDM-aware" and take advantage of network services offered at the optical layer. For the remainder of this section we will describe optical layer services and algorithms that can significantly improve the performance of the network and simplify the design of the electronic layer.

---

[1] In some instances, only one of either Frame Relay or ATM is used.

3.1.2 Protection and Restoration

Various failures can occur that disrupt network services, such as fiber cuts, line card and switch failures, and software failures [WU92]. Protection and restoration are two methods networks use to recover from these failures. Protection refers to hardware-based, pre-planned, fast failure recovery; restoration refers to software-based, dynamic, slow recovery. Protection is generally limited to simple topologies like rings or the interconnection of rings; restoration works on general mesh networks and is typically more bandwidth efficient. Recently, fast protection mechanisms at the optical layer have been proposed for general mesh networks [ES96, FMB97, FMB98], and for ring networks [GR97, MB96].

Failure recovery must be done at the electronic layers in order to recover from line card or electronic switch failures. Electronic recovery mechanisms, e.g., as is done in SONET, can also be used to protect against failures at the optical layer such as a fiber cut or a malfunctioning optical switch. However, in many cases, optical layer recovery is more natural and provides enhanced reliability. For instance, consider the case of 32 SONET rings being supported over a WDM ring network with 32 wavelengths. Without optical protection, each of the 32 SONET rings would need to individually recover from a single fiber cut, e.g., by loop-back in a SONET bi-directional ring network. On the other hand, the fiber cut can be optically restored with a simple 2x2 optical switch, thereby simultaneously restoring service to many electronic connections. A simple example of optical loop-back protection using a 2x2 switch is given below. Protection at the optical layer has the added advantage that the failure is transparent to SONET, allowing each SONET ring to individually respond to additional failures such as a line card failure.

For a simple example of optical loop-back protection, consider the two fiber bi-directional ring shown in Figure 4. On each ring half of the wavelengths are used for working traffic and the other half are reserved for protection against a cut in the fiber on the other ring. In the event of a fiber cut, the wavelengths from the cut fiber can be switched onto the uncut fiber, using a two-by-two switch at the node before the fiber cut. They can then be looped back to bypass the cut fiber and rejoin their original ring using another switch at the node immediately following the fiber cut.

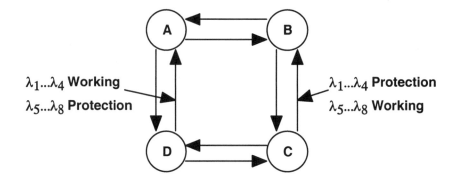

Figure 4. Protection in WDM ring networks

However, there are problems providing restoration at both the optical and electronic layers if the layers work independently of each other. For instance, if care is not taken, restoration will be duplicated at both the optical and electronic layers leading to a 75% loss in efficiency (assuming 50% efficiency for each layer of protection). In addition, differing time scales may lead to race conditions and topology flapping. Also, in the case of a SONET network, optical protection must somehow be completed before SONET starts its protection process. This may be difficult as SONET starts its protection process as soon as loss of power is detected. Hence, care must be taken to coordinate the protection mechanisms at the electronic and optical layers.

3.1.3   Virtual topology reconfiguration

One of the benefits of having a configurable WDM ring for the feeder network is that the virtual (electronic) topology of the network can be changed in response to changes in traffic conditions. The virtual topology, seen by the electronic layer, consists of a set of nodes interconnected by lightpaths (wavelengths). In this way, WDM networks provide a way to interconnect electronic switches with high bandwidth bit pipes without dedicating a fiber pair between each pair of switches. However, the configurable nature of WDM also allows the electronic topology to be dynamically optimized in response to changes in traffic conditions.

This is achieved by changing the lightpath connectivity between electronic switches and routers, thereby reconfiguring the electronic virtual topology. Lightpaths can be changed via tuning of the transmitter wavelengths in combination with frequency-selective-switches that can alter the route of a wavelength inside the network. For example, consider four nodes physically connected in a ring. Assume that each node has one port (i.e., single receiver and transmitter) and that the fiber supports two wavelengths, $\lambda 1$ and $\lambda 2$. A connected, fixed logical topology must take the form of a unidirectional ring, as pictured in Figure 5a. If a full-wavelength call is in progress from node 1 to node 3, and a call request arrives between nodes 2 and 4, then that request must be blocked. In a reconfigurable system, both calls can be supported as shown in Figure 5b.

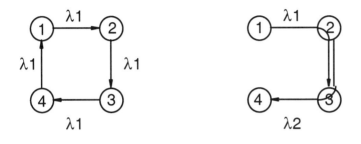

a) Fixed topology   b) reconfigurable topology

Figure 5. Using WDM to reconfigure the electronic topology

Preliminary studies on reconfiguration of a WDM ring show significant promise [SM00]. The work in [SM00] assumes that calls take up a full wavelength and cannot be rearranged or rerouted. The assumption about rerouting is made in order to eliminate the possibility of calls being adversely affected by reconfiguration. Shown in Figure 6 is the gain that can be achieved through reconfiguration. This gain is defined to be the ratio of the load that can be supported by a reconfigurable system to that of a bi-directional, fixed topology system at a given blocking probability (a blocking probability of 0.01 is used in Figure 6). As can be seen from the figure, the capacity gain due to reconfiguration is most significant when the ratio of wavelengths to ports per node (W/P) is large and the number of ports per node is small. The results from [SM00] indicate that when the number of wavelengths is much larger than the number of available ports per node, a capacity gain on the order of N can be obtained, where N is the number of nodes in the network. When the number of wavelengths approaches the number of ports per node, the benefits of reconfiguration are significantly diminished with approximately a factor of two gain in capacity. The concept of topology reconfiguration can similarly be applied to packet networks (e.g., IP) with the objective of reducing network queuing delays [NM00].

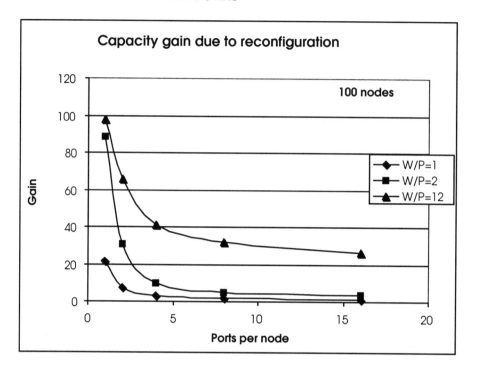

Figure 6. Performance of reconfiguration in a WDM ring network

3.1.4   Traffic grooming

One of the most important functions of the electronic layer of the network is multiplexing lower rate streams into higher rate channels or wavelengths. However, if calls are indiscriminately multiplexed on to wavelengths then each wavelength will have to be electronically processed at every node. Alternatively, if calls are *groomed* with foresight onto wavelengths, then the number of wavelengths that need to be processed at each node can be significantly reduced.

For example, when a SONET ring network is used to provide point-to-point OC-3 circuits between pairs of nodes, SONET Add/Drop Multiplexers (ADMs) are used to combine up to 16 OC-3 circuits into a single OC-48 that is carried on a wavelength. If a wavelength carries traffic that originates or terminates at a particular node, then that wavelength must be dropped at that node and terminated in a SONET ADM. In order to reduce the number of ADMs used, it is better to groom traffic such that all of the traffic to and from a node is carried on the minimum number of wavelengths (and not dispersed among the different OC-48's). Traffic grooming algorithms can be designed to minimize electronic costs while simultaneously making efficient use of wavelengths. Traffic grooming on SONET ring networks with uniform all-to-all traffic show a significant reduction in the number of required ADMs [MC98, CM98, SGS98,CM00].   Figure 7 shows the significant savings in the number of ADMs needed in a SONET ring network using an algorithm developed in [CM98]. The results from [CM98] assume the use of a unidirectional ring and uniform traffic. Similar results, however, were obtained in [SGS98] for a bi-directional ring. In [CM00] simple algorithms were developed for general, non-uniform traffic, that show similar savings in the number of ADMs. In [BM99] traffic grooming for dynamic traffic is considered, where the number of circuits between node pairs is allowed to change dynamically with the only restriction being that the total number of circuits entering or leaving a node is constrained. Further, while the discussion here focused on the use of SONET equipment, similar optimization can be done for other electronic technologies (e.g., ATM or IP) with the goal of minimizing the number of electronic ports used in the network.

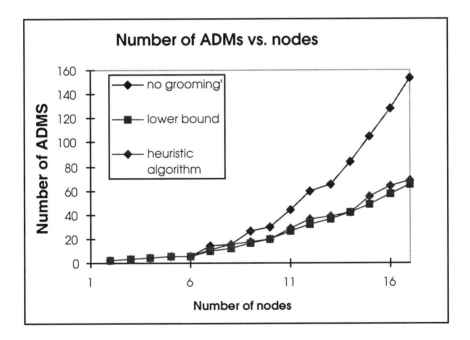

Figure 7. Performance of grooming in a WDM/SONET ring network

3.2   C/D Architecture

The C/D network is responsible for aggregating traffic from user locations to the Access Node (AN). With about 10 AN's per feeder network, each C/D network will cover an area of approximately 100 sq. km and support between 10 and 100 businesses with an aggregate transmission capacity of 100 Gbps. As stated earlier the C/D network should be passive (PON). Many PON architectures have been proposed in the past for use in the access network [LS89, MC89, FPS89, WL89]. Some alternative architectures are shown in Figure 8. The simplest architecture would use a dedicated fiber pair for each node. Of course, this approach can be costly because it may require a significant amount of fiber. In addition, a dedicated fiber architecture requires dedicated transceivers at the head-end (located at the AN). An alternative architecture, using a broadcast star at the head-end, would allow the lasers at the head-end to be shared among multiple users, but still requires the same amount of fiber as the dedicated fiber architecture. In order to reduce the cost of fiber, solutions that allow fiber to be shared have been proposed [LS89]. The simplest of which is the double star, with one broadcast star at the head-end and another at a remote location where a cluster of users is located. The most general form of a shared fiber architecture is the tree architecture also shown in Figure 8.

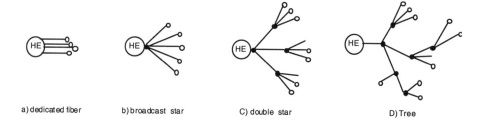

a) dedicated fiber   b) broadcast star   C) double star   D) Tree

Figure 8. PON architectures

One shortcoming of all of these shared fiber architectures is the power losses incurred due to the splitting of the fiber. These losses can be particularly significant in older fiber plants where the plant losses are already high. For this reason, many older architectures use dedicated fiber. However, in newer fiber plants, a shared fiber solution is more promising. In addition, recent improvements in passive components significantly reduce the amount of excess loss incurred in the splitters. Consequently, a tree architecture can be used to support more than 100 users, over a 10 km distance, at 155 Mbps [LS89, RS98]. Furthermore with the additional use of a fiber amplifier at the head-end even higher data rates and many more users can be supported.

An alternative architecture for a WDM PON based on a wavelength router at the head-end was proposed in [IFD95, FR94]. With this architecture each user would be communicating on a dedicated wavelength. This architecture has a number of advantages including lower power losses and wavelength isolation. However, it limits the number of users in the PON to the number of wavelengths and it also requires a dedicated fiber pair for each user. Alternatively, one could also design a wavelength routed tree PON where the fiber is shared but each user gets its own wavelength. With this approach, at each splitting point on the tree a WDM mux/demux is used to split off the proper wavelengths. This approach has a number of attractive features including wavelength isolation among the different users and reduced splitting losses. However, a wavelength routed PON has one significant disadvantage in that it cannot be used for end-to-end optical flows. In the absence of wavelength changers in the network, setting up end-to-end lightpaths requires the use of the same wavelength along the path. With a wavelength routed PON, setting up end-to-end lightpath between two users becomes rarely possible. For this reason, in our architecture we only consider broadcast PONs where each user can access any of the wavelengths.

In our proposed architecture, as shown in Figure 2, we use passive WDM broadcast rings and trees in the C/D network. As we will show next, a tree architecture provides greater scalability as it can achieve more efficient power splitting. On the other hand, the ring architecture is proposed for use in cases where diversity is needed for protection, as discussed later in this section.

3.2.1   Power budget

Using our assumptions on the size and capacity of an access network, a C/D network must support as many as 100 users, each transmitting at a minimum rate of 2.5 Gbps (OC-48) and traversing a distance of up to 10 km. These requirements impact the power budget in the C/D network and hence the topology of the network. To demonstrate this impact we provide a simple analysis of the power budget for a tree and ring C/D network. Consider a broadcast tap (i.e., coupler) at each of the splitting point in the C/D network, as shown in Figure 9. At each tap (on the tree or ring) a fraction, T, of the power will go to the left side and 1-T to the right side. In addition, each tap will incur in an additional <u>excess</u> power loss of α, for imperfection in the coupler and splicing of the coupler into the fiber.

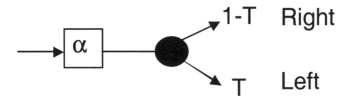

Figure 9. Power splitting at a C/D network tap

When the C/D network is a binary tree, T should be 0.5, because an equal fraction of the power should be sent along either side of the tree. With a binary tree network, each leaf node will be at depth $Log_2(N)^2$. When the C/D network is a ring, each leaf node will be at a different depth and hence nodes near the head of the ring will experience different losses than nodes further away. Ideally, one could set the tap values at each node on the ring so that all nodes receive the same amount of power [RS98]. However, designing a network where each node has a different tap is not practical. Instead, one can use a single tap value that would minimize the maximum loss on the ring. For a ring with N nodes the maximum loss is sustained by the last node and is equal to $L_{max} = T^{N-1}(1-T)$. In order to minimize this loss a value of T=(N-1)/N should be used. While the exact value of T depends of the size of the ring, for moderately sized rings (10-20 nodes) a value of T=0.95 can be used. Shown in Figure 10 is the power loss vs. number of nodes for a ring and tree networks where the excess loss in each tap is 0.3 and T=0.5 for a tree and 0.95 for the ring. After accounting for fiber plant losses (2.5 dB for 10 km), transmitter launch power (5 dBm), receiver sensitivity (-25 dBm at 2.5 Gbps and BER=10^{-12}), and a 5 dB link margin, a power budget of about 22dB is available. As can be seen from the Figure with this power budget a ring C/D network can support only about 10 to 20 nodes while a tree can support over 100 nodes. For this reason, when a large number of nodes must be supported on a single C/D network, a tree is the preferred choice. However, as we will show later in this section, a ring network can be used to provide passive protection and so, should be used in cases where protection is essential.

---

[2] We assume, for simplicity, that users are located only at the leaf nodes.

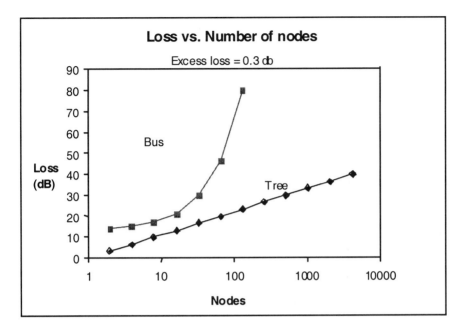

Figure 10. Power loss in the C/D network

3.2.2   Medium access control (MAC) protocol

Most existing WDM networks employ circuit switching, typically with one connection having exclusive use of an entire wavelength. This approach may not scale to the access network where the number of users may be much larger than the number of wavelengths. Furthermore, this approach is not well suited to bursty data traffic, where even partially aggregated traffic may require very low data rates during periods of inactivity and much higher rates at other times. An access mechanism is needed that provides both scalability and flexibility in provisioning bandwidth. In the C/D network, such an access mechanism is particularly important to allow users to share access to the IP router located at the AN.

There are a number of approaches that can be considered for providing scaleable access. One approach is to increase the number of available wavelengths. While present WDM technology provides tens of wavelengths, it is likely that over 100 wavelengths may soon be possible. Nonetheless, even with an increase in the number of wavelengths, it is likely that in certain locations, where fiber is precious, there would not be sufficient capacity to allocate dedicated wavelengths to users. An alternative approach that would allow efficient wavelength sharing is using electronic multiplexing equipment at the fiber merging points (e.g., on poles, pedestals, manholes, etc.). While this approach makes efficient use of the fiber, practical issues regarding the placement of electronic equipment as well as cost considerations and maintenance problems make it infeasible in many circumstances. We therefore propose the use of a Medium Access Control (MAC) protocols for sharing wavelengths in C/D networks.

Although many WDM MAC protocols for LANs have been proposed and studied in the literature, most of the proposed systems assume a synchronized and slotted system and many require multiple transceivers per node, contributing to their high cost and complexity [MUK92, SKG94, BS95]. Furthermore, most of these protocols were designed for a low latency LAN environment and would perform poorly in the access network where propagation delays are relatively high.

Our architecture uses a MAC protocol similar to that proposed in [MB98] that eliminates the need for slotting and synchronization, uses one tunable transceiver per node, yet results in efficient bandwidth utilization in high latency. The system is based on a simple (and potentially low cost) master/slave scheduler able to schedule transmissions efficiently and overcome the effects of propagation and transceiver tuning delays. As shown in Figure 11, a centralized scheduler is located at the AN and responsible for coordinating the transmissions. Users send their transmission requests to the scheduler on a shared control wavelength, λ_C, using a random access protocol, (e.g., Aloha). The scheduler, located at the AN, schedules the requests and informs the users on a separate wavelength, $\lambda_{C'}$, of their turn to transmit. Upon receiving their assignments, users immediately tune to their assigned wavelength and transmit. Hence users do not need to maintain any synchronization or timing information.

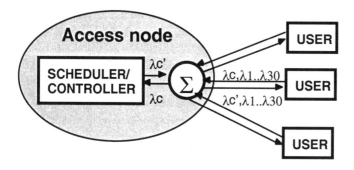

Figure 11. Scheduler based MAC protocol

The scheduler is able to overcome the effects of propagation delays by measuring the round-trip delay of each user to the AN and using that information to inform users of their turn to transmit in a timely manner. For example consider Figure 12. In order for user B's transmission to arrive at the AN at time T, the scheduler must send the assignment to user B at time T-2τ, where τ is user B's propagation delay to the AN. In this way the transmissions of different terminals can be scheduled back-to-back, with little dead time between transmissions. The operation of this MAC protocol, and in particular the ranging process, is somewhat similar to that of the proposed protocol for HFC networks [PG98].

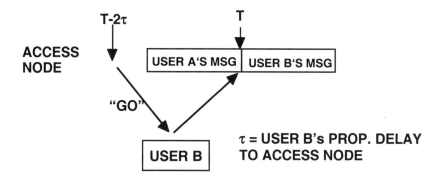

Figure 12. Use of ranging to overcome propagation delays.

An important and novel aspect of this system is the way in which ranging is accomplished. Unlike other systems where terminals need to range themselves to their hubs in order to maintain synchronization [KAM96], here we recognize that it is only the hub that needs to know this range information. Hence ranging can be accomplished in a straightforward manner. The scheduler ranges each terminal by sending a control message telling the terminal to tune to a particular wavelength and transmit. By measuring the time that it takes the terminal to respond to the request, the scheduler can obtain an estimate of the round trip delay for that terminal. This estimate will also include the tuning time delays. Furthermore the scheduler can repeatedly update this estimate to compensate for fiber inaccuracies. These measurements can also be made by simply monitoring the terminal's response to ordinary scheduling assignments. The significance of this approach is that terminals are not required to implement a ranging function, which simplifies their design.

Other important aspects of this MAC protocol include the control channel access mechanism and the scheduling algorithms used by the scheduler. These issues are addressed in [MB99]. Scheduling algorithms for transmitting multicast traffic in WDM broadcast-and-select networks are discussed in [MO98, MO99]. The performance of the protocol depends primarily on the scheduling algorithm used by the scheduler. In [MB98] simple scheduling algorithms are described that achieve nearly full utilization. This is a significant improvement over unscheduled WDM MAC protocols that achieve very low channel utilization [MUK92].

3.2.3   Passive protection and restoration

The passive C/D architecture has the advantage that it is less susceptible to failures because there are no active components in the network. The use of passive components also reduces the maintenance costs of the networks. However, one shortcoming of the passive C/D architecture is that protection or restoration from fiber cuts must be provided optically. Providing protection in the tree C/D network inevitably requires some diversity routing which may eliminate many of the cost benefits of the tree architecture. Therefore the tree architecture is a good choice for users that do not require rapid protection (e.g., homes, small businesses). When rapid protection is more critical, the passive C/D ring can be used. Recently, a number of approaches for providing protection in a passive WDM ring network have been proposed [WW92, GLA96]. A simple example of passive protection in a WDM ring network is shown in Figure 13. This passive ring network uses two fiber pairs, one as a primary (P) and one as a backup (B). Within each pair, one fiber is used for upstream traffic to the head end and the other is used for downstream traffic. Consider, first, the downstream operation. Downstream transmission takes place on both the primary and backup fibers. In the event of a fiber cut in either fiber, nodes will receive the transmission on the alternative fiber. If both fibers are cut, at the same location, those nodes before the cut will receive the transmission on the primary fiber and those after the cut will receive the transmission on the backup fiber. Upstream operation is similar, however, it also requires a mechanism to detect which fiber is cut so that it can be switched off (at the head end).

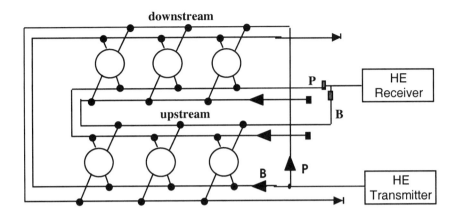

Figure 13. A four fiber passive protection ring

4. THE NGI ONRAMP TEST-BED

The concepts described in this chapter will be demonstrated in the ONRAMP test-bed. The ONRAMP test-bed will consist of a 5 node configurable WDM feeder ring, as shown in Figure 14. A node at the AT&T Boston hub will provide connectivity to backbone networks and other NGI test-beds. The ONRAMP feeder ring is a bi-directional, dual fiber ring with each fiber supporting 16 wavelengths. Eight wavelengths, in each direction, are used for working traffic and eight for loop-back protection. Bi-directionality is used to provide shortest path routing which can increase utilization by a factor of 2 over a uni-directional ring.

Attached to each AN, via a C/D network, are IP routers, workstations and other electronic devices (e.g., ATM switches). Two type of services will be provided; an IP service and a switched optical (wavelength) service. The IP service will be provided via an IP router located at the AN, as shown in Figure 16. Users requiring an IP service can connect to the IP router via the C/D network. Such a connection can be established using a dedicated wavelength, or using a MAC protocol to share wavelengths. In this way, workstations and routers can connect to the IP service at high-rates (e.g., 2.5 Gbps).

In addition to the IP service, ONRAMP will provide a switched optical service. This switched optical service can be used to transparently support other conventional electronic services. For example, it can be used to support a private ATM network, with the ATM switches located at the customer premises and connected via lightpaths. Similarly, SONET, Frame Relay and other legacy equipment can be supported transparently to the access network.

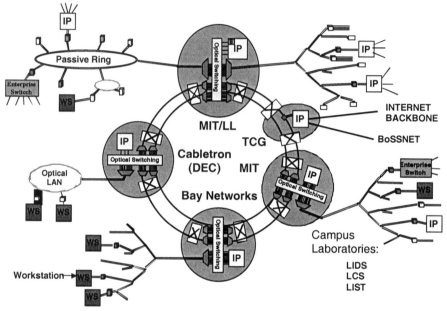

Figure 14. The Onramp feeder ring

The transparent optical service provided by ONRAMP will also enable optical flow switching (OFS). The basic idea behind OFS is very similar to IP switching or Tag switching that is used to bypass IP routing in the internet [NLM96, REK97]. With OFS, a lightpath can be established for large data transactions such as the transfer of large files or long duration and high bandwidth streams across the network. This optical flow will bypass all of the electronics in the network and be switched at the WDM layer as shown in Figure 15.

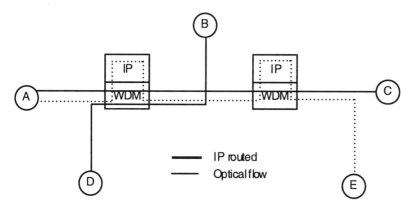

Figure 15. Optical Flow Switching

The AN, shown in Figure 16, has a combination of optical and electronic switching. Electronic switching is provided in the form of an IP router and optical switching in the form of a configurable WDM add/drop multiplexer. Since the feeder network is a configurable WDM ring, each node has the ability to add or drop any of the wavelengths from the ring to the local C/D network. The configurable WDM add/drop multiplexer is designed using a WDM mux and de-mux and a series of 2x2 switches (one per wavelength) that can be configured to either drop the wavelength at a node or bypass that node.

Protection is provided using optical loop-back using the dual fiber ring. In the clockwise direction a band of wavelengths (A) is used for working traffic and in the counter-clockwise direction band (B) is used for working traffic. The working bands are reserved for protection in the opposite direction. A pair of 2x2 switches at each node is used to provide loop-back protection in the event that a loss of power (light) is detected (due to a fiber cut or node failure). An A/B multiplexer at each node is used to allow the protection bands to bypass that node. This bypass helps reduce the cost of the node, because those wavelengths do not have to be demultiplexed further.

Wavelengths from the C/D network can either terminate at a port on the AN's IP router (for an IP service) or be switched onto the feeder ring to provide a switched optical service. The AN's IP router can access the feeder ring using a number of tunable interfaces (3 shown). These ports are used to connect the routers to one another and to the backbone network. Having tunable ports, combined with the configurable optical switches will allow for reconfiguration of the electronic topology of the network.

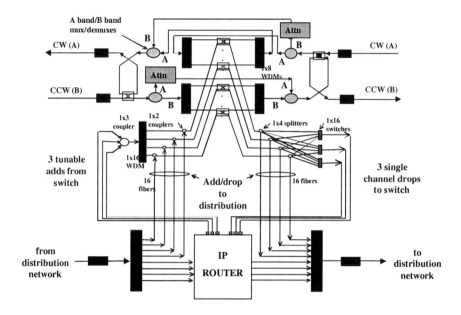

Figure 16. Feeder network AN

ACKNOWLEDGEMENTS

This work was supported by Darpa under the Next Generation Internet (NGI) initiative. The author wishes to acknowledge contributions from: Rick Barry, Vincent Chan, Angela Chiu, Nan Froberg, Mark Kuznetsov, Aradhana Narula, Kristin Rauschenbach , Adel Saleh, Brett Schein, Jane Simmons, and Eric Swanson.

REFERENCES

[BM99] Randy Berry and Eytan Modiano, "Minimizing Electronic Multiplexing Costs for Dynamic Traffic in Unidirectional SONET Ring Networks," ICC '99, Vancouver, CA, June 1999.

[BS95] Bo Li and K. M. Sivalingam, "Channel Access Protocols for High Speed LANs Using WDM: A Comparative Study", in Proc. SPIE Conference on All-Optical Communications Systems: Architecture, Control, and Network Issues, (Philadelphia, PA), pp. 283--294, Oct. 1995.

[CM98] A. L. Chiu and E. Modiano, "Reducing Electronic Multiplexing Costs in Unidirectional SONET/WDM Ring Networks via Efficient Traffic Grooming," Globecom '98, Sydney, Australia, November, 1998.

[CM00] Angela Chiu and Eytan Modiano, "Traffic Grooming Algorithms for Reducing Electronic Multiplexing Costs in WDM Ring Networks," *IEEE Journal on Lightwave Technology*, January, 2000.

[ES96] G. Ellinas and T.E. Stern, "Automatic Protection Switching for Link Failures in Optical Networks with Bi-directional Links," Globecom '96.

[FMB97] S.G. Finn, M. Medard, and R.A. Barry, ``A Novel Approach to Automatic Protection Switching Using Trees", ICC '97.

[FMB98] S.G. Finn, M. Medard, and R.A. Barry, "A New Algorithm for Bi-directional Link Self Healing for Arbitrary Redundant Networks," OFC '98.

[FPS89] D. W. Faulkner, et. al., "Optical Networks for Local Loop Applications," Journal of Lightwave Technology, November, 1989.

[FR94] N.J. Frigo, et. al., "A Wavelength Division Multiplexed passive optical network with cost shared components," IEEE Photonic Technology Letters, June, 1994.

[FR96] N. J. Frigo, " Local Access Optical Networks," IEEE Network, Dec., 1996.

[GLA96] B. Glance, et. al., "Novel optically restorable WDM ring network," OFC '96.

[GR97] O. Gerstel and R. Ramaswami, "Multiwavelength Ring Architectures and Protection Schemes," Photonic Networks, Advances in Optical Communications, pp. 42-51, Springer-Verlag, New York, 1997.

[IFD95] P.P. Iannone, N.J. Frigo, and T.E. Darcie, "WDM passive optical network architecture with bidirectional optical spectral slicing," OFC '95.

[KAM96] I.P. Kaminow, et. al., "A Wideband All-Optical WDM Network", IEEE JSAC, June, 1996.

[LS89] Y. M. Lin and D. R. Spears, "Passive Optical Subscriber Loops with Multiaccess," Journal of Lightwave Technology, November, 1989.

[LU96] X. Lu, et. al., "Mini-Fiber Node Hybrid Fiber/Coax Networks for Two way Broadband Access," OFC '96.

[LU98] Angelo Luvison, "The architecture of full service access network," OFC '98, San Jose, CA, March, 1998.

[MB96] J. Manchester and P. Bonenfant, "Fiber Optic Network Survivability: SONET/Optical Protection Layer Interworking," National Fiber Optic Engineers Conference 1996.

[MB99] E. Modiano and R. Barry , "Design and analysis of an asynchronous WDM local area network using a master/slave scheduler," Infocom '99, New York, NY, March, 1999.

[MC98] Eytan Modiano and Angela Chiu, "Traffic Grooming Algorithms for Minimizing Electronic Costs in Unidirectional SONET/WDM Ring Networks," CISS'98, Princeton, NJ, March, 1998.

[NM00] Aradhana Narula-Tam and Eytan Modiano, "Load Balancing Algorithms for WDM-based IP networks," Infocom, 2000, Tel Aviv, Israel, April, 2000.

[MC89] I. M. McGregor, "Implementation of a TDM Passive Optical Network for Subscriber Loop Applications," IEEE JLT, Nov., 1989.

[MO98] Eytan Modiano, "Unscheduled Multicasts in WDM Broadcast-and-Select Networks," Infocom '98, San Francisco, CA, March 1998.

[MO99] Eytan Modiano, "Random Algorithms for scheduling multicast traffic in WDM broadcast-and-select networks," IEEE/ACM Transactions on Networking, August, 1999.

[MUK92] B. Mukherjee, "WDM-Based Local Lightwave Networks Part I ," IEEE Network, May, 1992.

[NLM96] P. Newman, T. Lyon, and G. Minshall, ``Flow Labeled IP: A Connection Approach to ATM." IEEE Infocom, 1996.

[PB95] W. Puh and G. Boyer, "Broadband Access: Comapring alternatives," IEEE Communication Magazine, August, 1995.

[PG98] S. Perkins and A. Gatherer, "Two-way Broadband CATV-HFC Networks: State of the art and Future trends," Computer Networks, Vol. 31, No. 4, February, 1999.

[REK97] Y. Rekhter, *et. al.*, ``Cisco Systems' Tag Switching Architecture Overview," IETF RFC 2105, Feb. 1997.

[RS92] M.N. Ransom and D.R. Spears, "Applications of public Giga-bit networks," IEEE Network, March, 1992.

[RS96] R. Ramaswami and K. Sivarajan, "Design of logical Topologies for Wavelength Routed Optical Networks," IEEE JSAC, June, 1996.

[RS98] R. Ramaswami and K. Sivarajan, *Optical Networks*, Morgan Kaufmann, San Francisco, CA, 1998.

[SB97] S. Subramanian and R. Barry, "Wavelength Assignment in Fixed Routing WDM Networks,", Proc. of ICC '97.

[SGS98] Simmons, J.M., Goldstein, E.L., Saleh, A.A.M, "On The Value of Wavelength-Add/Drop in WDM Rings With Uniform Traffic," OFC'98, San Jose, CA, February, 1998.

[SKG94] G. N. M. Sudhakar, M. Kavehrad, N.D. Georganas, "Access Protocols for Passive Optical Star Networks," Computer Networks and ISDN Systems, pp. 913-930, 1994.

[SM00] Brett Schein and Eytan Modiano, "Quantifying the benefits of configurability in circuit-switched WDM ring networks," Infocom, 2000, Tel Aviv, Israel, April, 2000.

[WL89] S.S. Wagner and H. L. Lemberg, "Technology and System Issues for a WDM-Based Fiber Look Architecture," IEEE JLT, Nov., 1989.

[Wu92] T.-H. Wu, Fiber Network Service Survivability, Artech House, 1992.

[WW92] T.-H. Wu, W.I. Way, ``A Novel Passive Protected SONET Bidirectional Self-Healing Ring Architecture", IEEE Journal of Lightwave Technology, vol. 10, no. 9, September 1992.

Chapter 14

Optical Network Management

Alan McGuire
Core Transport, Internet & Data Networks, OP7, B29, Adastral Park, Martlesham Haeth, Suffolk, United Kingdom IP5 3RE. Email: alan.mcguire@bt.com

Abstract: Optical networks will play an essential role in meeting the demands for future communications bandwidth. With such large traffic volumes at risk network management is fundamental to the running of such networks. To achieve this on an industrial scale management solutions must be based on an unambiguous framework that describes the entities that need to be managed. This chapter describes such a framework and provides a high level summary of how it can be applied to the management of both simple and complex structures. Nevertheless management of this new technology is still at a very early stage and considerable effort is required within the industry before the vision of an optical transport network can be fully realized.

1. INTRODUCTION

The accelerating demand for bandwidth fuelled by growth in both the Internet and broadband services represents a major challenge to network operators. Globally operators are facing up to a shortage of fiber capacity in parts of their networks. Whereas this has been most acute in the US where growth in traffic is greatest it is rapidly becoming an international problem. Conventionally an operator would have increased capacity by deploying more fiber or introducing higher bit rate digital systems. Approximately three years ago a new option became commercially available, wavelength division multiplexing. This technology has emerged from the research labs and in a relatively short time frame has become an essential weapon in the transmission engineer's arsenal. Over 1000 of these systems have now been deployed globally. Depending on the number of channels utilized WDM increases the capacity of a fiber from between 2.5 – 10 Gbps with existing single channel systems to between 40 and 200 Gbps. With large traffic volumes at risk network management is an essential component of such systems. At the present time the management of these systems is relatively simple. In the next few years we can expect to see the introduction of more complex optical systems such as rings and cross-connect meshes. There has been a considerable amount of literature published regarding the transmission characteristics of such networks but very little on the network management challenge. Yet the lack of large scale industrial strength management systems represents a major barrier to the deployment of the optical network. This is particularly true in the existing competitive environment where automation of many of the tasks involved in running and maintaining the network will become a necessity to drive down operating costs and to provide faster provision of services.

In this chapter we shall examine some of the features of existing optical network management solutions and describe what will be required in order to manage the optical transport network of the future. But first we need to understand what it is that we want to manage.

2. OPTICAL TRANSPORT NETWORK ARCHITECTURE

In order to manage a communications network it is necessary to describe the entities that need to be managed in a rigorous and unambiguous manner. The integration of technology and function is now so great that it is impossible to accurately describe what functions a network element provides by means of a semi-formal technique. Furthermore, the achievement of successful scalable software systems is predicated on a clear definition of what is to be managed and its behaviour.

Transport networks can be described in terms of layer networks in accordance with the architectural principles cited in (ITU-T Recommendation G.805, 1995). Each layer network represents a set of inputs and outputs (or access points) that can be interconnected and the layer network is characterized by the information that is transported across it. This information, or characteristic information as it is termed, is a signal of characteristic rate, coding and format. Examples of layer networks include the VC and VP layers in ATM, the VC-12, VC-4, multiplex and the regenerator sections in SDH. A layer networks topology can be described in terms of subnetworks and links between them, as illustrated in Figure 1.

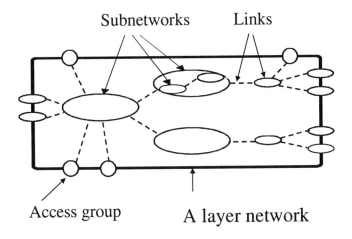

Figure 1: Components of a layer network

A subnetwork can be decomposed into smaller subnetworks interconnected by links, in other words it is recursive. This decomposition can, if required go from a global network right down to the smallest subnetwork that is equivalent to a single network elements cross-connect fabric. Connectivity in any layer network can be managed at the network level in the same manner independently of technology. In other words, once we know how to manage connectivity in one layer we should be able to do it for all layers. Managed objects that represent resources within a layer include connection, link, subnetwork, trail, network trail termination points (NW TTPs) and Network Connection Termination Points (NW CTPs). These managed objects represent an abstract view of the resource that can be manipulated by a network manager.

Layer networks have client/server relationships with each other, and in many cases one server layer may support different types of clients (Figure 2). An example is the VC-4 network layer which can support (is the server of) VC-3, VC-2-nc, VC-12, ATM VP etc. Whereas it can support these different clients the definition of its characteristic information is separate and distinct. In turn the VC-4 layer network is the client of the multiplex section. To get from one layer to another requires some processing to alter the characteristic information and this is provided by entities known as adaptation and termination functions. Adaptation functions provide functionality such as multiplexing/demultiplexing, frequency justification, timing recovery, alignment and soothing. Termination functions provide means of ensuring signal integrity supervision within a layer by means, for example, of error detection codes, trail trace identifiers, remote indicators and performance monitoring. The transfer of validated information is termed a trail.

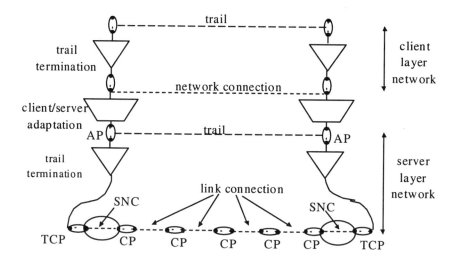

Figure 2: The client/server relationship
TCP – termination connection point, CP – connection point

In addition to providing a network view, the architecture can also be applied to network elements where adaptation and termination functions are combined to describe functionality as it can be observed from the inputs and outputs of the network element. The internal structure of the implementation (the equipment design) does not need to be identical to the structure of the functional model as long as the external observable behaviour is the same.

The management view of a network element is based upon an information model containing managed objects that can be manipulated by a management system. The definition of a managed object is derived from a specific part of the functional model. For example, generic trail termination point and connection termination point classes are defined generically such as CTP and TTP, from which technology dependent subclasses such as rsTTP and rsCTP (in SDH) can be developed using the object oriented principle of inheritance. These managed objects have attributes and behaviour that is manipulated from a management system; i.e., they can generate alarms or change their connectivity to other objects. The managed object effectively hides the implementation of the resource that it represents from the management systems and only provides information about aspects of that resource which are important from a network management view.

In essence, element managers manipulate managed objects and relationships between these managed objects within a network element, whereas network management is concerned with entities such as network connections, which may use resources from several network elements. The reader is warned however that this is a very simplified picture of the reality.

ITU-T Recommendation G.872 (1995), which is a technology dependent version of G.805, defines three optical layers, as shown in Figure 3, in the optical transport network (OTN):

- an optical channel (OCh) layer network that provides end-to-end networking of optical channels for transparently conveying digital client information of varying formats (e.g., SDH, PDH and ATM)
- an optical multiplex section (OMS) layer network that provides functionality for transport of a multi-wavelength optical signal
- an optical transmission section (OTS) layer network that provides functionality for transmission of optical signals on optical media

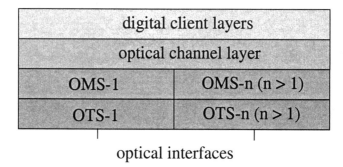

Figure 3: The Optical transport network layers

The optical transport network architecture of G.872 is extremely flexible and supports the following features (not all of which may appear in a single instance of a network):

– Unidirectional, bidirectional, and point-to-multipoint connections.
– Individual optical channels within a multiplex may support different client types.
– Cross-connection of optical channels can be accomplished by either wavelength assignment or wavelength interchange.
– Optical transport network functionality can be integrated with client functionality in the same equipment.
– Interworking between equipment with existing single-channel optical interfaces (ITU-T Recommendation G.957, 1999) and equipment containing optical transport network functionality.

The last two points are significant since they provide network operators with a degree of flexibility in the design of their networks, both now and in the future.

Each of these layer networks provides overhead for the operations administration and maintenance of its layer. OAM functions include continuity and connectivity supervision, defect indications (upstream and downstream), protection switching protocols and channels for transporting management information. Not all of these functions are found in each layer. Initially there was considerable debate with regard to the nature of the overhead at the optical channel level as some people viewed the optical channel as a transparent entity within the network, but such an entity, almost by definition cannot be managed. Instead of optical transparency the optical channel provides service transparency, and this is perhaps more central to the concept of optical networking as described by the ITU. The optical channel overhead is created within a digital frame that takes the client layer signal in the form of a continuous data stream and adds both the overhead and a forward error correction mechanism for improving system margin. This frame is known as a digital wrapper.

The concept of the digital wrapper can be viewed in one of two ways, firstly as an overhead that is carried end-to-end and so may be switched in optical channel subnetworks. However, this requires that the optical channel remain in the optical domain from beginning to end. Alternatively at intermediate points part, but not all, of the overhead may be processed. At such points the optical channel is regenerated, the FEC is computed and the relevant parts of the overhead are processed. There is however no need to obtain access to the payload. In contrast to all optical networks, optical transport networks utilize the strengths of both electronics and optics to produce much more scalable networks. Nevertheless there is considerable debate within the standards community as to the appropriate overheads and the choice of FEC.

The following figures (4, 5 and 6) show the relationships between the resources of the transport network and managed objects. The actual objects are subject to change in the standards' body and the ones provided here are for indication purposes. With these in hand we have the basic resources of the optical transport network that need to be managed.

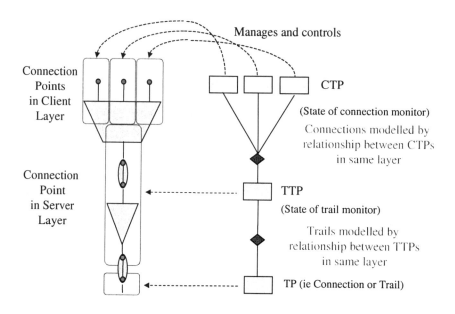

Figure 4: Relationship between equipment resources and managed objects. The adaptation and termination functions and their connectivity are managed and controlled via managed objects

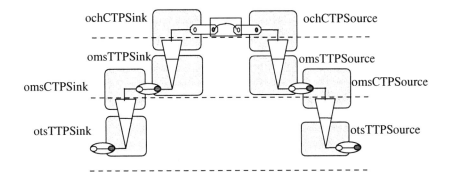

Figure 5: Relationship between optical transport resources and optical managed objects. The example shown can be considered as an optical channel cross-connect with two ports

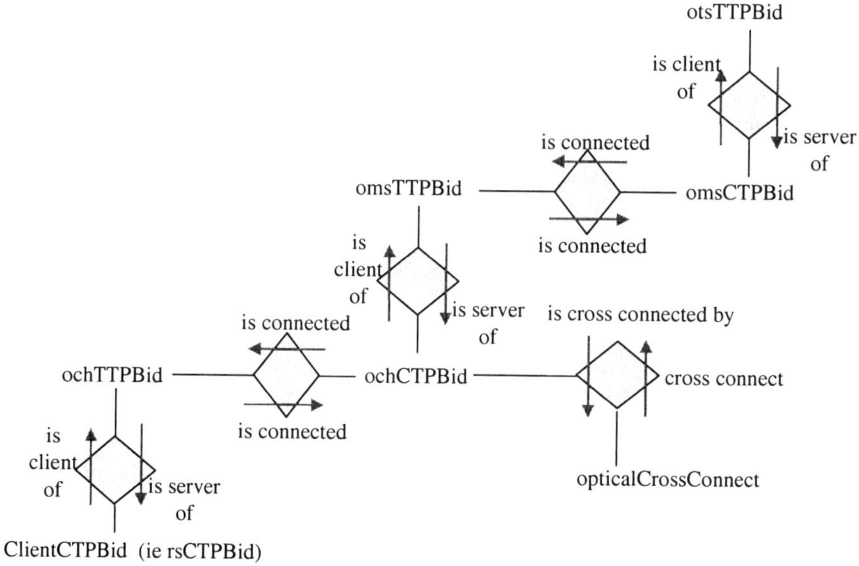

otsTTPBid

is client
of

is connected

omsTTPBid — omsCTPBid

is connected

is
client
of

is connected

is server
of

is cross connected by

ochTTPBid — ochCTPBid — cross connect

is connected

is
client
of   is server
of   opticalCrossConnect

ClientCTPBid (ie rsCTPBid)

Figure 6: Entity-Relationship between optical managed objects

3. OPTICAL NETWORK MANAGEMENT ARCHITECTURE

Network management is an activity that allows a network operator to administrate, plan, provision, install, maintain, monitor and operate a telecommunications network and its services. Within the ITU-T, the general architecture of the management network is described in terms of a Telecommunications Management Network (TMN) as described in (M.3010, 1996). The TMN concept can be applied to a variety of scenarios including public and private networks, exchanges, digital and analog transmission systems, ISDN, circuit and cell-based networks, operations systems, PBX's and signalling systems. It can also be applied to optical networking. Although there are some issues concerning the viability of TMN, the principles described below are very general and can be applied to any management architecture.

The basis for the management of the optical network and its network elements is the following simple rule:

The management of the optical layers must be separable from its client layers, or, put another way the management of the optical layers is not dependent on a particular client layer even if they are in the same box.

This rule is of fundamental importance; it ensures that the optical network can support both existing and future, unforeseen, clients. It does not necessarily mean that a link in an optical network simultaneously supports a wide variety of clients, rather it suggests that regardless of which protocol is supported in an instance of an optical network, the architecture and management is the same as in all other instances of an optical network. It allows WDM in a stand-alone platform connected to existing SDH equipment to be managed in the same way as WDM and SDH integrated within the same physical platform. This principle is also implicit in other technologies such as ATM.

An optical management network (OMN) is defined as a subset of a TMN that is responsible for managing optical network elements. An optical network element, ONE, is that part of a network element that contains entities from one or more OTN layer networks. According to this definition the functions of an ONE may be contained in a stand-alone physical entity (an equipment) that may or may not support other layer networks. Where non-OTN layer network entities are in the same equipment they should be considered as being administered separately from OTN entities. This separation is a logical separation and it indicates that the management of different technologies is not interdependent, and nor should it be. It does not prevent them from both being available on a single element manager albeit as separate applications.

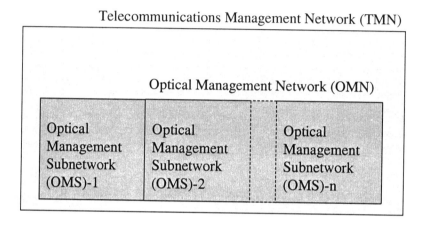

Figure 7: Architecture of management networks

An OMN may be subdivided into a set of optical management subnetworks, OMSNs (see Figure 7). In formal terms an OMSN can be defined as a separate set of optical transport network embedded communications channels and associated intrasite data communications links which have been interconnected to form an operations data communications control network within any given OTN transport topology. In other words an OMSN represents a local communications portion of a network. The entities contained within the OMSN and OMN are illustrated in Figure 8. The operating system may be an element manager, a subnetwork manager, or a network manager and there can be a hierarchy of such systems.

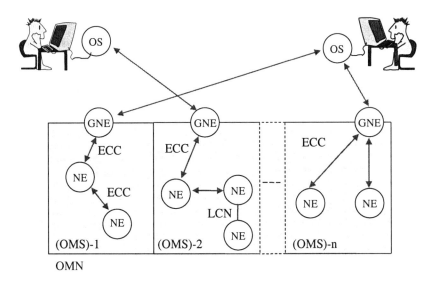

Figure 8 Components of the management system

The concepts described above can be understood with reference to a 1+1 optically protected line system, similar to those being deployed in the BT network, containing line terminals and optical line amplifiers, which is used to interconnect SDH line terminals.

The optical line terminals and optical line amplifiers are managed within a single OMSN. It should be noted that an OMSN might be applied to any network topology, not just to optical line systems. The SDH terminals are outside of the OMSN. Within the OMSN management communications is by means of a messaging channel known as an embedded communications channel (ECC) that is available to every network element. In the context of G.872 it is generated and terminated on an OTS basis. This messaging channel is only between ONEs. The ECC may be used to carry event notifications and alarms from network elements or to convey instructions/interrogations to the network element such as remote protection switching requests or channel status. The ECC is carried within an optical supervisory channel on a wavelength separate to all of the other optical channels. It is inserted and extracted separately from the other optical channels. For example in an optical line amplifier the optical supervisory channel is extracted, converted to a digital signal, processed then reinserted, whereas the other channels are amplified in the optical domain. A supervisory channel also carries other information for the purposes of operations administration and maintenance.

All of the network elements in the OMSN are managed by an operations system, OS, (in this case it is an element manager). The OS is directly connected to the two optical line terminals by means of a data communications network. This network does not extend to the optical line amplifiers. Instead the line terminals act as gateway network elements that allow the OS to communicate with the line amplifiers by means of the optical supervisory channel. The element manager is considered to be part of the optical management network and can be used to manage optical network elements in a number of OMSNs. Large networks will contain many OMSNs and a number of element managers. Management workstations are considered to be outside of the OMN but may be located with the OS or remotely.

An example of a possible management hierarchy for managing optical line systems is now considered. Optical line systems are in the lowest tier and have access to the element managers via diverse data communications links. If a data link or router is unavailable at one end of the line system the system can be managed from the other end. The element managers (EM) are located at a network administration and computing centre (NACC). This allows the element managers to be maintained in central locations where they can be maintained by specialized personnel.

The element managers are connected (again by way of diverse links) to workstations at a remote network operations unit (NOU) which would be responsible for managing not only the optical network but the SDH network as well. Such a management network should be designed to meet the following operational requirements:

- visibility of the optical network 24 hours a day, 365 days a year
- resilient network communications between workstations, element managers and optical line systems to ensure availability even in the event of management network failures
- provision of fallback facilities for the NOU

– management of the data-communication networks' routers, links and hubs
– provision of remote access facilities to allow supplier support under the control of the operations personnel
– support of multiple user accounts wishing concurrent access
– visibility of multiple network elements simultaneously
– support of real time event handling and remote configuration
– allowance of management of optical line systems in the same way as other technologies without the need for being an optical transmission expert (It is essential that the personnel required to manage the optical network elements are also skilled in managing multiple technologies. A common approach to management is therefore required).

4. THE ROLE OF THE ELEMENT MANAGER

The element manager can be thought of as a go-between for the user located at the workstation and the network element. Its role is to process information for the purpose of monitoring/coordinating and/or controlling network element functions. In TMN terms the relationship between the managing entities and the managed entities is described by means of a manager/agent relationship. A manager represents the managing process whereas an agent represents the managed process, both of which are software applications. In essence the manager software provides the management functions and management services whereas the agent applications provide access to the management information related to the managed resources or managed objects described earlier. The manager/agent model is hierarchical with the manager sending requests and receiving replies. The operations system or element manager contains the manager and the agent is contained within a network element. One manager is normally associated with many agents. The element manager can be used to retrieve information from network elements such as:
– network element id and location
– inventory, build number
– configuration
– current alarms or historical alarms from logs
– performance monitoring threshold levels
– active wavelengths, power levels, supported client signals, etc.
– ECC status

It can also be used to set information in the network element such as:
– initialization of network element

- optical protection state
- latching alarms, inhibiting alarm reporting
- initiation of 15 minute and 24 hour performance monitoring bins
- network element Id, Card Id etc.
- network element time
- single channel power levels and aggregate power levels

From the network element the element manager receives the following

- current alarms (these may be traffic or non-traffic affecting)
- threshold crossing alerts
- event notifications such as protection switching events, insertion/removal of cards
- local logging, remote logging, user access information

The element manager also provides facilities such as security, in the form of passwords and user profiles which determine the privileges of a user, (read only, expert, administrator etc.), back-up facilities, network element synchronization, and other administrative features.

5. FAULT MANAGEMENT IN STAND-ALONE SYSTEMS

A stand-alone optical line system is one that can be deployed without having to modify the SDH systems connected to it. From the perspective of the SDH systems the optical line system is "invisible" and can be considered as virtual fiber. The network elements of the line system contain optical layer network entities which are managed as part of an OMSN and layer networks which are necessary for interworking with the legacy systems (these layer network entities provide part of the functions of a transponder). Although this interworking function within the line system, is not part of the OMSN it is managed by the same element manager as the rest of the line system. It should also be noted that different interworking functions (transponders) could be used to support a variety of systems, not just SDH. The layer networks of the SDH systems belong to a SDH management subnetwork (SMS) and may be managed by a separate element manager. Management communications is available between elements within the optical line system and between the SDH systems but not between the line systems and the SDH systems.

Where stand-alone systems are used, it is highly likely that operations personnel will initially rely upon the element managers of the SDH systems for identifying network problems. This is due to the size of the installed SDH base. The number of SDH circuits is likely to be considerably larger than the number of optical line systems, and personnel will not automatically assume that a number of circuit failures occurring at the same time correspond to failure of a multi-wavelength optical line system. The natural assumption will be that there is a cable or SDH equipment failure.

Consider the case of a fiber failure that occurs within the optical line system. This will result in downstream loss of signal that generates alarms on the SDH network manager. Upon receipt of these alarms the SDH manager will indicate the affected circuits and it will be assumed that there is a multiple fiber failure within a duct. Upon checking network databases it will become apparent that one or more of the underlying fibers supports a WDM system. The operations personnel will then check the optical element manager thereby determining the affected optical transmission section. If the working route in the protected optical line system is affected, a loss of line alarm and a protection switch notification will both be present. Otherwise channel failures and possibly equipment failures are likely causes. As a side note, if no alarms are detected on the optical element manager then the fault can be assumed to be between the optical line system and the SDH system receiver. Once the nature of a fault has been determined and/or localized field personnel are dispatched to repair it under the supervision of the NOU.

6. FAULT MANAGEMENT IN INTEGRATED SYSTEMS

The majority of optical line systems currently being deployed globally are stand-alone systems that are being used to support legacy systems. Where new build is being considered, there are a number of advantages to integrating the optical transport layer networks into SDH equipment. The principal advantage is the removal of optical transponders, which represent a significant part of the cost of optical line systems. In this example every network element is also an optical network element and the optical layers are managed within an OMSN whereas the client layer network entities within the end terminals are managed within a SMSN. Although these two management subnetworks are logically separate they can be managed from a single element manager. In contrast to the stand-alone case, management communications is available between all of the network elements, and network faults can be diagnosed in a single step manual process via a single element manager. A further advantage is that the data communications network structure is more straightforward than the stand-alone case.

7. MANAGEMENT OF OPTICAL RINGS

The management of optical line systems is relatively straightforward. In addition to element managers the network operator should hold one or more configuration databases that provide a record of the relationships between the client circuits, the optical layers, fiber, cable and duct and the location and type of optical network elements. In addition, for a network containing a large number of optical line systems the operator may consider introducing an interface between the optical element managers and a network level fault manager for the purpose of fault correlation across technologies and identifying the root cause of the fault. This will also minimize the number of alarms that have to be processed by the user.

Optical rings, although only slightly more complex than optical line systems, require far more network management capabilities to be provided by the supplier. The major difference is that the optical channel is now reconfigurable. In addition to simple element management, it is a requirement that the optical connections between network elements are visible. This requires some level of network management of links, trails and subnetwork connections. This ability to reconfigure means that it is no longer acceptable to have a configuration database recording a static relationship between an optical channel and its server layers. Manual updating of this relationship is not reliable and it will not be possible to keep up with configuration changes due, for example, to protection switching. In case of a discrepancy, which is correct the entry in the database or the connection in the network? For this reason it is better to automate the configuration process and ensure synchronization with the network by means of an application programming interface between the database and the element managers:

The network management system for a ring also requires the following capabilities
- an electronic interface between the element managers and the subnetwork manager to allow event messages to be passed to the subnet manager for surveillance and configuration messages to be passed to the element managers
- capability of detecting and notifying mis-connections. The optical channel overhead provides such a capability by means of a trail trace,

which compares the incoming label with the expected label. Trail trace is not required at the optical channel level in an optical line system, but is in any system where optical channels can be reconfigured

- ability to suppress alarms that are associated with configuration changes and filtering to reduce consequential alarms
- indication of circuit status, e.g. reserved, in-service, failed routings, etc.
- support for monitoring repair work and to provide a log of repair activities until the fault is resolved
- support for concurrent users involved in planning, configuration, surveillance and repair
- planning tools. In contrast to digital networks 3R regeneration is not available in every network element for every channel. Therefore, when designing the network it will be necessary to measure fiber parameters such as loss and polarization mode dispersion to calculate if the network elements can be placed in the desired locations and that any and all paths can be configured within the ring for both working and protection modes. It is not acceptable to put the equipment into the ground and then discover it will not work when an attempt is made to set-up the first or indeed the last circuit. This represents a major challenge for optical networking. The greater the number of optical network elements the greater the problem. This indicates why the digital wrapper is likely to be more successful than any all-optical counterpart.

As the number of optical rings increases the operator will eventually want to interconnect traffic between them. There are essentially two ways in which this can be achieved. The simplest is to manually connect a tributary from an OADM in one ring to a tributary on an OADM in another ring. The result of traffic churn is that such a connection may need to be altered many times. For some network operators this may not be seen as a problem, particularly if it is a small network. However, for larger operators there is considerable operational saving to be made by providing automatic provisioning. This can be achieved by means of an optical cross-connect between the rings.

How does the network operator set up such a connection automatically? Again there are two ways. The first is that personnel in the network management centre plan the route using a network database and then remotely configure a connection through each network element (having to set up each network element separately) in the end-to-end path. This is time consuming, laborious and prone to error. If the supplier provides only element management, the network operator may have no other choice. However, the number of circuits that need to be set-up and torn down each day will increase with network size, and this method will not scale.

The second way is to provide auto-routing capabilities. This requires a network database that contains a model of the network, including not only the equipment contained within it but also the connectivity of all of the links and subnetworks in each of the layer networks. It is important to reiterate that the information stored in this configuration database should be obtained from the network (via the management interfaces to the network elements) and not from manually keying it in. This ensures that the database reflects what is in the network. Auto-routing tools would use this database to calculate the appropriate path through the network, based on algorithms and planning rules. Once the path is determined the network management system provides configuration commands to the network elements. Such an approach allows large numbers of circuits to be configured per day. Instead of drilling down into each network element one after the other, the user need now only specify the two end points. Whereas some network operators currently have such a capability for SDH and can configure the network with a small number of people it is of little use if the underlying optical layer still requires manual configuration between its subnetworks and its connections to the SDH network. Unfortunately this may be the case for some time to come.

At the present time nearly all optical network management is based upon element managers, and first generation subnetwork managers are likely to manage small subnetworks such as rings. These may be provided by suppliers or built by the network operator. If we assume the former it is obvious that if there is more than one supplier in the network it will be necessary to provide a network manager that can manage both subnetworks and set-up connections across them. At this point it should be obvious that we are now talking about very large systems and the major problems are volumetrics and scalability.

8. MANAGING INTERCONNECT

Up until now the discussion has focussed on network management within a single administrative domain but there will also be a need to network optical channels between administrative domains. There are some differences between inter- and intra-domain network management, and these can be understood by means of reference to a simple interconnect regime – an optical line system connecting two operators. Such an environment is likely to be controlled by national or regional regulators. A consequence of this is that interconnection will probably be across a transversely compatible interface that allows network operators to have different vendors equipment at each end. The boundary between operators in this system can be considered as an accessible point on a fiber (e.g., via a footway box), known as the point of interconnection (POI). The optical channel link connection between administrative domains will be regenerated at both ends (to ensure that the ingress and egress signals are of the highest possible quality) and tandem connection monitoring will be employed. However, no network management information will be transferred between network operators using the embedded communications channel. This is unsurprising; network operators are unlikely, except by mutual consent, to allow other operators management access to their network. Instead the operators will rely on a single ended-maintenance strategy that allows management of the network element within their own administrative domain and the use of defect indicators to detect a far end problem in the other operators domain. Where problems are identified on the fiber, each operator will dispatch maintenance personnel to test each end of the link and identify which side of the POI it occurs. Manual processes are therefore required between the operators in order to co-ordinate the management of this link. A similar management strategy can be envisaged for international links.

9. SUMMARY

This chapter has provided a brief overview of the network management domain. It should be recognized that optical networks will be extremely difficult to operate without large scale management systems. The current trend of providing hardware products well in advance of systems that can manage them constitutes a major threat to successful widespread deployment.

References
1. ITU-T Recommendation G.805, "Generic Functional Architecture of Transport Networks."
2. ITU-T Recommendation G.872, "Architecture of Optical Transport Networks."
3. ITU-T Recommendation G.957, "Optical Interfaces for Equipment and Systems Relating to the Synchronous Digital Hierarchy."
4. ITU-T Recommendation M.3010, "Principles for a Telecommunications Management Network."

Chapter 15

HITLESS SWITCHING IN A DATA-CENTRIC WDM OPTICAL NETWORK

Georgios Ellinas
Telcordia Technologies
georgios@ctr.columbia.edu

Krishna Bala
Tellium, Inc.
kbala@tellium.com

Chien-Ming Yu
Columbia University

Abstract   This paper proposes a methodology for reconfiguring a Wavelength Division Multiplexed (WDM) optical network to adapt to changing traffic requirements at the Asynchronous Transfer Mode (ATM) layer. The ATM network layer in turn is reconfigured to adapt to changes in traffic patterns. These changing ATM network topologies, or connectivities, are known and are used to design the network. The proposed method sizes the ATM switches and then assigns wavelengths between pairs of ports at the switches in order to support the required ATM network topologies and their reconfiguration in a hitless manner. The ATM switches are sized at each node so that they can support the union of an ATM topology and the next topology to which it reconfigures. Upper and lower bounds on the number of wavelengths needed at the WDM layer for such a hitless reconfiguration are found and two wavelength assignment heuristic algorithms are proposed. These heuristic algorithms are then implemented and their results are compared with the lower and upper bounds.

1.   INTRODUCTION

Wavelength routed optical networks using WDM cross-connects to provide wavelength re-use have been proposed [Acampora, 1994; Alexander et al., 1993; Brackett et al., 1993; Hill et al., 1993]. This work considers end users attached to access stations equipped with ATM switches, communicating with each other using fixed packets or cells. The topology representing the inter-connection pattern among the ATM switches is called the ATM layer topology. Interconnection "links" between two ATM switches at the ATM layer are es-tablished by selecting a path between them in the WDM network, assigning a wavelength for that path and setting the cross-connect states of the optical switching elements for the corresponding wavelength in an appropriate man-ner [Acampora, 1994; Alexander et al., 1993; Bala et al., 1995; Brackett et al., 1993; Hill et al., 1993; Ramaswami and Sivarajan, 1995]. It is envisioned that wavelengths in the network will be used to establish optical connections among access stations such that the connections remain in place for durations of time that are large compared to individual packet or cell transmission times.

Reconfiguration within the WDM network can be used to allow the ATM layer to adapt its connectivity to changing inter-nodal traffic [Acampora, 1994]. This work considers a particular scenario in which the changes in traffic patterns and the corresponding ATM layer configurations are known in advance. The interconnectivity at the ATM layer that best matches a given traffic pattern is determined by some other method (ATM network design) which is not the subject of this work. It should be noted that the de-sign of the ATM topologies is not independent of the wavelength assignment problem. Since there are many ways to design a topology to meet the ATM traffic demand, the ATM topologies could be designed to maximize the overlap between subsequent topologies and thus minimizing the number of required wavelengths. However, since the problem is already complex in its current form, this work treats the wavelength assignment problem independent of the ATM topology design problem.

The paper studies the reconfiguration of the WDM optical layer due to chang-ing traffics at the ATM layer, resulting in a requirement for changing topologies at the ATM layer. ATM networks are already designed to change in response to a failure and have extensive automatic topology discovery features. In this case, the ATM network operates on a changing physical topology as follows : the old topology stays in place until the new topology has been discovered and routing does not begin until the new topology has been established. In general, the ATM layer network topology (T) may be required to change a number of times in the order T_0 to T_1, T_1 to T_2, ..., T_i to T_{i+1}, ..., T_{N-2} to T_{N-1} and T_{N-1} back to T_0 that starts the process over. For a complete cycle of topologies the order of configurations and the configuration patterns are fixed. Nevertheless, since the system is cyclical and there are N possible ways to name the topologies in the cyclic direction (assuming the ATM layer network topology is required to change N times per cycle), no topology is treated different from the others.

There might be several practical applications of a system that evolves cyclically with a fixed order of configurations. For example, a network interconnecting financial institutions may have topology T_0 in the morning where the data traffic is relatively heavy, topology T_1 in the early afternoon where the traffic subsides, and topology T_2 at night where the institutions exchange data for the day's transactions and the data traffic is very heavy.

The proposed method sizes the ATM switches and then assigns wavelengths between pairs of ports at the switches to support the required ATM network topologies and their reconfiguration in a hitless manner. Hitless reconfiguration is defined as the reconfiguration process without the loss of even a single cell as a result of this process. The goal of this work is to ensure service continuation throughout the reconfiguration process, thereby ensuring that a customer does not lose any information during that process.

Previous reconfiguration methods discussed in [Auerbach and Segall, 1995; Gerstel and Segall, 1995; Labourdette et al., 1994; Lee and Li, 1996] did not address the hitless aspect of the problem shown in this work. In those cases, connections had to be torn down and re-established during reconfiguration, which in turn introduced some type of service degradation in the network, including packet loss and de-sequencing.

Section 2 describes the general network architecture and the ATM network layer. Section 3 explains the transition process from one topology to the next and presents a sufficient sizing of the ATM switches in order to provide for hitless reconfiguration of the ATM layer. This is followed in Section 4 by bounds on the number of wavelengths required in the network for the purpose of achieving hitless reconfiguration when dynamic and fixed wavelength assignments are used. An exact solution is given for a single topology and lower and upper bounds are found when the network evolves from any topology T_i to topology T_{i+1} in a hitless manner. Heuristics for a dynamic and fixed wavelength assignment as well as results from their implementation are presented in Section 5. Conclusions follow in Section 6.

2.   NETWORK ARCHITECTURE

Network access stations transmit and receive optical signals to and from the WDM network. The ports of the ATM switch are attached to a laser (receiver) array of W wavelengths so as to transmit (receive) optical signals to (from) the WDM network. In this case, cell by cell wavelength translation functionality is provided using the electronic ATM switch. As an example, Fig. 15.1 shows a cell entering the ATM switch on wavelength λ_1 and leaving on wavelength λ_w (i.e., wavelength λ_1 is translated to wavelength λ_w).

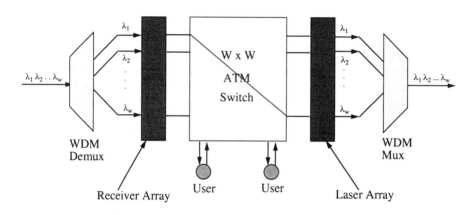

Figure 15.1   Network Access Station (NAS) architecture.

The complete network architecture is shown in Fig. 15.2(a). It demonstrates an ATM network layer over a WDM network. The WDM cross-connects used in the WDM network are $K \times K$ optical switching elements that can be dynamically reconfigured to achieve any permutation of input to output ports independently for each of the W wavelengths, $\lambda_1, \lambda_2, ..., \lambda_w$. Fig. 15.2(b) shows the resulting interconnectivity at the ATM layer as a result of wavelength routing at the optical layer shown in Fig. 15.2(a).

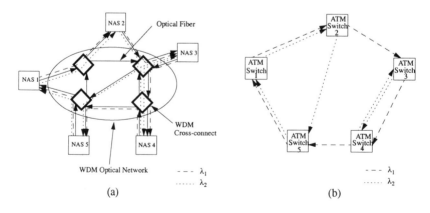

Figure 15.2   (a) ATM network layer over a WDM optical network (b) ATM network layer.

3. SIZING OF ATM SWITCHES FOR HITLESS RECONFIGURATION

A sufficient theoretical condition for the reconfiguration process to be hitless is developed in this section. Although in practice there could be a loss depending on the specific instance, the proposed scheme has a very good chance of allowing hitless reconfiguration under the conditions specified. The assumptions underlying this work are:

- Each ATM network topology is required to be in place for a long period of time (orders of magnitude longer than individual cell times).

- Even though a change of traffic pattern (requiring a transition from topology T_i to T_{i+1}) will not occur abruptly but over a period of time, no change in traffic patterns requiring a transition from topology T_{i+1} to T_{i+2} takes place before the change from topology T_i to T_{i+1} is completed.

- The ATM layer is embedded on an optical layer that is non-blocking, i.e., the WDM network can always be set-up to allow a wavelength path between any pair of ATM switches.

- Any wavelength routing performed is subject to the Color Clash (CC) constraint. This constraint states that optical signals simultaneously sharing a single fiber are required to have different wavelengths. Note that this constraint also applies to the input and output fibers of the access station ATM switches. Since the third assumption states that the optical layer always allows any wavelength path, special care has to be taken to ensure that the CC constraint is satisfied at the fibers through which the ATM switches access the WDM network.

- The frequency of topology changes is lower than the lifetime of the ATM connections.

An example requiring three changes in the ATM topology is shown in Fig. 15.3. Initially, traffic conditions require that the topology at the ATM layer be T_0. After a long enough period of time (orders of magnitude longer than individual cell transmission time), the traffic pattern changes so that topology T_1 is required at the ATM layer instead of T_0. From a practical standpoint, the change in traffic pattern (from T_0 to T_1) is not instantaneous but it occurs over a finite, although short, period of time. Similarly, the traffic pattern changes require a transition from topology T_1 to T_2 and from topology T_2 to T_0 and the process starts all over again.

Definitions :

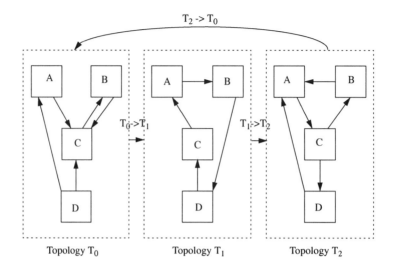

Figure 15.3   Reconfiguration of the ATM network layer for changing traffics.

The ATM network topologies are modeled by digraphs whose vertices v represent the ATM switching nodes and directed edges e represent the links interconnecting the ATM switches. V now represents the set of the ATM switching nodes and E_i represents the set of the directed edges interconnecting the vertices corresponding to the ith ATM network topology ($T_i = G(V, E_i)$). Topology $T_{i,j}$ ($T_{i,j} = T_i \bigcup T_j = G(V, E_i \cup E_j)$) represents the union of the ith and j^{th} ATM network topologies and $d^+\{T_i(u_k)\}$, $d^-\{T_i(u_k)\}$ represent the outdegree and indegree respectively of node u_k for a given ATM network topology T_i.

Observation 1: Given a configuration cycle $T_0, T_1, ..., T_{N-1}, T_0...$, sizing the ATM switches as shown below is sufficient to provide for hitless reconfiguration of the ATM layer under the conditions specified in this work.

$$p^+(u_k) = \max_{i=1}^{N} d^+\{T_{(i-1),(i)modN}(u_k)\}$$

$$p^-(u_k) = \max_{i=1}^{N} d^-\{T_{(i-1),(i)modN}(u_k)\} \qquad (15.1)$$

where $p^+(u_k)$ is the number of output ports on the ATM switch and $p^-(u_k)$ is the number of input ports on the ATM switch represented by node u_k. For simplicity, the switches are chosen to be symmetrical (same number of input and output ports), e.g., having size $p \times p$ where $p = \max\{p^+(u_k), p^-(u_k)\}$.

Equation 15.1 states that the switch size at each node should support the union of all "adjacent" topologies T_i and T_{i+1} at that node. Suppose the ATM layer topology is required to change from T_0 to T_1 and then later to T_2. The transition from topology T_0 to T_1 is achieved by first allocating all the links of

T_1 without removing the links from T_0. Thus, the links from the union of both topologies T_0 and T_1 are supported simultaneously. Then, the links of T_0 are "disabled" i.e., no new traffic is allocated to them and all new traffic is carried on the links of T_1. The assumption is that the previously allocated traffic on the links of T_0 will leave the network after a finite period of time. As this happens, the links of T_0 are removed from service and the network completes its transition from topology T_0 to T_1 in a hitless manner. The same process is repeated for all transitions from one topology to the next. According to equation 15.1, nodes A, B, C and D in Fig. 15.4 require 4×4, 3×3, 2×2 and 3×3 ATM switches respectively in order to achieve hitless reconfiguration.

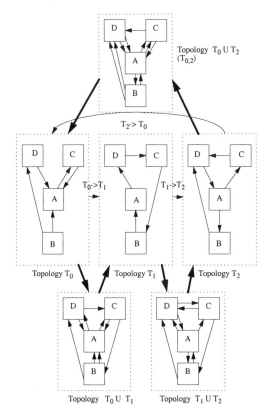

Figure 15.4 Union of adjacent ATM network topologies.

This section proposed a method to size the ATM switches appropriately (independent of the WDM layer) so as to provide hitless reconfiguration. Since the ATM layer is supported over a reconfigurable WDM optical layer, the focus of the following sections is the WDM layer. In particular, the next section presents bounds on the number of wavelengths required in the network to achieve hitless reconfiguration and proposes wavelength allocation algorithms that assign wavelengths between pairs of ATM ports that require an ATM layer link between them.

4.   WAVELENGTH ASSIGNMENT TECHNIQUES

This section attempts to find the number of wavelengths required in the network for the purpose of achieving hitless reconfiguration under the conditions specified here. One of the assumptions states that the WDM layer can support the required ATM layer topologies. Thus, only blocking due to the CC condition on the edges of the network, where the access stations transmit and receive optical signals on multiple wavelengths, is considered.

Lemma 1 :   *For a single ATM network topology T_i with M ATM switches, wavelengths can always be assigned to the ports of the ATM switches such that W, the number of wavelengths required in the WDM network, is given as follows:*

$$W = \max_{j=1}^{M}(d^+\{T_i(u_j)\}, d^-\{T_i(u_j)\}) \tag{15.2}$$

For the access station architectures shown in Fig. 15.1, each port of the ATM switch is associated with a single (distinct) wavelength and these wavelengths are multiplexed onto a single fiber before entering the WDM network. Clearly, based on the CC constraint it follows that :

$$W \geq \max_{j=1}^{M}(d^+\{T_i(u_j)\}, d^-\{T_i(u_j)\})$$

Wavelengths should be assigned to the edges of each ATM network topology T_i such that no two incoming or outgoing edges at a node have the same wavelength. Such an assignment of wavelengths (colors) to the edges of topology T_i will satisfy the CC constraint at the access stations. This problem is similar to the problem of *digraph edge coloring*. A significant difference between the two problems is that for digraph edge coloring, no more than one incoming or outgoing edge can use the same color, whereas in the current problem, an incoming and an outgoing edge can have the same color. The problem can be reduced to that of *bipartite digraph edge coloring* by using the construction described below :

Each node u_k in the network is divided into two nodes u'_k and u''_k with u'_k containing all the outgoing ports (no incoming ports) and u''_k containing all the incoming ports (no outgoing ports). For every (k, j) edge in the original network, an edge from u'_k to u''_j is created [Bala et al., 1996] (Fig. 15.5(a)). Clearly, the constructed network is bipartite. The edge coloring of the bipartite graph is done such that no two edges incident at a node have the same color.

Thus, edge coloring the bipartite graph will produce the desired wavelength assignment in the original network without violating the CC constraint. It has been shown that the edge chromatic number (number of colors required to edge-color a graph) of a bipartite graph is equal to the maximum among the degrees of all nodes [Thulasiraman and Swamy, 1980], which is exactly what equation 15.2 states. Bipartite graph edge coloring algorithms based on bipartite matching can be found in the literature [Papadimitriou and Steiglitz, 1982]. The wavelength assignment corresponding to the example of Fig. 15.5(a) is shown in Fig. 15.5(b). Note that the number of wavelengths required is equal to the largest degree in the network.

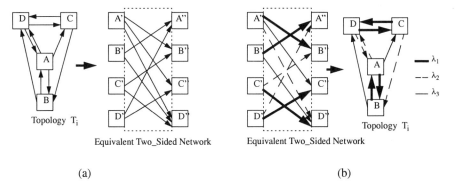

(a)   (b)

Figure 15.5   (a) Two sided network construction (b) Wavelength assignment in a two sided network.

For the scenario considered here, say the ATM layer topology is required to change from topology T_0 to T_1 and then later to T_2. As explained in the discussion following Observation 1, the transition from topology T_0 to T_1 is effected by first allocating all the links of T_1. Thus, the links from the union of both topologies T_0 and T_1 are supported simultaneously. Based on Lemma 1, to color the edges of the union topology $T_{0,1}$, a number of wavelengths equal to the largest size switch (node degree) in the union topology $T_{0,1}$ is required. As explained earlier, the links of T_0 are "disabled" and they are gradually removed from service. For the transition from topology T_1 to T_2, the network should support the union of the two topologies T_1 and T_2. If wavelengths could be re-assigned to the links of T_1 the union topology $T_{1,2}$ could be colored with a number of wavelengths equal to the largest size switch (node degree) in the union $T_{1,2}$. However, this will not result in a hitless transition. For a hitless transition, the new links of topology T_2 should be assigned to wavelengths such that no CC occurs at the access links with the wavelengths previously assigned to T_1 during the transition from T_0 to T_1. Clearly, this problem is more complicated than the one presented in Lemma 1.

4.1.   LOWER BOUND ON NUMBER OF WAVELENGTHS REQUIRED

Definitions :

M is the number of nodes in all topologies considered.

N is the number of times per cycle the ATM layer topology changes.

$d^+\{T_{i,(i+1)}(u_j)\}$ = outdegree of node u_j (ATM switch) in the union of adjacent network topologies T_i and $T_{(i+1)}$.

$d^-\{T_{i,(i+1)}(u_j)\}$ = indegree of node u_j (ATM switch) in the union of adjacent network topologies T_i and $T_{(i+1)}$.

$\max_{j=1}^M(\max_{k=1,k\neq j}^M(d^+\{T_{i,(i+1)}(u_j)\} + d^-\{T_{i,(i+1)}(u_k)\})) = max_{sum2}(T_{i,(i+1)})$ is the maximum of the summation between indegree and outdegree for all node combinations in topology $T_{i,(i+1)}$.

$K_{(u,v)}(T_{i,(i+1)})$ is defined as the number of edges that originate from node u and terminate at node v in topology $T_{i,(i+1)}$.

$\max_{j=1}^M(\max_{k=1,k\neq j}^M((d^+\{T_{(i-1),i,(i+1)}(u_j)\}) + (d^-\{T_{(i-1),i,(i+1)}(u_k)\}))) = max_{sum3}(T_{(i-1),i,(i+1)})$ is the maximum of the summation between indegree and outdegree for all node combinations in topology $T_{(i-1),i,(i+1)}$.

$d_{max}(T_i) = \max_{j=1}^M(d^+\{T_i(u_j)\}, d^-\{T_i(u_j)\})$ is defined as the maximum indegree or outdegree of topology T_i.

$d^+\{T_{(i-1),i,(i+1)}(u_j)\} = d^+_{3adj(T_i)}(u_j)$ is the outdegree of node u_j in the union of three adjacent topologies $T_{(i-1),i,(i+1)}$.

$d^-\{T_{(i-1),i,(i+1)}(u_j)\} = d^-_{3adj(T_i)}(u_j)$ is the indegree of node u_j in the union of three adjacent topologies $T_{(i-1),i,(i+1)}$.

$K_{(j,k)}(T_{(i-1),i,(i+1)}) = K_{(j,k)3adj(T_i)}$ is the number of edges that originate from node j and terminate at node k in the union of three adjacent topologies $T_{(i-1),i,(i+1)}$.

Lemma 2 : *The minimum number of wavelengths W required in the WDM network in order to perform hitless reconfiguration is given by :*

$$W_{min} = \max_{i=1}^N(\max_{j=1}^M(d^+\{T_{(i-1),(i)modN}(u_j)\}, d^-\{T_{(i-1),(i)modN}(u_j)\}))$$

Enough wavelengths are required to support the union of all adjacent topologies. The maximum degree (indegree or outdegree) from all the nodes is found for each union of adjacent topologies. The maximum of this set results in the minimum number of wavelengths needed to support a hitless reconfiguration.

4.2.   UPPER BOUND ON NUMBER OF WAVELENGTHS REQUIRED

The need for more wavelengths than the minimum number shown in Lemma 2 arises from the fact that for a hitless transition from topology T_i to topology T_{i+1}, the wavelengths previously allocated to the links of topology T_i cannot be changed. The new links of topology T_{i+1} should be assigned to wavelengths such that no CC occurs at the access links with the wavelengths previously assigned to topology T_i. If the wavelengths assigned to the links of topology $T_{i,(i+1)}$ could be re-assigned, the number of wavelengths required would be equal to W_{min}. However, this will not result in a hitless transition.

Lemma 3 : *A very loose upper bound that can be imposed on the number of wavelengths required (wavelength assignment algorithm independent) is given by :*

$$W_{loose}^u = |\bigcup_{i=1}^{N} E_i|$$

W_{loose}^u is sufficient to support the union of all edges for all topologies and thus it will be sufficient to support the union of all adjacent topologies. Stricter upper bounds on the number of wavelengths required in the WDM network for hitless reconfiguration are discussed below.

Dynamic Wavelength Assignment

Dynamic wavelength assignment is defined as the wavelength assignment process that assigns different wavelengths to topology T_i every time it returns to T_i after a reconfiguration cycle.

Lemma 4 : *Assuming a general wavelength re-use algorithm that tries to re-use the wavelengths whenever possible, a stricter upper bound for the dynamic wavelength assignment, W_u^d, can be found as follows :*

$$W^d = \max_{i=1}^{N}(\max_{j=1}^{M}(\max_{k=1,k\neq j}^{M} (d^+\{T_{(i-1),imodN}(u_j)\}$$
$$+d^-\{T_{(i-1),imodN}(u_k)\})))$$

$$W_u^d = \max_{i=1}^{N}(\max_{j=1}^{M}(\max_{k=1,k\neq j}^{M} (d^+\{T_{i-1,imodN}(u_j)\}$$
$$+d^-\{T_{i-1,imodN}(u_k)\}) - K_{(j,k)}(T_{i-1,imodN})))$$

W^d calculates the maximum of the summation between indegree and outdegree for all node combinations for all unions of adjacent topologies. The cases of interest are the ones where the number of wavelengths required is greater than the maximum indegree or outdegree for a union of adjacent topologies $T_{i,(i+1)}$ because of previous wavelength assignments in topology T_i. In those cases, the maximum number of wavelengths required will be equal to the maximum of the summation between indegree and outdegree for all node combinations in topology $T_{i,(i+1)}$ (defined as $max_{sum2}(T_{i,(i+1)})$). This way, regardless of the wavelength assignments in the previous topology T_i, there will always be enough wavelengths available for assignment in topology $T_{(i+1)}$. If $max_{sum2}(T_{i,(i+1)})$ is found for all different unions of adjacent topologies, and the maximum of this set is taken, a loose upper bound on the number of wavelengths can be obtained. To tighten the bound, the edges that originate at the node with the highest outdegree and terminate at the node with the highest indegree have to be taken into account. The tighter bound W_u^d calculates the maximum of the summation between indegree and outdegree for all node combinations, while subtracting the common number of edges for each node combination, for all unions of adjacent topologies.

Fixed Wavelength Assignment

Fixed wavelength assignment is defined as the wavelength assignment process which does not assign different wavelengths to topology T_i every time it returns to T_i. It assigns wavelengths to each topology only once at the time of the network design and this assignment always remains the same. For fixed wavelength assignment, the wrap-around from topology T_{N-1} to topology T_0 introduces some additional complexity. If topologies T_{N-2} and T_0 are using all the available wavelengths (for inbound or outbound edges for a single node), then topology T_{N-1} needs to use different wavelengths for its edges for that specific node, thus going over the maximum (indegree or outdegree of union of adjacent topologies) value.

Lemma 5 : *Assuming a general wavelength re-use algorithm that tries to re-use the wavelengths whenever possible, a stricter upper bound for the fixed wavelength assignment, W_u^f, is given as follows:*

$$W_u^f = \max[(\min_{i=1}^{N}(\max_{j=1}^{M}(\max_{k=1,k\neq j}^{M}(d_{3adj(T_i)}^+(u_j) + d_{3adj(T_i)}^-(u_j)$$
$$-K_{(j,k)_{3adj(T_i)}})))), (W_{u1}^d)] \qquad (15.3)$$

Given that the system is cyclical, there are N possible ways to name the topologies. Since for the fixed problem the algorithm is run "off-line", the coloring can start from any topology T_i and the wavelength assignment for all topologies can be obtained, i.e., any topology can be the starting one. The maximum of the summation between indegree and outdegree for all node combinations in topology $T_{(i-1),i,(i+1)}$ (defined as $max_{sum3}(T_{(i-1),i,(i+1)})$) is calculated, while subtracting the common number of edges for each node combination, for all unions of three adjacent topologies. A tighter upper bound can be obtained by taking the minimum of all the W_u^f's calculated from all N possible namings. Note that for the calculation of W_u^f an extra operation is needed. If the minimum number of the summation of highest degrees of three adjacent topologies subtracting the "double-counted" edges, is less than the maximum number of the summation of highest degrees of two adjacent topologies subtracting the "double-counted" edges, $W_u^f = W_u^d$ wavelengths have to be used.

5. HEURISTIC ALGORITHMS FOR WAVELENGTH ASSIGNMENT

5.1. DYNAMIC WAVELENGTH ASSIGNMENT

This section presents a dynamic wavelength allocation heuristic algorithm that assigns wavelengths for each edge between a pair of ATM ports for a given topology T_{i-1}. The assignments can be stored in a file and then used when a change in topology from T_{i-1} to $T_{(i)modN}$ is required.

This is a greedy algorithm that begins by allocating wavelengths to the ports of the ATM nodes in T_0. As explained in Lemma 1, this is done using a bipartite edge coloring algorithm [Papadimitriou and Steiglitz, 1982] on topology T_0. The resulting wavelength allocation is stored in a file. Say, w wavelengths $(\lambda_1, \lambda_2.....\lambda_w)$ are required for the edge coloring of T_0. To effect a transition from T_0 to T_1 the network must support the union of the two topologies $T_{0,1}$. The algorithm considers each edge e_j from topology T_1 and tries to allocate to the edge e_j a wavelength, starting with λ_1. An attempt is made to allocate wavelength λ_1 to edge e_j while checking whether it violates the CC condition with the previously allocated wavelengths to the edges of topology T_0. If λ_1 fails the CC condition test then the next wavelength λ_2 is chosen and tested. This process is continued until a wavelength is found that satisfies the Color Clash constraint. If no wavelength from λ_1 through λ_w satisfies the CC constraint, a new wavelength λ_{w+1} is created which is added to the wavelength set. The resulting wavelength allocation for the edges of topology T_1 is again stored in a file. The algorithm repeats this process for all the transition topologies T_{i-1} to $T_{(i)modN}$. Each time the results of the wavelength allocation are stored in a file. Since this is a dynamic wavelength assignment, when the process cycles back to T_i a new assignment is made.

An example of the dynamic wavelength heuristic applied to a network with three transitional topologies (T_0, T_1, T_2) is demonstrated in Fig. 15.6. This figure shows the wavelength assignments after the first two transitions (T_0 to T_1

and T_1 to T_2). The edge coloring of topology T_0 requires two colors. To color topology T_1, the heuristic allocates wavelength λ_1 to the edge from vertex C to vertex A and wavelength λ_3 to the edge from vertex D to vertex A (CC prevents that edge to be allocated wavelengths λ_1 or λ_2). For the transition from T_1 to T_2, the algorithm assigns wavelengths to the edges of T_2, while making sure that no CC exists with the previous wavelength assignmentin T_1. Thus, the edge from vertex D to vertex C is allocated λ_1. The same process is repeated for each new transition.

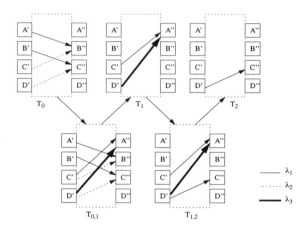

Figure 15.6   Example of the wavelength assignment heuristic after two transitions (T_0 to T_1, T_1 to T_2) (dynamic wavelength assignment).

5.2.   FIXED WAVELENGTH ASSIGNMENT

The fixed wavelength assignmentalgorithm is again a greedy algorithm that begins by allocating wavelengths to the ports of the ATM nodes in T_0. The algorithm repeats the same procedure as outlined in Section 5.1 for all transitions from topology T_{i-1} to $T_{(i)mod N}$ until it reaches topology T_{N-1}. For the last transition from T_{N-2} to T_{N-1}, the algorithm is required to check the wavelength assignments in both T_{N-2} and T_0 before allocating a wavelength to an edge $e_j \in E_{(N-1)}$.

Since it is desirable to minimize the number of wavelengths required, the coloring should start from the topology that will require W_u^f wavelengths. This topology is re-named as T_0 (and all other topologies accordingly), and the algorithm to find the wavelength assignment starts from that topology.

An example for the fixed wavelength heuristic applied to a network with three transitional topologies (T_0, T_1, T_2) is demonstrated in Fig. 15.7. The edge coloring for T_0 requires two colors. The heuristic then allocates wavelength λ_1 to an edge from vertex C to vertex A in topology T_1. The CC constraint prevents the edge from D to A to be allocated wavelengths λ_1 and λ_2, so it is allocated wavelength λ_3. For the transition from T_1 to T_2, the edge from vertex D to vertex C in T_2 cannot be allocated wavelengths λ_1 and λ_2 (CC violation with the previously allocated wavelengths in T_0) or wavelength λ_3 (CC violation with the edges of topology T_1). Hence, it is allocated wavelength λ_4.

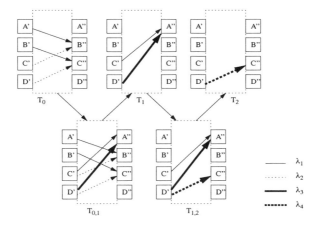

Figure 15.7 Example of the wavelength assignment heuristic (greedy algorithm) for fixed wavelength assignment.

5.3.   IMPLEMENTATION OF THE HEURISTIC ALGORITHMS

Both dynamic and fixed wavelength assignment heuristic algorithms were implemented for a number of networks and the number of wavelengths found was compared to the upper and lower bounds calculated in the analysis. In each implementation of the algorithms, one of three variables (average node outdegree, number of topologies in the network and total number of nodes in each topology) was varied while the other two remained constant. The upper bounds used were W_u^d and W_u^f for dynamic and fixed wavelength assignment respectively and the lower bound was always given by W_{min}.

Fig. 15.8 shows a plot of the number of wavelengths for a varying average node outdegree, for both fixed and dynamic wavelength assignments. The number of topologies and the number of nodes in each topology were kept constant. For both heuristics, the number of wavelengths found falls between its respective upper and lower bounds and it is closer to the theoretically established lower bound than to the upper bound. This indicates that the simple greedy approach can color all topologies with a number of wavelengths not significantly larger than the theoretical lower bound. The plot also indicates that while the lower bound is the same for both the fixed and dynamic approaches, the upper bound for the fixed method is significantly higher than the upper bound for the dynamic method for networks where the average node outdegree is large. This is consistent with the analysis, since for the fixed approach the union of three topologies is taken into consideration, compared to the union of two topologies for the dynamic approach. The implementation of the two heuristics also indicates the same trend. In general, a larger number of wavelengths is required for the fixed than for the dynamic approach. The difference appears greater for networks with large average node outdegrees since in that region the topologies become more dense.

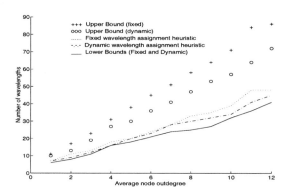

Figure 15.8   Number of wavelengths vs. average node outdegree (fixed and dynamic cases).

Plots of the average number of wavelengths for a varying number of topologies using fixed and dynamic wavelength assignments are shown in Fig. 15.9(a) and 15.9(b) respectively. Once again, the number of wavelengths found falls between its respective upper and lower bounds and it is closer to the theoretically established lower bound than to the upper bound. The implementation of the fixed algorithm shows an interesting trend in the number of wavelengths required to color the topologies. This number is always higher for an odd number of topologies than for the previous even number of topologies. This is consistent with the analysis, since for even number of topologies, when the fixed wavelength assignment is used, the last topology (T_{N-1}) can potentially re-use some of the wavelengths used in topologies T_0 and T_{N-2}. In the dynamic algorithm, the number of wavelengths is independent of the number of topologies, since the wavelength assignment depends only on the assignment in the previous topology. In general, a larger number of wavelengths is required for the fixed approach than for the dynamic approach. The difference becomes negligible though for a large number of topologies because in that region the wavelengths can be more easily re-used for the fixed case and will approach the number of wavelengths required for the dynamic case.

Plots of the number of wavelengths for a varying number of nodes in each topology are shown in Fig. 15.10(a) and 15.10(b) for the fixed and dynamic wavelength assignments respectively. Once again, the number of wavelengths follows closely the established lower bound. For the fixed approach, the four points on the plot indicate that the number of wavelengths remains approximately constant as the number of nodes doubles ($8, 16, 32$ and 64 nodes in this case). This is consistent with the analysis, since when the number of nodes doubles and the average node outdegree remains constant, the number of wavelengths required to color the topologies will also remain approximately constant.

6. CONCLUSIONS

This work proposed a methodology for reconfiguring a WDM optical network to adapt to changing traffic requirements at the ATM layer. It assumed that changing traffic patterns result in the requirement for changing ATM network topologies in a cyclic order and that these topologies, their ordering and configuration patterns are known in advance. The focus of this work was on the edges of the network where the ATM switches access the WDM layer. Sizes for the ATM switches and bounds on the number of wavelengths needed to support the required hitless reconfiguration were also discussed. Two heuristic algorithms that assign wavelengths between pairs of ports at the ATM switches to support the required ATM network topologies in a hitless manner were proposed. These algorithms, even though greedy, exhibited polynomial computational complexities and were easy to implement. It was shown that the number of wavelengths found using these two algorithms falls within the proposed lower and upper bounds and it is closer to the established lower bound than to the upper bound. This indicates that the algorithms proposed are efficient in finding a small number of wavelengths and that more complex allocation algorithms will produce only marginally better results.

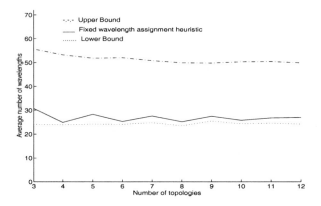

(a)

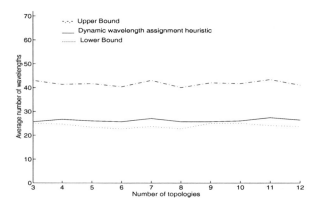

(b)

Figure 15.9   Average number of wavelengths vs. number of topologies (a) Fixed and (b) Dynamic case.

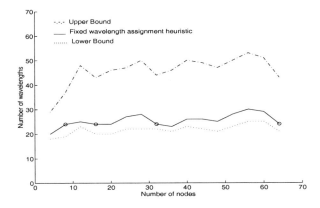

(a)

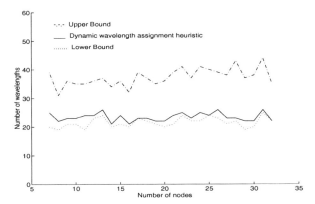

(b)

Figure 15.10   Number of wavelengths vs. number of nodes in each topology (a) Fixed and (b) Dynamic case.

Acknowledgments

We would like to thank Michael Post, Chien-Chung Shen, and John Wei from Telcordia Technologies and Neophytos Antoniades from Corning Inc. for their technical contributions to different parts of the work.

References

Acampora, A. (1994). The scalable lightwave network. *IEEE Communications Magazine*, pages 36–42.

Alexander, S. B., Bondurant, R. S., Byrne, D., Chan, V. W. S., Finn, S. G., Gallager, R., Glance, B. S., Haus, H. A., Humblet, P., Jain, R., Kaminow, I. P., Karol, M., Kennedy, R. S., Kirby, A., Le, H. Q., Saleh, A. A. M., Schofield, B. A., Shapiro, J. H., Shankaranarayanan, N. K., Thomas, R. E., Williamson, R. C., and Wilson, R. W. (1993). A precompetitive consortium on wide-band all-optical networks. *IEEE/OSA J. Lightwave Tech.*, 11(5/6):714–735.

Auerbach, J. and Segall, A. (1995). Use of delegated tuning and forwarding in wavelength division multiple access networks. *IEEE Trans. Comm.*, 43(1):52–63.

Bala, K., Chung, F. R. K., and Brackett, C. A. (1996). Optical wavelength routing, translation and packet/cell switched networks. *IEEE/OSA J. Lightwave Tech.*, 14(3):336–343.

Bala, K., Stern, T. E., Simchi-Levi, D., and Bala, K. (1995). Routing in linear lightwave networks. *IEEE/ACM Trans. Networking*, 3(4):459–469.

Brackett, C. A. et al. (1993). A scalable multiwavelength multihop optical network: A proposal for research on all-optical networks. *IEEE/OSA J. Lightwave Tech.*, 11(5/6):736–753.

Gerstel, O. and Segall, A. (1995). Dynamic maintenance of the virtual path layout. In *Proc. INFOCOM*, pages 330–337.

Hill, G. R. et al. (1993). Multi-wavelength transport network: A transport layer based on optical network elements. *IEEE/OSA J. Lightwave Tech.*, 11(5/6):667–679.

Labourdette, J. P., Hart, G. W., and Acampora, A. S. (1994). Branch-exchange sequences for reconfiguration of lightwave networks. *IEEE Trans. Comm.*, 42(10):2822–2832.

Lee, K. C. and Li, V. O. K. (1996). A wavelength rerouting algorithm in wide-area all-optical networks. *IEEE/OSA J. Lightwave Tech.*, 14(6): 1218–1229.

Papadimitriou, C. and Steiglitz, K. (1982). *Combinatorial Optimization: Algorithms and Complexity*. Prentice Hall.

Ramaswami, R. and Sivarajan, K. N. (1995). Routing and wavelength assignment in all-optical networks. *IEEE/ACM Trans. Networking*, 3(5):489–500.

Thulasiraman, K. and Swamy, M. N. S. (1980). *Advanced Graph Theory*. Prentice Hall.

Chapter 16

OPTICAL WDM NETWORKS: FUTURE VISION

Imrich Chlamtac
University of Texas at Dallas
chlamtac@utdallas.edu

Jason P. Jue
University of Texas at Dallas
jjue@utdallas.edu

Introduction

Advances in optical WDM communications have created the potential of virtually unlimited bandwidth [Chlamtac, 2000]. WDM systems capable of supporting 40 OC-48 channels are commercially available, while products capable of 128 OC-192 (10 Gbps) channels, or a total of 1.28 Tbps, are expected soon. Similar progress is being made in the advancement of optical components which will compose the solution for optical WDM networking, and these advances are fueling the evolution of a new optical layer which will provide capabilities such as optical reconfiguration, wavelength provisioning, and all-optical protection and restoration. New optical network architectures in the local access, metro, and wide area environments will utilize this emerging optical layer, and will be poised to provide an efficient optical network, that will, ultimately, provide end-to-end optical wavelength-based services.

This vision of an all-optical, end-to-end WDM network is already driving significant research efforts of many transport and optical networking companies. The concept of IP packets being transmitted directly over optics is a hot research issue, potentially leading to the exploration of an entirely new IP-routing layer. The ongoing integration of optical networking concepts and the Internet, along with the Internet's phenomenal growth, will continue to provide technological challenges and opportunities to drive optical networking research, development, and commercialization. In fact, the indisputable role of Internet as a mainstay of the global marketplace guarantees that the role of the accompanying optical networking technology will remain in the commercial focus for years to come.

In this article, we will discuss some of the most promising trends in WDM optical networks and describe some of the challenges which will need to be met in order to fulfill the promise of WDM as the lead technology for high bandwidth communication in the first decade of the 21st century.

1.   ISSUES IN WIDE AREA WDM COMMUNICATIONS

Telecommunication systems are rapidly evolving to meet the changing demands of network users. The rapid increase in computer processing power in the last decade has encouraged the development and deployment of many high speed applications. Combined with the sudden surge in Internet growth over the past few years, these applications have pushed the demand for network bandwidth to a new level. Fortunately, in the same time period, the bandwidth of carries has been revolutionized. Advances in optical communication technology and DWDM systems have raised the bandwidth capacity of a single fiber link to over 100 Gbps. However, in addition to bandwidth, emerging applications will also require fast switch and router architectures which allow data to be moved across the network via intermediate switching nodes at speeds comparable to the fiber link speeds. The expected increase in multimedia applications will also require future networks to provide increased performance and reliability, stricter response times, and guaranteed delivery of data.

To better understand the types of services demanded of future networks, we need to examine the requirements of emerging applications. Examples of possible high-bandwidth applications include video telephony, video conferencing, video distribution, and data transfer services. Currently, these applications already exist in some form or another over the Internet or the World Wide Web; therefore, it is reasonable to expect that, as the network infrastructure continues to improve, these applications will become more prominent in the future. Most of these applications can be characterized by their requirements in terms of bandwidth, latency, error tolerance, multicasting, and security. Real-time interactive applications, such as video telephony, will have strict latency requirements and will require bandwidth on the order of hundreds of Kbps to several Mbps in each direction, depending on the desired quality of the transmission. Video conferencing applications will have the same latency requirements as video telephony, but will require a greater amount of bandwidth, since the data must be transmitted to several destinations. Conferencing applications may also require some level of multicast support from the underlying network in order to provide a more efficient use of the network bandwidth. Video distribution applications, such as video-on-demand or TV broadcast distribution, will require bandwidth ranging from around 1.5 to 100 Mbps for TV-quality video to around 400 Mbps for the transmission of uncompressed HDTV streaming video. Although video distribution applications will not have the same strict latency requirements as real-time interactive video, there will still be requirements with respect to jitter or frame loss. Video-on-demand and video broadcast applications may also require some level of multicasting support if a large number of users request the same video streams. Finally, data applications, such as software distribution or the transfer of text or images, are not as sensitive to latency as video applications, but may tend to be less tolerant of errors or data losses. As we can see, this wide range of traffic types necessitates a network design which can offer a variety of services as well as different levels of service.

One of the main objectives of a network architecture is to optimize cost and performance by trading resources to supply the performance requirements of an application at minimal cost. Until recently, the scarce resource in wide area communication networks has been bandwidth. Existing network solutions, therefore, still stress elaborate electronic multiplexing and switching of data to reduce the cost of transmission by efficiently sharing link bandwidth among multiple users. Current store-and-forward architectures are therefore protocol processing intensive, allowing bandwidth to be shared at the expense of additional processing, buffering, and switching at intermediate nodes.

WDM has been deployed commercially in long-distance networks, with systems capable of operating up to 40 channels on each optical fiber link. These systems provide a straightforward replacement of links in conventional WAN networks, offering point-to-point WDM connections which require the signal on each optical channel to be converted to electronic form for switching and multiplexing at each intermediate node. The electronic processing and switching costs in such a system can potentially be very high, which may possibly lead to severe performance bottlenecks at the intermediate switching nodes and impede the delivery of the optical link bandwidth to the end user. In order to provide effective rates on the order of Tbps and to limit electronic bottlenecks and delays at intermediate switching nodes, the upcoming generation of wide area, as well as local networks, needs to be molded by the characteristics of maturing optical technologies.

Emerging WDM systems, which contain optical switching and interconnection elements, will allow wavelengths to bypass intermediate nodes entirely in the optical domain, thus reducing the amount of electronic processing required at a node. Such systems include ring networks which utilize optical add-drop multiplexers (ADMs) [Gerstel et al., 1998] and wavelength-routed mesh networks, introduced in [Chlamtac et al., 1989], in which all-optical lightpaths are set up on specific wavelengths between pairs of nodes [Banerjee and Mukherjee, 1996; Chlamtac et al., 1992; Chlamtac et al., 1996; Mokhtar and Azizoglu,

1998; Mukherjee et al., 1996; Ramaswami and Sivarajan, 1995]. Currently, commercial optical ADM systems exist and are being considered for deployment in metropolitan area networks, while wavelength routedmesh networks exist only as testbeds. In order for wavelength-routed WDM networks to become commercially viable, a number of issues still need to be resolved. The most basic issues include developing mechanisms for the control and management of lightpaths, and providing protection and restoration for lightpaths at the optical layer. Emerging WDM systems must also provide the appropriate services for higher-layer protocols, such as IP, ATM, SONET, and Ethernet. At the very least, wavelength-routed network should have both the capability to set up static lightpaths to serve as links in electronic circuit-switched or packet-switched networks, and the capability to set up and take down lightpaths dynamically to accommodate all-optical circuit-switched connections. More advanced WDM networks may have the capability to provide enhanced services such as optical multicasting and optical packet switching.

Initially, the WDM layer would be transparent to the higher-layer protocols, providing basic optical transport services over lightpaths, as well as protection for the lightpaths. As the network evolves, new applications and protocols may be developed which are aware of the optical layer and which can take full advantage of additional services provided by the optical layer. These applications and protocols may be able to request lightpaths on demand, allowing for fully dynamic reconfiguration of logical network topologies as well as all-optical end-to-end connections for specialized classes of traffic.

1.1.   IP OVER WDM

The Internet is growing at a rapid rate, and IP is expected to be the dominant source of traffic in the backbone in the next decade. As such, it is especially important for the optical layer to be capable of supporting services for the IP layer, and for the IP layer to be capable of taking advantage of these services provided by the optical layer.

As the Internet continues to expand, it will be expected to support a growing number of applications, such as Internet telephony, video conferencing, and video distribution. To support these applications, the network must be capable of not only providing the required bandwidth, but also providing guarantees with respect to quality of service, security, and fault tolerance.

In many existing telecommunication networks, SONET is used to provide services such as transmission, multiplexing, and protection; thus, we may envision an IP layer implemented over SONET links, and SONET implemented over WDM lightpaths. However, an alternate approach which has been gaining much attention recently is the implementation of IP directly over a WDM optical layer, with a minimal layer of electronics between these two layers. In this scenario, much of the SONET functionality, such as protection, may be shifted to the IP layer or WDM layer, while only SONET framing is maintained. IP routers are then interconnected via wavelengths or lightpaths. Many important issues, such as determining a logical topology for the IP network, routing the lightpaths, and assigning wavelengths to the lightpaths have been studied in the literature; however, there are a number of other issues which also deserve attention. In order to realize an IP over WDM network environment, we need to define the services and the functionality to be offered by each layer of the network, and we need to ensure that the various layers complement each other rather than conflict with each other. For example, we may wish to address such issues as the interaction between lightpath reconfiguration and IP routing, or the interaction between IP restoration and optical layer protection. It may even be beneficial to consider the possibility of modifying the IP layer to take advantage of the services provided by the WDM layer. For example, if the IP layer can provide information with regard to traffic type and QoS requirements of packets, then the WDM layer may be able to establish lightpaths on-demand for specific high-volume flows of IP traffic, or otherwise provide differentiated services for different classes of traffic.

Another emerging possibility for packet over WDM networks is to deploy 10 Gb Ethernet in the wide-area backbone, either over SONET OC-192 links, or directly over WDM lightpaths. The benefit of this approach is that it provides a low-cost end-to-end electronic packet-switching layer. Currently, work is being done to develop the standards for 10 Gigabit Ethernet. Additional work is required to investigate the networking issues involved in deploying Ethernet over a wide-area environment, and to develop WDM architectures and protocols suited for Ethernet traffic.

2.   WDM IN LOCAL AREA AND ACCESS NETWORKS

Local area network (LAN) architectures of today have evolved rapidly in terms of the bandwidth offered to the end-user, with current and emerging LAN standards such as Gigabit Ethernet and 10 Gbps Ethernet utilizing high-speed optical fiber links. As the need for bandwidth in the local area increases with the emergence of high speed applications, new LAN architectures which are able to provide high transmissions rates in a cost-effective and scalable manner will need to be developed. We expect that WDM-based optical technologies will be able to take the rapid evolution of LANs to the next level.

In addition to providing higher bandwidth, these new architectures will also need to provide a greater variety of services to the end user. These services may include multicasting, security, connectionless packet-switched services, connection-oriented services, and priority-based services for different types of applications. Finally, it may be desirable for next-generation LANs be compatible with legacy LAN equipment, so as to minimize upgrade costs.

An architecture for WDM LANs which has received much attention in the literature is the broadcast-and-select architecture based on the passive-star coupler. Because the star coupler is a passive component, it is relatively inexpensive. Also, the broadcast nature of the star coupler offers the possibility of implementing multicasting services with relative ease. The coupler is scalable with respect to the number of wavelengths, due to its wavelength-insensitive property; however, it may not be scalable with respect to the number of nodes connected to the coupler, due to the additional splitting losses incurred.

A number of MAC protocols have been proposed for the broadcast and select architecture [Mukherjee, 1992a; Mukherjee, 1992b]. While none of these WDM LANs has been commercially deployed, a few have been implemented in testbeds [Chen et al., 1990; Goodman et al., 1990; Dowd et al., 1996].

The most viable local-area application of WDM in the near future is in the local access network. Many current local access networks are broadcast in nature and many also contain optical fiber; thus, it seems natural to extend this environment to include WDM broadcast and select networks. In the future, we may also envision WDM to the home or desktop, with users sharing the bandwidth of multiple wavelengths. Some wavelengths may be allocated for the distribution of broadcast video signals, while other may be used for data or high-bandwidth interactive applications, such as video conferencing. The most significant factors limiting the deployment of broadcast-and-select WDM networks in the desktop LAN environment are the high cost of network interfaces and the limited compatibility with existing legacy LAN equipment. Work must be done to incorporate traditional protocols, such as Ethernet, into WDM networks in order to ensure compatibility with existing infrastructure. Such compatibility is essential for situations in which a large investment has already been made in legacy LAN equipment. Also, WDM-based LANs will need to provide significant benefits over existing LAN architectures in order to be commercially viable.

3.   OPTICAL PACKET SWITCHING

An emerging alternative to both the wavelength-routed WDM network in the wide area and the broadcast and select network in the local area is a WDM network based on photonic packet switching. In a photonic packet-switched network, a packet is routed from source to destination entirely in the optical domain, thereby reducing delays and providing data transparency.

A significant issue which needs to be resolved for such networks is the issue of packet contention. Since it is more difficult to buffer packets in the optical domain than in the electronic domain, it is important to either carefully design optical buffers so as to minimize dropped packet, or to develop other approaches for handling contention. A number of variations of optical packet buffers have been proposed, although much additional work is needed to study the properties of these buffers. An alternative to buffering is deflection routing in which contention is resolved by intentionally misrouting a packet. In this case, the links in the network itself serve as buffers.

Another significant issue in optical packet switching is synchronization. Packets coming into a node on different input fibers must be perfectly aligned so that they may traverse the optical switching element at the same time. In order to achieve this packet alignment, we may require adjustable delay elements at each fiber input, as well as the appropriate control mechanisms.

A promising all-optical packet-switched architecture for WDM access networks is one based on the concept of photonic slot routing [Chlamtac et al., 1999]. In photonic slot routing, data is transmitted in the form of photonic slots which are fixed in length and span all wavelengths in the network. Each wavelength in the photonic slot may contain a single packet, and all packets in the photonic slot are destined to the same node. All packets in a slot must have the same destination; thus, the photonic slot may be routed as a single integrated unit without the need for demultiplexing individual wavelengths. Also, since photonic slot routing does not require wavelength-sensitive components, it results in less complexity, faster routing, and a greater degree of scalability.

4.   CONCLUSIONS

The amount of traffic being transmitted over existing networks is rising at an unprecedented rate, and telecommunications and data communications companies are racing to provide the means for meeting these demands. Many companies are turning to wavelength-division multiplexing to provide increased bandwidth; however, WDM has the potential to provide many additional benefits in terms of cost and performance. By developing architectures and protocols which fully utilize the benefits of WDM, we can ensure that this tremendous growth will continue well into the future.

References

Banerjee, D. and Mukherjee, B. (1996). A practical approach for routing and wavelength assignment in large wavelength-routed optical networks. *IEEE Journal on Selected Areas in Communications*, 14(5):903–908.

Chen, M.-S., Dono, N. R., and Ramaswami, R. (1990). A media access protocol for packet-switched wavelength division multiaccess metropolitan area networks. *IEEE Journal on Selected Areas in Communications*, 8(6):1048–1057.

Chlamtac, I. (2000). Optical networking - editorial. *SPIE/Baltzer Optical Networks Magazine*, 1(1).

Chlamtac, I., Elek, V., Fumagalli, A., and Szabo, C. (1999). Scalable WDM access network architecture based on photonic slot routing. *IEEE/ACM Transactions on Networking*, 7(1):1–9.

Chlamtac, I., Farago, A., and Zhang, T. (1996). Lightpath (wavelength) routing in large WDM networks. *IEEE Journal on Selected Areas in Communications*, 14(5):909–913.

Chlamtac, I., Ganz, A., and Karmi, G. (1989). Purely optical networks for terabit communication. In *Proceedings, IEEE INFOCOM '89*, Waterloo, Canada.

Chlamtac, I., Ganz, A., and Karmi, G. (1992). Lightpath communications: A novel approach to high bandwidth optical WANs. *IEEE Transactions on Communications*, 40(7).

Dowd, P., Perreault, J., Chu, J., Hoffmeister, D., Minnich, R., Hady, F., Chen, Y.-J., Dagenais, M., and Stone, D. (1996). Lightning network and systems architecture. *IEEE/OSA Journal of Lightwave Technology*, 14(6):1371–1387.

Gerstel, O., Ramaswami, R., and Sasaki, G. (1998). Cost effective traffic grooming in WDM rings. In *Proceedings, IEEE INFOCOM '98*, pages 69–77, San Francisco, CA.

Goodman, M. S., Gimlett, J. L., Kobrinski, H., Vecchi, M. P., and Bulley, R. M. (1990). The LAMBDANET multiwavelength network: Architecture, applications, and demonstrations. *IEEE Journal on Selected Areas in Communications*, 8(6):995–1004.

Mokhtar, A. and Azizoglu, M. (1998). Adaptive wavelength routing in all-optical networks. *IEEE/ACM Transactions on Networking*, 6(2):197–206.

Mukherjee, B. (1992a). WDM-based local lightwave networks – part I: Single-hop systems. *IEEE Network Magazine*, 6(3):12–27.

Mukherjee, B. (1992b). WDM-based local lightwave networks – part II: Multihop systems. *IEEE Network Magazine*, 6(4):20–32.

Mukherjee, B., Banerjee, D., Ramamurthy, S., and Mukherjee, A. (1996). Some principles for designing a wide-area WDM optical network. *IEEE/ACM Transactions on Networking*, 4(5):684–696.

Ramaswami, R. and Sivarajan, K. N. (1995). Routing and wavelength assignment in all-optical networks. *IEEE/ACM Transactions on Networking*, 3(5):489–500.

Index